PURE AND APPLIED MATHEMATICS

SCHUMAKER—Spline Functions: Basic Theory
SENDOV and POPOV—The Averaged Moduli of Smoothness
*SIEGEL—Topics in Complex Function Theory
    Volume 1—Elliptic Functions and Uniformization Theory
    Volume 2—Automorphic Functions and Abelian Integrals
    Volume 3—Abelian Functions and Modular Functions of Several Variables
STAKGOLD—Green's Functions and Boundary Value Problems
*STOKER—Differential Geometry
STOKER—Nonlinear Vibrations in Mechanical and Electrical Systems
TURÁN—On a New Method of Analysis and Its Applications
WHITHAM—Linear and Nonlinear Waves
ZAUDERER—Partial Differential Equations of Applied Mathematics, 2nd Edition

*Now available in a lower priced paperback edition in the Wiley Classics Library.

TWO-DIMENSIONAL GEOMETRIC VARIATIONAL PROBLEMS

TWO-DIMENSIONAL GEOMETRIC VARIATIONAL PROBLEMS

Jürgen Jost
Ruhr-Universität Bochum
Fakultät und Institut für Mathematik
Germany

A Wiley-Interscience Publication

JOHN WILEY & SONS
Chichester · New York · Brisbane · Toronto · Singapore

Other Wiley Editorial Offices

John Wiley & Sons, Inc., 605 Third Avenue,
New York, NY 10158–0012, USA

Jacaranda Wiley Ltd, G.P.O. Box 859, Brisbane,
Queensland 4001, Australia

John Wiley & Sons (Canada) Ltd, 22 Worcester Road,
Rexdale, Ontario M9W 1L1, Canada

John Wiley & Sons (SEA) Pte Ltd, 37 Jalan Pemimpin 05–04.
Block B, Union Industrial Building, Singapore 2057

Library of Congress Cataloging-in-Publication Data:

Jost, Jürgen, 1956–
Two-dimensional geometric variational problems / Jürgen Jost.
p. cm.—(Pure and applied mathematics)
Includes bibliographical references and index.
1. Harmonic maps. 2. Variational inequalities (Mathematics)
3. Riemannian manifolds. I. Title. II. Series: Pure and applied mathematics (John Wiley & Sons)
QA614.73.J67 1990
514′.74—dc20 90-12622
CIP

British Library Cataloguing in Publication Data:

Jost, Jürgen
Two-dimensional geometric variational problems.
1. Boundary value problems. Solution. Variational methods
I. Title
515.62

ISBN 0 471 92839 9

Typeset by Thomson Press (India) Ltd, New Delhi
Printed in Great Britain by Courier International Ltd, Tiptree, Essex

CONTENTS

INTRODUCTION

In this monograph, we treat variational problems for mappings from a surface equipped with a conformal structure into Euclidean space or a Riemannian manifold. We assume that the variational problems are invariant under conformal reparametrizations of the domain. Solutions to such variational problems consist of conformal mappings between surfaces, minimal surfaces in Riemannian manifolds, harmonic maps from a surface into a Riemannian manifold, and solutions of prescribed mean curvature equations.

We present here a general theory of such variational problems, proving existence and regularity theorems with particular conceptual emphasis on the geometric aspects of the theory and thoroughly investigating the connections with complex analysis. Our approach is purely parametric, and, consequently, we do not address the question of geometric regularity of the solutions (immersion and embeddedness properties); some references to the relevant literature can be found at the end of Section 4.4, however. We treat the existence of closed solutions as well as of solutions to various boundary value problems (Dirichlet, Plateau-type and free boundary conditions). Usually, we assume that the boundary configuration is of class C^2, and we make no systematic attempt to weaken this assumption, although usually this can rather easily be achieved by approximation arguments (some examples of more general assumptions can be found in Sections 4.7 and 5.1, however). In the same spirit, we always assume that the target space is a Riemannian manifold of bounded geometry (see the Appendix on notation and terminology for a discussion of this concept).

In Chapter 1, which is mainly of an introductory nature, we first discuss Plateau's problem as the prototype of a two-dimensional geometric variational problem. This allows us to exhibit some useful and important methods of reasoning at a (nowadays) trivial example. We then introduce the relevant definitions and concepts of this book, and we treat some elementary relations of our variational problems with holomorphic quadratic differentials (Gauss maps of minimal surfaces, Hopf's Theorem on the non-existence of a sphere of constant mean curvature in $\mathbb{R}^3$ other than the standard one and related results, Liebmann's Theorem, etc.).

In Chapter 2, which is partly of an expository nature, we deal with the regularity theory of solutions of our problems. Whenever possible, we refer to the author's monograph *Harmonic Mappings Between Riemannian Manifolds* (item [J6] of the bibliography) for proofs where the general regularity theory of harmonic maps between Riemannian manifolds is covered[1]. For all other results needed in this book, we give detailed proofs. Apart from some results on free boundary regularity, a topic not covered in [J6], these are results that are particular to the situation of a two-dimensional domain. We present continuity results for weak solutions (higher regularity then follows from [J6]), and also the result of Sacks and Uhlenbeck on the non-existence of isolated singularities. We then prove a version of the Hartman–Wintner lemma and discuss some of its consequences, in particular the asymptotic expansion of solutions near branch points. Finally, we present estimates from below for the functional determinant of univalent harmonic mappings. Such estimates were first obtained in the fundamental work of Heinz and were further developed by Jost and Karcher. Here, we have achieved considerable simplifications of the original work.

In Chapter 3, we present a variational approach for obtaining various conformal representation theorems. The results are of course known from other methods, but our approach fits naturally into the scheme of the present book, and many of the arguments will again be useful later on. For the conformal representation of surfaces of higher genus, we make use of the collar lemma and of Mumford's compactness theorem; elementary proofs of these are included.

In Chapter 4, we prove a general existence theorem for harmonic mappings. It was discovered by Sacks and Uhlenbeck [SkU1] and in a different context independently by Wente [W2] that even an energy-minimizing sequence may be non-compact because of the possible splitting off of minimal spheres. This kind of phenomenon has *a fortiori* to be taken into account if one seeks saddle points, i.e. unstable solutions. Our method to deal with these problems is reminiscent of the curve-shortening process used for obtaining closed geodesics in Riemannian manifolds as well as of the alternating method of Schwarz and Neumann and the balayage method of Poincaré. The method depends on a local existence and uniqueness result and proceeds by simultaneous replacement on small balls of a continuous family of maps by harmonic maps. As these balls overlap, the number of times each point is affected by a local replacement has to be controlled by a covering argument. Controlled blowing-up at points where the modulus of continuity goes to infinity in the course of iterated applications of this replacement procedure then detects the splitting off of minimal spheres. A further possible loss of compactness which could not be handled by previous

[1] References to the original papers on regularity theory can also be found in [J6]. In this respect, we also mention the survey articles [Hi3] by Hildebrandt and [S] by Schoen on the analytic aspects of harmonic maps.

approaches is treated by suitable rescalings on annuli. Our method applies to closed solutions as well as to solutions of Dirichlet, Plateau-type, or free boundary value problems, and in Section 4.4 we present all kinds of existence results for harmonic maps and minimal surfaces that follow from our procedure, demonstrating its generality.

In Section 4.5 we present some instructive non-existence results and contrast them by showing the existence of maps with holomorphic quadratic differentials, a condition which has occasionally been used as an alternative definition of harmonic maps (e.g. in [GR]). This definition should, however, be abandoned since the existence theory with this definition becomes quite different.

In Section 4.6, then, we present a simplified variant of our general method to give another proof of the existence of unstable minimal surfaces. Finally, in Section 4.7, we treat the Plateau–Douglas problem which consists in showing the existence of minimal surfaces of higher genus under appropriate conditions.

We omit existence results for surfaces of prescribed mean curvature, although our method is applicable here as well, because this topic is already treated by a different method in [St5].

Chapter 5 is devoted to harmonic maps between surfaces. We prove the existence of harmonic diffeomorphisms between surfaces, in particular the result of Jost and Schoen [JS]; this depends on the estimates of Heinz for the functional determinant presented in Section 2.7. We also discuss some local computations, some applications (including Kneser's theorem) and the question when harmonic maps are branched coverings.

The last chapter then gives a new approach to Teichmüller theory via harmonic maps. Given two surfaces with hyperbolic structures, a harmonic map between them gives rise to a holomorphic quadratic differential on the domain which can be studied in its dependence on domain and image structure. The effect of variations of the image structure was computed by Wolf [Wf] while the author found a way to handle variations of the domain structure. With the help of these computations, we are able to obtain all the basic structures of Teichmüller space, namely the topological (cf. [Wf]), differentiable, complex, metric (Weil–Petersson metric) and Kählerian ones, and to compute the curvature tensor of the Weil–Petersson metric. Our approach has the principal advantage that quasiconformal maps, the standard analytic tool for studying Teichmüller space, are replaced by harmonic maps. While quasiconformal maps are defined by a pointwise variational principle, harmonic maps arise from an integral variational problem and are therefore analytically much better controlled than quasiconformal maps.

The present book in particular is meant to supersede the author's book [J4], apart from some aspects of the regularity theory which, however, are covered in a more general setting in [J6]. We should also mention some survey articles concerned with topics treated here. For minimal surfaces, we refer to [J13] by the author, for the geometric aspects of harmonic maps to [EL1], [EL4] and [EL5] by Eells and Lemaire and [J14] by the author. In the bibliography, we only list papers mentioned in the text. No attempt has been made at completeness and we refer to the survey articles just cited for further information on the literature.

The main results of this monograph were described in the author's address [J15] at the International Congress of Mathematicians in Berkeley [1986].

In this monograph, we utilize the standard concepts of Riemannian geometry (metric, curvature, covariant differentiation, etc.) without further comment. The required geometric background can be found in any text on Riemannian geometry, e.g. [Kl] or [GKM]. In many places, however, we give both a coordinate-free treatment and computation in local coordinates.

We also use regularity results for solutions of linear elliptic partial differential equations (potential and Schauder theory) without proof; references are [BJS] and [GT].

Except for some elementary aspects, we develop all necessary results about Riemann surfaces in the text. Nevertheless, a certain acquaintance with Riemann surfaces may be useful in order to provide a perspective on certain parts, such as Chapter 6 of the present book.

The sections are of varying technical difficulty. While 2.2, the end of 2.3, 2.4–2.7, 4.2, 4.3, 5.1, and 6.3–6.5 are probably more difficult to read, I have tried to explain many of the essential ideas in an easier setting in 1.1, 1.3, 1.4, the beginning of 2.3, parts of 3.1 and 3.3, 4.1, 4.5, 4.6, 5.2, and 5.3, and these sections should be more accessible.

The research, the results of which are presented in this monograph, was supported by SFB 72 (Deutsche Forschungsgemeinschaft) at the University of Bonn and Stiftung Volkswagenwerk. I want to thank Stefan Hildebrandt for his continuous support and advice over many years. I am also grateful to Michael Wolf and Rugang Ye for stimulating discussions. I want to thank Xiaowei Peng and Tilmann Wurzbacher for some help in proofreading and Ursula Rupprecht for her careful and accurate typing of my manuscript.

1 EXAMPLES, DEFINITIONS, AND ELEMENTARY RESULTS

1.1. Plateau's problem

The prototype of a two-dimensional geometric variational problem is Plateau's problem, namely to find a minimal surface $\Sigma \subset \mathbb{R}^3$ bounded by a given Jordan curve $\gamma \subset \mathbb{R}^3$. Of course, here one needs to specify the class of admissible surfaces among which one looks for a minimal one. Around 1930, Douglas and Radó independently arrived at a formulation of the problem for which they could obtain the following solution ([D1, R2, R3]).

Theorem 1.1.1: *Let γ be a closed Jordan curve in $\mathbb{R}^d$ $(d \geqslant 2)$. Then there exists a continuous map $f: D \to \mathbb{R}^d$ $(D = \{x + iy \in \mathbb{C}: x^2 + y^2 \leqslant 1\}$ is the closed unit disk) mapping ∂D monotonically onto γ and satisfying*

$$\Delta f = 0$$
$$f_x^2 - f_y^2 - 2i f_x \cdot f_y = 0$$

in the interior of D.

f, being harmonic and conformal, thus yields a minimal surface $f(D)$, *a priori* possibly with branch points. Interior branch points for this particular solution (to be constructed below) could later on be ruled out by Osserman [0], Gulliver [G], and Alt [Alt]. Except for a real analytic γ, treated in [GL], the question of whether this solution may have boundary branch points (for smooth γ) is still open.

Since f is harmonic, it is in particular real analytic in the interior of D. The question of boundary regularity again could only be solved much later by Hildebrandt [Hi1], with modifications or extensions obtained in [Ni1], [Ki],

[H6], [HT], [HH]. The result is that f is as regular as γ permits ($\gamma \in C^{k,\alpha}$, $k \geq 1$, $0 < \alpha < 1 \Rightarrow f \in C^{k,\alpha}$).

While the original arguments of Douglas and Radó were quite complicated, rather simple proofs later were given by Courant [C1] and McShane [Mc], each of them using (different) arguments originally invented by Lebesgue.

We define the *energy* of f as

$$E(f) := \tfrac{1}{2} \int_D |df|^2 \qquad (\text{for } f \in H^{1,2}(D, \mathbb{R}^d)).$$

We shall need the so-called Courant–Lebesgue lemma:

Lemma 1.1.1: *Let $f \in H^{1,2}(D, \mathbb{R}^d)$, $E(f) \leq K$, $\delta < 1$, $p \in D$. Then there exists some $r \in (\delta, \sqrt{\delta})$ for which $f|_{\partial B(p,r) \cap D}$ is absolutely continuous and*

$$|f(x_1) - f(x_2)| \leq (8\pi K)^{1/2} \left(\log \frac{1}{\delta} \right)^{-1/2} \tag{1.1.1}$$

for all $x_1, x_2 \in \partial B(p,r) \cap D$.

Here, we use the notation $B(p,r) := \{q : \text{dist}(p,q) \leq r\}$.

Proof: We use polar coordinates (ρ, φ) centred at p. Since f is a Sobolev function, for almost all r, $f|_{\partial B(p,r) \cap D}$ is absolutely continuous, and for all $x_1, x_2 \in \partial B(p,r) \cap D$

$$\begin{aligned} |f(x_1) - f(x_2)| &\leq \int_{(\rho,\varphi) \in \partial B(p,r) \cap D} |f_\varphi(x)| \, d\varphi \\ &\leq (2\pi)^{1/2} \left(\int |f_\varphi(x)|^2 \, d\varphi \right)^{1/2}. \end{aligned} \tag{1.1.2}$$

Now the energy of f on $B(p,r) \cap D$ is

$$E(f; B(p,r) \cap D) = \tfrac{1}{2} \int_{B(p,r) \cap D} \left(|f_\rho|^2 + \frac{1}{\rho^2} |f_\varphi|^2 \right) \rho \, d\rho \, d\varphi.$$

Hence, there exists some $r \in (\delta, \sqrt{\delta})$ with

$$\tfrac{1}{2} \int_{(\rho,\varphi) \in \delta B(p,r) \cap D} |f_\varphi|^2 \, d\varphi \leq \frac{K}{\int_\delta^{\sqrt{\delta}} \rho^{-1} \, d\rho} = \frac{2K}{\log(\delta^{-1})}. \tag{1.1.3}$$

Equation (1.1.1) then follows from (1.1.2) and (1.1.3).

q.e.d.

Proof of Theorem 1.1.1: We fix three distinct points x_1, x_2, $x_3 \in \partial D$ and three distinct points $p_1, p_2, p_3 \in \gamma$. Let

$$S_\gamma := \{f \in H^{1,2} \cap C^0(D, \mathbb{R}^d)\colon f \text{ maps } \partial D \text{ monotonically onto } \gamma, \\ f(x_i) = p_i \ (i=1,2,3)\}.$$

For simplicity we assume that S_γ is not empty, i.e. that γ bounds a surface of finite energy. (The general case can be handled by approximation; the argument is not difficult but not of interest for our present purpose.)

We let $(f_n)_{n\in N} \subset S_\gamma$ be a minimizing sequence for the energy E. We assume $E(f_n) \leqslant K$. After selection of a subsequence, f_n converges weakly in $H^{1,2}$ to a map f.

Nevertheless, the continuity properties of the f_n may be quite bad so that f_n need not converge uniformly. In order to overcome this difficulty we construct a new minimizing sequence by improving each f_n without increasing its energy. We actually want to discuss two different procedures to achieve this:

(1) We replace f_n by the *harmonic* map $u_n\colon D \to \mathbb{R}^d$ with $u_n|_{\partial D} = f_n|_{\partial D}$. Then $E(u_n) \leqslant E(f_n)$.

(2) Let $f \in S_\gamma$. We write $f = (f^1, \ldots, f^d)$. For each $i \in \{1, \ldots, d\}$, we perform the following construction. Let $(\sigma_k)_{k\in N}$ be dense in $\mathbb{R}$. For each component Ω_1 of $\{x \in D\colon f^i(x) \geqslant \sigma_1\}$ with $\Omega_1 \cap \partial D = \emptyset$, we replace f^i by σ_1 on Ω_1, and likewise for each component Ω_1' of $\{x \in D\colon f^i(x) \leqslant \sigma_1\}$ with $\Omega_1' \cap \partial D = \emptyset$. We obtain a map g_1^i. We repeat the procedure with σ_2 instead of σ_1 and g_1^i instead of f^i. Iteratively, we obtain maps g_k^i, with

$$E(g_k) \leqslant E(f)$$

where $g_k = (g_k^1, \ldots, g_k^d)$.

Also, as $k \to \infty$, g_k converges uniformly to a map g, with the following property: for each $B \subset D$, and each $i \in \{1, \ldots, d\}$

$$\min_B g^i = \min_{\partial B} g^i \qquad \text{and} \qquad \max_B g^i = \max_{\partial B} g^i. \tag{1.1.4}$$

Moreover $g|_{\partial D} = f|_{\partial D}$.

We thus perform this construction for each element f_n of our original minimizing sequence to get maps $u_n \in S_\gamma$ with

$$E(u_n) \leqslant E(f_n). \tag{1.1.5}$$

If we then replace f_n by u_n, constructed from either (1) or (2), we get an equicontinuous minimizing sequence. This is seen as follows. Let $\eta \in (0,1)$ be so small that each $B(x,\eta) \cap D$ $(x \in D)$ contains at most one of the three points x_1, x_2, x_3 fixed in the definition of S_γ. For each $x_0 \in D$ and $\delta \in (0, \eta^2)$, by Lemma 1.1.1, we can find $r_n \in (\delta, \sqrt{\delta})$, for which u_n $(\partial B(x_0, r_n) \cap D)$ is contained in a ball of radius $(8\pi K)^{1/2}[\log(\delta^{-1})]^{-1/2}$.

In the case $B(x_0, r_n) \cap \partial D \neq \emptyset$, the monotonicity of $u_n|_{\partial D}$ and the normalization $u_n(x_i) = p_i$ imply that u_n maps $\partial D \cap B(x_0, r_n)$ into the 'smaller' one of the two subarcs into which γ is divided by $u_n(y_1)$ and $u_n(y_2)$, where

$\partial D \cap \partial B(x_0, r_n) = \{y_1, y_2\}$, namely into the one that contains at most one of the points p_1, p_2, p_3.

Thus we can control the image of $\partial(D \cap B(x_0, r_n)$ under u_n and hence by the properties of u_n (the maximum principle for harmonic u_n resp. (1.1.4) for u_n constructed by (2)) we then control also the size of $u_n\,(B(x_0, r_n) \cap D)$, and this size goes to zero as $\delta \to 0$ (cf. Lemma 1.1.1). Since $r_n \geqslant \delta$, this implies equicontinuity of the u_n. After selection of a subsequence, u_n then converges weakly in $H^{1,2}$ and uniformly to a map $f \in S_\gamma$, and because of (1.1.5), and the lower semicontinuity of E under weak $H^{1,2}$ convergence, f minimizes E in the class S_γ.

Therefore, f is automatically harmonic. In order to show that f is also conformal, we proceed as follows.

First of all, we observe that f minimizes the energy among all maps g mapping ∂D monotonically onto γ, not necessarily satisfying the three-point normalization; namely, for each such g, there is a conformal automorphism $\tau: D \to D$, for which $g \circ \tau$ satisfies the three-point condition, and

$$E(g \circ \tau) = E(g) \qquad \text{(cf. Section 1.3 below).} \tag{1.1.6}$$

If now $\sigma_t: D \to D$ is a family of diffeomorphisms, depending differentiably on t, with $\sigma_0 = \mathrm{id}$, then $f \circ \sigma_t$ also maps ∂D monotonically onto γ; hence because of the minimizing property of f,

$$\frac{\mathrm{d}}{\mathrm{d}t} E(f \circ \sigma_t^{-1})|_{t=0} = 0. \tag{1.1.7}$$

Putting

$$\sigma_t = \xi + \mathrm{i}\eta \qquad \left.\frac{\partial \sigma_t}{\partial t}\right|_{t=0} = v + \mathrm{i}\omega$$

we have

$$E(f) = \tfrac{1}{2} \int_D (f_x^2 + f_y^2)\,\mathrm{d}x\,\mathrm{d}y$$

and

$$E(f \circ \sigma_t^{-1}) = \tfrac{1}{2} \int_D \{f_x^2(\xi_y^2 + \eta_y^2) - 2 f_x f_y (\xi_x \xi_y + \eta_x \eta_y) + f_y^2(\xi_x^2 + \eta_x^2)\}(\xi_x \eta_y - \xi_y \eta_x)^{-1}\,\mathrm{d}x\,\mathrm{d}y.$$

Equation (1.1.7) then yields (noting $\sigma_0 = x + \mathrm{i}y$, i.e. $\xi_x = \eta_y = 1$, $\xi_y = \eta_x = 0$ for $t = 0$)

$$\int_D \{(f_x^2 - f_y^2)(v_x - \omega_y) + 2 f_x \cdot f_y (v_y + \omega_x)\}\,\mathrm{d}x\,\mathrm{d}y = 0. \tag{1.1.8}$$

With

$$\varphi := f_x^2 - f_y^2 - 2\mathrm{i} f_x f_y$$

this is

$$\mathrm{Re} \int_D \varphi (v + \mathrm{i}\omega)_{\bar{z}}\,\mathrm{d}x\,\mathrm{d}y = 0. \tag{1.1.9}$$

Then, since in particular all v and ω with compact support in the interior of D are admissible, we conclude

$$\varphi_{\bar{z}} \equiv 0 \tag{1.1.10}$$

i.e. φ is holomorphic (this also follows from the fact that f is harmonic).

We choose a holomorphic function ψ on D with

$$\psi''(z) = \varphi(z).$$

Since $\varphi \in L^1(D)$, $\psi \in H^{2,1}(D)$; in particular, ψ is continuous on $\bar{D}$ by the Sobolev embedding theorem. We put

$$\gamma_r := \{|z| = r\} \qquad (r \leqslant 1).$$

Integrating (1.1.8) by parts, we obtain

$$\lim_{r\to 1} \int_{\gamma_r} \{(f_x - f_y)(v\,\mathrm{d}y + \omega\,\mathrm{d}x) + 2f_x\cdot f_y(\omega\,\mathrm{d}y - v\,\mathrm{d}x)\} = 0. \tag{1.1.11}$$

We then choose polar coordinates (ρ, φ) centred at 0, and near $\gamma_1 = \partial D$

$$v + \mathrm{i}\omega = \Lambda(\varphi)\,\mathrm{i}\,\mathrm{e}^{\mathrm{i}\varphi}.$$

Equation (1.1.11) gives

$$\lim_{r\to 1} \int_{\gamma_r} \Lambda(\varphi)\,\mathrm{Im}(\psi''((1-r)\,\mathrm{e}^{\mathrm{i}\varphi})\,\mathrm{e}^{2\mathrm{i}\varphi})\,\mathrm{d}\varphi = 0.$$

Since $\psi' \in H^{1,1}(D)$, ψ' has boundary values in $L^1(\partial B)$; hence we can put $r = 1$ after an integration by parts; thus (noting $\psi' = (\partial/\partial z)\psi$, but $\Lambda' = (\mathrm{d}/\mathrm{d}\varphi)\Lambda$)

$$\begin{aligned}
0 &= \mathrm{Im}\left(\int_{\gamma_1} \psi'(\mathrm{e}^{\mathrm{i}\varphi})\{\mathrm{i}\mathrm{e}^{\mathrm{i}\varphi}\Lambda'(\varphi) - \mathrm{e}^{\mathrm{i}\varphi}\Lambda(\varphi)\}\,\mathrm{d}\varphi\right)\\
&= \mathrm{Im}\left(\int_{\gamma_1} \psi'(\mathrm{e}^{\mathrm{i}\varphi})\,\mathrm{i}\mathrm{e}^{\mathrm{i}\varphi}\Lambda'(\varphi)\,\mathrm{d}\varphi\right)\\
&\quad - \mathrm{Im}\left(\int_{\gamma_1} \psi'(\mathrm{e}^{\mathrm{i}\theta})\,\mathrm{e}^{\mathrm{i}\theta}\left\{\Lambda(0) + \int_0^\theta \Lambda'(\varphi)\,\mathrm{d}\varphi\right\}\mathrm{d}\theta\right)\\
&= \mathrm{Im}\left(\int_0^{2\pi} \Lambda'(\varphi)\{\mathrm{i}\mathrm{e}^{\mathrm{i}\varphi}\psi'(\mathrm{e}^{\mathrm{i}\varphi}) - \mathrm{i}\psi(\mathrm{e}^{\mathrm{i}\varphi})\}\,\mathrm{d}\varphi\right)\\
&\quad + \mathrm{Im}\left(\mathrm{i}\Lambda(0)\int_0^{2\pi} \frac{\mathrm{d}}{\mathrm{d}\theta}\psi(\mathrm{e}^{\mathrm{i}\theta})\,\mathrm{d}\theta\right).
\end{aligned}$$

The last integral vanishes; since $\Lambda(\varphi)$ was arbitrary,

$$\mathrm{Im}(\mathrm{i}z\psi'(z) - \mathrm{i}\psi(z)) = \text{const. on } \partial D. \tag{1.1.12}$$

Consequently, $z\psi' - \psi$ is holomorphic across ∂D, and thus ψ is smooth across

∂D. Differentiating (1.1.12) w.r.t. φ, we get on ∂D

$$0 = \operatorname{Im}\left(iz\frac{d}{dz}(iz\psi' - i\psi) \right) = -\operatorname{Im}(z^2\varphi(z)).$$

Therefore, $z^2\varphi(z)$ can be reflected across ∂D to obtain a holomorphic function on S^2, which then is constant, and hence identically zero as it vanishes at $0 \in D$. Consequently, $\varphi \equiv 0$ which means that f is conformal. This completes the proof of Theorem 1.1.1.

q.e.d.

Let us discuss the principle arguments of the proof because similar ideas will be important for our subsequent presentation.

The Courant–Lebesgue lemma implies that for each element of a minimizing sequence we can always find circles of a controlled size, the images of which are arbitrary small. If we then improve the minimizing sequence, either by harmonic replacement or by the projections of (2), so that each element now satisfies a maximum principle, we obtain an equicontinuous minimizing sequence. A limit then, in particular, satisfies the imposed boundary condition. In our case, we have a Plateau boundary condition, meaning that ∂D can be mapped in an arbitrary (monotonic) way onto γ. This can be interpreted as a special case of a free boundary condition, and a limit f then is critical for the energy also w.r.t. composition with diffeomorphisms of D. Using suitable such diffeomorphisms, one concludes that for

$$\varphi := f_x^2 - f_y^2 - 2i f_x \cdot f_y$$

$z^2\varphi(z)$ is real on ∂D, hence identically zero, so that f is not only harmonic, but also conformal. Actually, because of the transformation properties, one should look at $\varphi\, dz^2$; this then is a holomophic quadratic differential which is real on ∂D, hence identically zero. We notice that this argument is no longer valid for a domain of higher genus or connectivity because a holomophic quadratic differential which is real on the boundary need not then vanish.

Moreover, for the proof of equicontinuity of an improved minimizing sequence, we also needed a three-point normalization on the boundary. This normalization did not impose any restriction on admissible variations of the limit f as it could always be achieved by composition with a conformal automorphisms of D that leaves the energy $E(f)$ invariant. On the other hand, without such a normalization, even an improved minimizing sequence need not be compact as the group of conformal automorphismus of D, namely $PL(2, \mathbb{R})$ is not compact. Later on, we shall encounter situations where such a normalization will no longer be possible, and then the conformal invariance of the variational problem together with the non-compactness of the conformal group will cause difficulties and actually lead to the generation of non-trivial solutions concentrated (in a sense to be made precise later on) at a point.

1.2. Two-dimensional conformally invariant variational problems

In order to clarify the scope of our subsequent results, we start with the following (easy) classification of two-dimensional conformally invariant variational problems, due to Grüter [Gr2][1]:

Theorem 1.2.1: *Let $F:\mathbb{R}^d \times \mathbb{R}^{2d} \to \mathbb{R}$ be of class C^1 w.r.t. the first d variables and of class C^2 w.r.t. the remaining $2d$ variables. Suppose*

$$\lambda|p|^2 \leqslant F(u,p) \leqslant \Lambda|p|^2 \qquad 0<\lambda\leqslant\Lambda<\infty \tag{1.2.1}$$

for all $(u,p)\in\mathbb{R}^d \times \mathbb{R}^{2d}$.

For $u\in H^{1,2}(\Omega,\mathbb{R}^d)$, Ω open in $\mathbb{R}^2$, we define

$$I(u):=\tfrac{1}{2}\int_\Omega F(u(x),Du(x))\,\mathrm{d}x$$

and we assume that I is conformally invariant, i.e.

$$I(u)=I(u\circ\varphi) \tag{1.2.2}$$

for every u and any conformal diffeomorphism $\varphi:\Omega'\to\Omega$. Then

$$F(u,p)=g_{ik}(u)p^ip^k + b_{ik}(u)\det(p^i,p^k)$$

with positive definite symmetric (g_{ik}) and skew-symmetric (b_{ik}) and

$$I(u)=\tfrac{1}{2}\int\{g_{ik}(u)\nabla u^i\cdot\nabla u^k + b_{ik}(u)\det(\nabla u^i,\nabla u^k)\}.$$

Proof: If $\varphi:\Omega'\to\Omega$ is a conformal diffeomorphism, we compute from (1.2.2)

$$\begin{aligned}\int F(u(\varphi(y)),Du(\varphi(y))\cdot D\varphi(y))\,\mathrm{d}y &= \int F(u(x),Du(x)\,\mathrm{d}x\\ &= \int F(u(\varphi(y)),Du(\varphi(y))\det(D\varphi)\,\mathrm{d}y\end{aligned}$$

and since this holds for all u,

$$F(u,Du\cdot D\varphi(y))=\det(D\varphi(y))F(u,Du). \tag{1.2.3}$$

Using homotheties $\varphi(x)=\mu x$, $\mu\neq 0$, thus

$$F(u,\mu p)=\mu^2F(u,p) \qquad (\text{with } p=Du).$$

Hence, since F is C^2 w.r.t. p, from Euler's theorem

$$F(u,p)=\tfrac{1}{2}F_{p^i_\alpha p^k_\beta}(u,0)p^i_\alpha p^k_\beta. \tag{1.2.4}$$

[1] A more general result was obtained by Kilpeläinen [Kp].

Let

$$A_{ik}^{\alpha\beta}(u) := F_{p_\alpha^i p_\beta^k}(u, 0) \qquad A_{ik} = (A_{ik}^{\alpha\beta})_{\alpha,\beta=1,2}$$

with

$$A_{ki}^{\beta\alpha} = A_{ik}^{\alpha\beta} \qquad \text{i.e. } A_{ki} = {}^t A_{ik}. \tag{1.2.5}$$

Then

$$I(u) = \tfrac{1}{2} \int {}^t\nabla^i u A_{ik}(u) \nabla^k u.$$

Using rotations $\varphi(x) = S(\theta)x$,

$$S(\theta) = \begin{pmatrix} \cos\theta & \sin\theta \\ -\sin\theta & \cos\theta \end{pmatrix}$$

from (1.2.3), (1.2.4)

$${}^t p^i A_{ik} p^k = {}^t p^i ({}^t S(\theta) A_{ik} S(\theta))^t p^k$$

and because of (1.2.5)

$$A_{ik}(u) = {}^t S(\theta) A_{ik}(u) S(\theta).$$

For $\theta = \pi/2$, we conclude that

$$A_{ik} = \begin{pmatrix} g_{ik} & b_{ik} \\ -b_{ik} & g_{ik} \end{pmatrix}$$

with (g_{ik}) symmetric, (b_{ik}) skew-symmetric. Thus

$$I(u) = \tfrac{1}{2} \int \{ g_{ik}(u) \nabla u^i \cdot \nabla u^k + b_{ik}(u) \det(\nabla u^i, \nabla u^k) \}. \tag{1.2.6}$$

Equation (1.2.1) implies that (g_{ik}) is positive definite.

q.e.d.

Remark: From the proof, we see that it already suffices to assume that I is invariant under homotheties and rotations.

Lemma 1.2.1: *The Euler–Lagrange equations of $I(u)$, given by* (1.2.6), *are*

$$\Delta u^i + \Gamma_{kl}^i \nabla u^k \cdot \nabla u^l = g^{im}(b_{mk,l} + b_{kl,m} + b_{lm,k}) \det(\nabla u^k, \nabla u^l) \tag{1.2.7}$$

with

$$(g^{ik}) := (g_{ik})^{-1} \qquad g_{kl,m} := \partial_m g_{kl} \qquad \left(\partial_m = \frac{\partial}{\partial u^m}\right)$$

$$\Gamma_{kl}^i := \tfrac{1}{2} g^{im}(g_{km,l} + g_{lm,k} - g_{lk,m}).$$

Proof: In general, the Euler–Lagrange equations are $(i = 1, \ldots, d)$:

$$D_1(F_{p_1^i}(u, Du)) + D_2(F_{p_2^i}(u, Du)) = F_{u^i}(u, Du).$$

In our case

$$2g_{ik}\Delta u^k + (\partial_l g_{ik} - \partial_i g_{kl} + \partial_k g_{li})\nabla u^k \cdot \nabla u^l = (\partial_l b_{ik} + \partial_i b_{kl} + \partial_k b_{li})\det(\nabla u^k, \nabla u^l)$$

and (1.2.7) follows.

q.e.d.

Let us now discuss examples of variational integrals of the form (1.2.6). We first treat the case where $b_{ik} = 0$.

Let Σ be a Riemann surface, i.e. a two-dimensional differentiable manifold with a (local) conformal structure; for the sake of generality, we also include non-orientable surfaces; also possibly $\partial\Sigma \neq \varnothing$.

Let $\lambda^2 \, dz \, d\bar{z}$ be a conformal metric on Σ. Let N be a complete Riemannian manifold of dimension d; let its metric in local coordinates be given by (g_{ik}), with Christoffel symbols Γ^i_{kl} as above.

Definition 1.2.1: For $u \in H^{1,2}(\Sigma, N)$, we define the energy density as

$$e(u)(z) = \frac{1}{\lambda^2} g_{ik}(u^i_x u^k_x + u^i_y u^k_y) \tag{1.2.8}$$

and the energy as

$$E(u) = \tfrac{1}{2}\int e(u)\lambda^2 \, dz \, d\bar{z}$$

$$= \tfrac{1}{2}\int g_{ik}(u^i_x u^k_x + u^i_y u^k_y) \, dx \, dy. \tag{1.2.9}$$

A solution of the corresponding Euler–Lagrange equations

$$\Delta u^i + \Gamma^i_{kl}(u^k_x u^l_x + u^k_y u^l_y) = 0 \qquad (i = 1, \ldots, d) \tag{1.2.10}$$

is called harmonic.

If u is in addition conformal, i.e. if

$$g_{jk}(u^j_x u^k_x - u^j_y u^k_y - 2i u^j_x u^k_y) = 0 \tag{1.2.11}$$

then it is called a (parametric) minimal surface in N.

Of course, the energy is a generalization of the Dirichlet integral

$$D(u) = \tfrac{1}{2}\int |\nabla u|^2 \qquad \text{for } u\colon \Omega \to \mathbb{R}^d, \Omega \subset \mathbb{R}^2.$$

Definition 1.2.2: The H-surface functional is given by

$$I(u) = \tfrac{1}{2}\int \{|\nabla u|^2 + Q(u)(D_1 u \wedge D_2 u)\} \qquad (u\colon \Omega \to \mathbb{R}^3, \Omega \subset \mathbb{R}^2) \tag{1.2.12}$$

where

$$\operatorname{div} Q(u) = 4H(u)$$

and in our previous notation

$$Q^1(u) = 2b_{23}(u) \qquad Q^2(u) = -2b_{13}(u) \qquad Q^3(u) = 2b_{12}(u).$$

A conformal solution of the Euler equations, namely

$$\Delta u = 2H(u)(D_1 u \wedge D_2 u) \tag{1.2.13}$$

is called a (parametric) H-surface (in $\mathbb{R}^3$).

More generally, an H-surface in the three-dimensional Riemannian manifold N with metric (g_{ik}) is a conformal solution of

$$\Delta u^i + \Gamma^i_{kl} \nabla u^k \nabla u^l = 2H\sqrt{g}\, g^{ik}(D_1 u \wedge D_2 u)^k \tag{1.2.14}$$

with $g = \det(g_{ik})$ and

$$H(u) = \left(\frac{1}{4g(u)}\right)^{1/2} (\partial_1 b_{23}(u) + \partial_2 b_{31}(u) + \partial_3 b_{12}(u)). \tag{1.2.15}$$

Remark: Actually, the general equation (1.2.7) can also be interpreted as an H-equation in a Riemannian manifold N with metric (g_{ik}). We define a 2-form ω on N via

$$\omega(u) := \tfrac{1}{2} b_{ik}(u)\, du^i \wedge du^k$$

and H as the unique $(2, 1)$ tensor satisfying for arbitrary vectors U, V, W

$$d\omega(U, V, W) = g_{ik} U^i H^k(V, W).$$

In this notation,

$$I(u) = E(u) + 2\int \omega(D_1 u, D_2 u) \tag{1.2.16}$$

and (1.2.7) becomes

$$\nabla u^i + \Gamma^i_{kl} \nabla u^k \nabla u^l = 2H^i(D_1 u, D_2 u). \tag{1.2.17}$$

In our previous notation,

$$H^i = \tfrac{1}{4} g^{ij}(\partial_l b_{jk} + \partial_j b_{kl} + \partial_k b_{lj})\, du^k \wedge du^l.$$

We note that for a conformal solution of (1.2.13), (1.2.14), or (1.2.17), $H(u)$ is the mean curvature of the surface described by u (see (1.4.11)). If u is not conformal, this is no longer true, however.

Thus, two-dimensional conformally invariant variational problems give rise to:

- conformal maps between surfaces
- (parametric) minimal surfaces in Riemannian manifolds
- harmonic maps from a surface into a Riemannian manifold
- solutions of the H-surface equations in Riemannian manifolds

the last possibility being the most general case that can occur.

Definition 1.2.3: A two-dimensional geometric variational problem is a two-dimensional conformally invariant variational problem (satisfying the assumptions of Theorem 1.2.1).

Lemma 1.2.2: *Suppose u is a solution of* (1.2.7). *Then*

$$\varphi(z)\,\mathrm{d}z^2 := g_{jk}(u)(u_x^j u_x^k - u_y^j u_y^k - 2\mathrm{i}u_x^j u_y^k)\,\mathrm{d}z^2$$

is a holomorphic quadratic differential. $\varphi \equiv 0 \Leftrightarrow u$ *is conformal.*

Proof: We compute (using symmetries and renaming indices in between)

$$\begin{aligned}
&\left(\frac{\partial}{\partial x} + \mathrm{i}\frac{\partial}{\partial y}\right)(g_{jk}(u)(u_x^j u_x^k - u_y^j u_y^k - 2\mathrm{i}u_x^j u_y^k)) \\
&= 2g_{jk}u_x^k(u_{xx}^j + u_{yy}^j) + g_{jk,l}u_x^k u_x^j u_x^l - g_{jl,k}u_x^k u_y^j u_y^l + 2g_{lk,j}u_x^k u_y^l u_y^j \\
&\quad - \mathrm{i}[2g_{jk}u_y^k(u_{xx}^j + u_{yy}^j) + 2g_{jk,l}u_y^k u_x^l u_y^j - g_{jl,k}u_y^k u_x^j u_x^l + g_{jl,k}u_y^k u_y^j u_y^l] \\
&= 2g_{jk}(u_x^k - \mathrm{i}u_y^k)\{\Delta u^j + \tfrac{1}{2}\Gamma_{lm}^j(u_x^l u_x^m + u_y^l u_y^m)\} \\
&= 2g_{jk}(u_x^k - \mathrm{i}u_y^k)\{g^{jm}(b_{mn,l} + b_{nl,m} + b_{lm,n})(u_x^n u_y^l - u_y^n u_x^l)\} \qquad \text{because of (1.2.7)} \\
&= 2(u_x^k - \mathrm{i}u_y^k)(b_{kn,l} + b_{nl,k} + b_{lk,n})(u_x^n u_y^l - u_y^n u_x^l) \\
&= 0 \qquad \text{because of the skew-symmetry of } (b_{ik}).
\end{aligned}$$

q.e.d.

We shall also need the concept of a critical point of $I(u)$, or of a weak solution of (1.2.7).

We find it preferable to use an intrinsic formulation, and for this we first state some geometric terminology. Let $\Omega \subset \mathbb{R}^2$, $z = (z^1, z^2) \in \Omega$. Let N^d again be a Riemannian manifold, with Levi–Civita connection ∇ and scalar product $\langle \cdot, \cdot \rangle$ in the tangent bundle TN.

For $u \in H^{1,2}(\Omega, N)$, du is a 1-form with values in $u^{-1}(TN)$, defined almost everywhere on Ω.

Since the cotangent bundle $T^*\Omega$ is Euclidean, we shall also use the notation $\langle \cdot, \cdot \rangle$ for the scalar product in $T^*\Omega \otimes u^{-1}(TN)$. The energy of u is then

$$E(u) = \tfrac{1}{2}\int_\Omega \langle \mathrm{d}u, \mathrm{d}u \rangle.$$

We let φ be a vector field along u, i.e. a section of $u^{-1}(TN)$. In local coordinates

$$\varphi = \varphi^i(z)\frac{\partial}{\partial u^i}.$$

and

$$\mathrm{d}\varphi = \nabla_{\partial/\partial z^\alpha}\left(\varphi^i \frac{\partial}{\partial u^i}\right)\mathrm{d}z^\alpha \tag{1.2.18}$$

$$= \frac{\partial \varphi^i}{\partial z^\alpha}\frac{\partial}{\partial u^i} \otimes \mathrm{d}z^\alpha + \varphi^i \Gamma_{ij}^k \frac{\partial u^j}{\partial z^\alpha}\frac{\partial}{\partial u^k} \otimes \mathrm{d}z^\alpha$$

is then a section of $T^*\Omega \otimes u^{-1}(TN)$.

We also need the product of φ with the right-hand side of (1.2.7) (for simplicity, $d=3$; cf (1.2.14)):

$$\varphi(u_{z_1}\wedge u_{z_2})=\sqrt{g}\varphi^k(u_{z_1}\wedge u_{z_2})^k$$

in local coordinates.

We shall now assume that φ has compact support in Ω, that φ is bounded and that

$$\int\langle \mathrm{d}\varphi,\mathrm{d}\varphi\rangle<\infty. \tag{1.2.19}$$

(In symbols: $\varphi\in H_0^{1,2}\cap L^\infty(\Omega,u^{-1}(TN))$.)

In all applications, $\varphi(z)$ will be of the form

$$\eta(z)\lambda(u(z)) \tag{1.2.20}$$

where $\eta\in\mathrm{Lip}_0(\Omega,\mathbb{R})=\{$compactly supported Lipschitz functions on $\Omega\}$, and $\lambda(u)$ is a C^1-section of TN. φ then induces a variation

$$u_t(z):=\exp_{u(z)}(t\varphi(z)) \tag{1.2.21}$$

of u. We compute

$$\frac{\mathrm{d}}{\mathrm{d}t}E(u_t)|_{t=0}=\int_\Omega\langle \mathrm{d}u,\mathrm{d}\varphi\rangle. \tag{1.2.22}$$

Definition 1.2.4: u is a weak solution of (1.2.10) (or a weakly harmonic map), if

$$\int_\Omega\langle \mathrm{d}u,\mathrm{d}\varphi\rangle=0 \tag{1.2.23}$$

for all $\varphi\in H_0^{1,2}\cap L^\infty(\Omega,u^{-1}(TN))$.

Likewise, u is a weak solution of (1.2.14) if

$$\int_\Omega\langle \mathrm{d}u,\mathrm{d}\varphi\rangle=-\int 2H(u)\varphi\cdot(u_{z_1}\wedge u_{z_2}) \tag{1.2.24}$$

for all such φ.

More generally, a weak solution of (1.2.17) has to satisfy for all such φ

$$\int_\Omega\langle \mathrm{d}u,\mathrm{d}\varphi\rangle=-\int 2H(u_{z_1},u_{z_2})\cdot\varphi. \tag{1.2.25}$$

A weakly harmonic non-constant u which is also weakly conformal, i.e.

$$\langle u_x,u_x\rangle-\langle u_y,u_y\rangle-2\mathrm{i}\langle u_x,u_y\rangle=0$$

almost everywhere in Ω (note that since $u\in H^{1,2}$, this expression is only defined almost everywhere), is called a weak minimal surface.

A non-constant weakly conformal solution of (1.2.25) is called a weak H-surface.

Let us relate this definition to the standard formulation of weak solution of the Euler–Lagrange equations w.r.t. local coordinates:

Lemma 1.2.3: *Equation* (1.2.23) *is equivalent to*

$$\int_\Omega (u^i_{z^\alpha}\psi^i_{z^\alpha} - \Gamma^i_{jk}u^j_{z^\alpha}u^k_{z^\alpha}\psi^i) = 0 \tag{1.2.26}$$

for $\psi^i = g_{ij}\varphi^j \in H^{1,2}_0 \cap L^\infty(\Omega)$. *Likewise,* (1.2.24) *is equivalent to*

$$\int_\Omega (u^i_{z^\alpha}\psi^i_{z^\alpha} - \Gamma^i_{jk}u^j_{z^\alpha}u^k_{z^\alpha}\psi^i) = -\int 2H(u)\sqrt{g}g^{ik}\psi^i(u_{z_1}\wedge u_{z_2})^k. \tag{1.2.27}$$

Proof: With $\phi = \phi^i(\partial/\partial u^i)$ and

$$\mathrm{d}\phi = \nabla_{\partial/\partial z^\alpha}\left[\phi^i\frac{\partial}{\partial u^i}\right]\mathrm{d}z^\alpha = \frac{\partial\phi^i}{\partial z^\alpha}\frac{\partial}{\partial u^i} + \phi^i\Gamma^k_{ij}\frac{\partial u^j}{\partial z^\alpha}\frac{\partial}{\partial u^k}$$

we compute

$$\langle \mathrm{d}u, \mathrm{d}\phi\rangle = g_{ij}\gamma^{\alpha\beta}\frac{\partial u^i}{\partial z^\alpha}\frac{\partial\phi^j}{\partial z^\beta} + \gamma^{\alpha\beta}g_{ik}\phi^l\Gamma^k_{lj}\frac{\partial u^j}{\partial z^\beta}\frac{\partial u^i}{\partial z^\alpha}. \tag{1.2.28}$$

On the other hand, we choose $\psi^i = g_{ij}\phi^j$ as a test vector in (1.2.26). Then the integrand of (1.2.26) becomes

$$\gamma^{\alpha\beta}g_{ij}\frac{\partial u^i}{\partial z^\alpha}\frac{\partial\phi^j}{\partial z^\beta} + \gamma^{\alpha\beta}g_{kj,l}\frac{\partial u^l}{\partial z^\beta}\frac{\partial u^k}{\partial z^\alpha}\phi^j - \gamma^{\alpha\beta}g_{ij}\Gamma^i_{kl}\frac{\partial u^k}{\partial z^\alpha}\frac{\partial u^l}{\partial z^\beta}\phi^j$$

$$= \gamma^{\alpha\beta}g_{ij}\frac{\partial u^i}{\partial z^\alpha}\frac{\partial\phi^j}{\partial z^\beta} + \gamma^{\alpha\beta}\tfrac{1}{2}(g_{kj,l} + g_{kl,j} - g_{jl,k})\frac{\partial u^l}{\partial z^\beta}\frac{\partial u^k}{\partial z^\alpha}\phi^j$$

which after changing some indices, is the same as (1.2.28). The computation for (1.2.27) is similar.

q.e.d.

We can also use the preceding formalism to write the Euler–Lagrange equation for harmonic maps in intrinsic notation; the result is that u is harmonic if

$$\operatorname{trace}\nabla\,\mathrm{d}u = 0$$

where ∇ is the covariant derivative in $T^*\Omega\otimes u^{-1}(TN)$. This is seen as follows, writing $z = x^1 + \mathrm{i}x^2$ as a coordinate on Ω and summing Greek indices from 1 to 2:

$$\nabla_{\partial/\partial x^\beta}(\mathrm{d}u) = \nabla_{\partial/\partial x^\beta}\left(\frac{\partial u^i}{\partial x^\alpha}\frac{\partial}{\partial u^i}\otimes\mathrm{d}x^\alpha\right)$$

$$= \frac{\partial}{\partial x^\beta}\left(\frac{\partial u^i}{\partial x^\alpha}\right)\frac{\partial}{\partial u^i}\otimes\mathrm{d}x^\alpha + \left(\nabla^{u^{-1}(TN)}_{\partial/\partial x^\beta}\frac{\partial}{\partial u^i}\right)\frac{\partial u^i}{\partial x^\alpha}\otimes\mathrm{d}x^\alpha$$

(since $T^*\Omega$ is Euclidean, hence $\nabla^{T^*\Omega}_{\partial/\partial x} dx^\alpha = 0$) and thus:

$$= \frac{\partial^2 u^i}{\partial x^\alpha \partial x^\beta} \frac{\partial}{\partial u^i} \otimes dx^\alpha + \Gamma^k_{ij} \frac{\partial u^j}{\partial x^\beta} \frac{\partial u^i}{\partial x^\alpha} \frac{\partial}{\partial u^k} \otimes dx^\alpha$$

and thus

$$(\text{trace}\, \nabla\, du)^i = \Delta u^i + \Gamma^i_{jk} \frac{\partial u^j}{\partial x^\alpha} \frac{\partial u^k}{\partial x^\alpha}.$$

With the preceding notations, we can also calculate the Hessian of a harmonic map u for vector fields v, w along u (i.e. v and w are sections of $u^{-1}(TN)$). For this purpose, we consider a smooth two-parameter variation u_{st} with $u_{00} = u$ and compact support:

$$v := \left.\frac{\partial u_{st}}{\partial s}\right|_{s,t=0} \qquad w := \left.\frac{\partial u_{st}}{\partial t}\right|_{s,t=0}.$$

We then want to calculate

$$H_u(v, w) := \left.\frac{\partial^2 E(u_{st})}{\partial s \partial t}\right|_{s,t=0}.$$

We have, writing u instead of u_{st}, and taking scalar products $\langle \cdot, \cdot \rangle$ in $T^*\Omega \otimes u^{-1}(TN)$, if not otherwise indicated,

$$\begin{aligned}
&\frac{\partial}{\partial t} \frac{\partial}{\partial s} \frac{1}{2} \left\langle \frac{\partial u}{\partial x^\alpha} dx^\alpha, \frac{\partial u}{\partial x^\beta} dx^\beta \right\rangle \\
&= \frac{\partial}{\partial t} \left\langle \nabla_{\partial/\partial s} \frac{\partial u}{\partial x^\alpha} dx^\alpha, \frac{\partial u}{\partial x^\beta} dx^\beta \right\rangle \\
&= \frac{\partial}{\partial t} \left\langle \nabla^{u^{-1}(TN)}_{\partial/\partial x^\alpha} \left(\frac{\partial u}{\partial s}\right) dx^\alpha, \frac{\partial u}{\partial x^\beta} dx^\beta \right\rangle \\
&= \left\langle \nabla_{\partial/\partial t} \nabla^{u^{-1}(TN)}_{\partial/\partial x^\alpha} \left(\frac{\partial u}{\partial s}\right) dx^\alpha, \frac{\partial u}{\partial x^\beta} dx^\beta \right\rangle \\
&\quad + \left\langle \nabla^{u^{-1}(TN)}_{\partial/\partial x^\alpha} \left(\frac{\partial u}{\partial s}\right) dx^\alpha, \nabla^{u^{-1}(TN)}_{\partial/\partial x^\beta} \left(\frac{\partial u}{\partial t}\right) dx^\beta \right\rangle \\
&= \left\langle \nabla^{u^{-1}(TN)}_{\partial/\partial x^\alpha} \nabla_{\partial/\partial t} \left(\frac{\partial u}{\partial s}\right) dx^\alpha, \frac{\partial u}{\partial x^\beta} dx^\beta \right\rangle \\
&\quad + \left\langle R^N \left[\frac{\partial u}{\partial t}, \frac{\partial u}{\partial x^\alpha} dx^\alpha\right] \frac{\partial u}{\partial s}, \frac{\partial u}{\partial x^\beta} dx^\beta \right\rangle \\
&\quad + \left\langle \nabla^{u^{-1}(TN)}_{\partial/\partial x^\alpha} v \, dx^\alpha, \nabla^{u^{-1}(TN)}_{\partial/\partial x^\beta} w \, dx^\beta \right\rangle. \qquad (1.2.29)
\end{aligned}$$

(R^N denotes the curvature tensor of N.) Now:

$$\int_\Omega \left\langle \nabla^{u^{-1}(TN)}_{\partial/\partial x^\alpha} \nabla_{\partial/\partial t} \frac{\partial u}{\partial s} \mathrm{d}x^\alpha, \frac{\partial u}{\partial x^\beta} \mathrm{d}x^\beta \right\rangle$$

$$= \int \frac{\partial}{\partial x^\alpha}\left(\left\langle \nabla_{\partial/\partial t} \frac{\partial u}{\partial s}, \frac{\partial u}{\partial x^\alpha} \right\rangle_{u^{-1}(TN)}\right) \mathrm{d}x^1\,\mathrm{d}x^2$$

$$- \int \left\langle \nabla_{\partial/\partial t} \frac{\partial u}{\partial s} \mathrm{d}x^\alpha, \nabla_{\partial/\partial x^\alpha} \frac{\partial u}{\partial x^\beta} \mathrm{d}x^\beta \right\rangle$$

$$= - \int \left\langle \nabla_{\partial/\partial t} \frac{\partial u}{\partial s}, \nabla_{\partial/\partial x^\alpha} \frac{\partial u}{\partial x^\alpha} \right\rangle_{u^{-1}(TN)} \qquad \text{by Stokes' Theorem}$$

$$= 0 \qquad \text{since } \nabla_{\partial/\partial x^\alpha}(\partial u/\partial x^\alpha) = \operatorname{trace} \nabla\,\mathrm{d}u = 0, \text{ as } u \text{ is harmonic.}$$

Thus

$$H_u(v,w) = \int_\Omega \left\langle \nabla^{u^{-1}(TN)}_{\partial/\partial x^\alpha} v, \nabla^{u^{-1}(TN)}_{\partial/\partial x^\beta} w \right\rangle_{u^{-1}(TN)} - \int_\Omega \left\langle R^N\left(v, \frac{\partial u}{\partial x^\alpha}\right)\frac{\partial u}{\partial x^\alpha}, w \right\rangle_{u^{-1}(TN)}$$

$$= \int_\Omega \left\langle \nabla^{u^{-1}(TN)} v, \nabla^{u^{-1}(TN)} w \right\rangle_{u^{-1}(TN)} - \int_\Omega \operatorname{trace} \left\langle R^N(v, \mathrm{d}u)\,\mathrm{d}u, w \right\rangle_{u^{-1}(TN)}.$$

In particular H_u is positive definite if N has negative curvature.

We shall also need the following definition:

Definition 1.2.5: $u \in H^{1,2}(\Omega, N)$, $\Omega \subset \mathbb{R}^2$ is called stationary w.r.t. variations of the independent variables for $I(u)$, if

$$\frac{\mathrm{d}}{\mathrm{d}t} I(u \circ \sigma_t)\bigg|_{t=0} = 0 \qquad (1.2.30)$$

for every family of diffeomorphisms σ_t: $\Omega \to \Omega$, depending differentiably on t, with $\sigma_0 = \mathrm{id}$,

$$\sigma_t(z) = \mathrm{id} \qquad \text{for } z \in \Omega \setminus K,\ K \subset \Omega \text{ compact.} \qquad (1.2.31)$$

We note that (1.2.30) is always satisfied if u is a (classical) solution of (1.2.7); it is not clear whether this also holds for weak solutions.

The proof of the next lemma, which generalizes Lemma 1.2.2, will also show that the derivative in (1.2.30) always exists.

Lemma 1.2.4: *Let u be stationary w.r.t. variations of the independent variables for $I(u)$. Then*

$$(\langle u_x, u_x\rangle - \langle u_y, u_y\rangle - 2\mathrm{i}\langle u_x, u_y\rangle)\,\mathrm{d}z^2$$
$$(= g_{jk}(u^j_x u^k_x - u^j_y u^k_y - 2\mathrm{i} u^j_x u^k_y)\,\mathrm{d}z^2)$$

is holomorphic ($z = x + \mathrm{i}y \in \Omega$).

Proof: We shall only treat the case of the energy integral (1.2.9) since an additional H-term can be handled in the same way as in the proof of Lemma 1.2.2.

As in the proof of Theorem 1.1.1, we put

$$\sigma_t = \xi + i\eta \qquad \left.\frac{\partial \sigma_t}{\partial t}\right|_{t=0} = v + i\omega.$$

From

$$E(u) = \tfrac{1}{2}\int_\Omega (\langle u_x, u_x\rangle + \langle u_y, u_y\rangle)\,dx\,dy$$

we compute

$$E(u\circ\sigma_t^{-1}) = \tfrac{1}{2}\int_\Omega \{\langle u_x, u_x\rangle(\xi_y^2+\eta_y^2) - 2\langle u_x, u_y\rangle(\xi_x\xi_y+\eta_x\eta_y)$$
$$+\langle u_y, u_y\rangle(\xi_x^2+\eta_x^2)\}(\xi_x\eta_y-\xi_y\eta_x)^{-1}\,dx\,dy. \tag{1.2.32}$$

This expression is differentiable w.r.t. t, and (1.2.30) implies

$$(\xi_x = \eta_y = 1,\ \xi_y = \eta_x = 0 \text{ for } t=0, \text{ as } \sigma_0 = x + iy)$$

that

$$\int_\Omega \{(\langle u_x, u_x\rangle - \langle u_y, u_y\rangle)(v_x - \omega_y) + 2\langle u_x, u_y\rangle(v_y+\omega_x)\}\,dx\,dy = 0. \tag{1.2.33}$$

With

$$\varphi := \langle u_x, u_x\rangle - \langle u_y, u_y\rangle - 2i\langle u_x, u_y\rangle.$$

(1.2.33) reads

$$\operatorname{Re}\int_\Omega \varphi(v+i\omega)_{\bar z}\,dx\,dy = 0. \tag{1.2.34}$$

Since all v and ω with compact support in the interior of Ω are admissible, we conclude (from Weyl's lemma) that φ is holomorphic:

$$\varphi_{\bar z} \equiv 0.$$

q.e.d.

Moreover, we shall need the following global result:

Lemma 1.2.5: *Let S be a compact surface with smooth boundary ∂S, u: $S \to N$, and suppose*

$$\frac{d}{dt} I(u\circ\sigma_t)|_{t=0} = 0 \tag{1.2.35}$$

for all families of diffeomorphismus σ_t: $S \to S$, depending differentiably on t, with $\sigma_0 = \mathrm{id}$ (but not necessarily satisfying (1.2.31)).

Then ($z = x + iy$ *a local conformal parameter on* S)

$$\varphi\,dz^2 := \langle u_x, u_x\rangle - \langle u_y, u_y\rangle - 2i\langle u_x, u_y\rangle$$

is a holomorphic quadratic differential which is real on ∂S. (*In a particular,* φ *is smooth up to the boundary* ∂S). *If under these assumptions,* $S = D$ *or* $S = S^2$, *then* $\varphi \equiv 0$, *i.e.* u *is (weakly) conformal.*

Proof: Again, we shall only treat the case of the energy integral $E(u)$. We shall continue to use the notations of the preceding proof. Lemma 1.2.4 implies that $\varphi\,dz^2$ is a holomorphic quadratic differential.

We represent a neighbourhood of ∂S conformally as an annulus A in the plane, with outer boundary $\{(x, y)\in\mathbb{R}^2 : x^2 + y^2 = 1\}$ corresponding to ∂S. Let

$$v + i\omega = \Lambda(\theta)ie^{i\theta} \qquad \text{near}\,\{x^2 + y^2 = 1\}\ (\text{writing } x + iy = re^{i\theta}) \tag{1.2.36}$$

and let it vanish near the other boundary curve of A. We put $\sigma_0(x, y) = x + iy$ and

$$\frac{d}{dt}\sigma_t(x, y) = (v + i\omega).$$

By assumption

$$\frac{d}{dt}E(u\circ\sigma_t^{-1})|_{t=0} = 0.$$

By the proof of Lemma 1.2.4, this implies

$$\mathrm{Re}\int_\Omega \varphi(v + i\omega)_{\bar z}\,dx\,dy = 0 \tag{1.2.37}$$

or after integrating by parts, putting $\gamma_r := \{x^2 + y^2 = r^2\}$

$$\lim_{r\to 1}\int_{\gamma_r}\{(\mathrm{Re}\,\varphi)(v\,dy + \omega\,dx) - (\mathrm{Im}\,\varphi)(\omega\,dy - v\,dx)\} = 0. \tag{1.2.38}$$

We can find $b, c\in\mathbb{C}$ with

$$\int_{\gamma_r}\left(\varphi + \frac{b}{z} + \frac{c}{z^2}\right)dz = 0 \tag{1.2.39}$$

and

$$\int_{\gamma_r} z\left(\varphi + \frac{b}{z} + \frac{c}{z^2}\right) = 0 \quad (\text{for } r \text{ close to } 1).$$

Hence, there exists an analytic function ψ on A with

$$\psi''(z) = \varphi(z) + \frac{b}{z} + \frac{c}{z^2}.$$

Since $\varphi \in L^1(B)$, ψ is continuous on $\bar{A}$. Equation (1.2.39) with $v + i\omega$ as in (1.2.36) yields

$$0 = \lim_{r\to 1} \int_{\gamma_r} \Lambda(\theta)[\mathrm{Im}(\psi''(re^{i\theta})e^{2i\theta}) + \mathrm{Im}(-be^{i\theta}) + \mathrm{Im}(-c)]\,d\theta.$$

Since $\varphi \in L^1(B)$, ψ' has boundary values in $L^1(\partial B)$, so that we can put $r = 1$ after integrating by parts. We obtain (with $\psi' = (\partial/\partial z)\psi$, but $\Lambda' := (d/d\theta)\Lambda$)

$$\begin{aligned}
&-\mathrm{Im}\int_{\gamma_1} \Lambda'(\theta) ibe^{i\theta}\,d\theta - \mathrm{Im}\int_{\gamma_1} \Lambda'(\theta) ic\log(e^{i\theta})\,d\theta \\
&= \mathrm{Im}\int_{\gamma_1} \psi'(ie^{i\theta}\Lambda'(\theta) - e^{i\theta}\Lambda(\theta))\,d\theta \\
&= \mathrm{Im}\int_{\gamma_1} \psi' ie^{i\theta}\Lambda'(\theta)\,d\theta - \mathrm{Im}\int_{\gamma_1} \psi' e^{i\lambda}\left(\Lambda(0) + \int_0^{\lambda} \Lambda'(\theta)\,d\theta\right) d\lambda \\
&= \mathrm{Im}\int_0^{2\pi} \Lambda'(\theta)(ie^{i\theta}\psi'(e^{i\theta}) - i\psi(e^{i\theta}))\,d\theta + \mathrm{Im}\left(i\Lambda(0)\int_0^{2\pi} \frac{d}{d\lambda}\psi(e^{i\lambda})\,d\lambda\right)
\end{aligned}$$

in which the last integral vanishes. Since $\Lambda(\theta)$ was arbitrary, we conclude

$$\mathrm{Im}(iz\psi'(z) - i\psi(z) + ibz + ic\log z) = \text{const.} \quad \text{on } \gamma_1. \tag{1.2.40}$$

Hence $iz\psi'(z) - i\psi(z) - bz$ is analytic along γ_1, and hence ψ is smooth along γ_1. Differentiating (1.2.40) w.r.t. θ, we obtain on γ_1

$$\begin{aligned}
0 &= \mathrm{Im}\left(iz\frac{d}{dz}(iz\psi' - i\psi + ibz + ic\log z)\right) \\
&= \mathrm{Im}(-z^2\varphi(z)) \quad \text{on } \gamma_1.
\end{aligned}$$

Therefore, $z^2\varphi(z)$ can be analytically continued across γ_1 as a holomorphic function. This implies that φ is smooth up to the boundary.

Also, $\varphi\, dz^2$ is real on ∂S. The last claim follows from Lemma 1.4.1 below.

q.e.d

Finally we want to define weak solutions of free boundary problems.

Definition 1.2.6: Let M be a submanifold of N, $u \in H^{1,2}(\Omega, N)$, Ω compact in $\mathbb{R}^2$. Then u is called a weakly harmonic map satisfying the free boundary condition defined by M, or shortly a weakly harmonic map with free boundary M, if

$$\int \langle du, d\varphi \rangle = 0 \tag{1.2.41}$$

whenever $\varphi \in H^{1,2} \cap L^\infty(\Omega, u^{-1}TN)$ satisfies

$$\varphi(z) \in T_{u(z)}M \subset T_{u(z)}N \tag{1.2.42}$$

for (almost all [2]) $z \in \partial\Omega$ with $u(z) \in M$ [3]. (Note that φ need not have compact support anymore.)

Likewise, if u satisfies (1.2.25) for such φ, it is called a weak solution of the H-system with free boundary M. If u satisfies (1.2.41) (for φ as before) and is in addition weakly conformal, i.e.

$$\langle u_x, u_x \rangle - \langle u_y, u_y \rangle - 2i\langle u_x, u_y \rangle = 0 \tag{1.2.43}$$

almost everywhere in Ω, it is called a weak minimal surface with free boundary M. A weak H-surface with free boundary M is defined similarly.

The following result is easily checked.

Lemma 1.2.6: *Let M be a C^1-submanifold of N, $u \in C^1(\bar{\Omega}, N)$ (Ω compact in $\mathbb{R}^2$) a nonconstant weakly harmonic map with free boundary M. Then*

$$u(\partial\Omega) \subset M$$

and if $\tau\colon T_vN \to T_vM$ ($v \in M$) denotes orthogonal projection, then

$$\tau(\mathrm{d}u(z)) = 0 \quad \text{for all } z \in \partial\Omega.$$

A more general result was proved by Jäger [Jä].

We also have:

Corollary 1.2.1: *If $u \in H^{1,2}(D, N)$ (D = unit disk in $\mathbb{R}^2$) is a weakly harmonic map with free boundary M and satisfies* (1.2.35), *then it is a weak minimal surface with free boundary M.*

Proof: This follows from Lemma 1.2.5.

q.e.d.

Concerning Definition 1.2.6, we point out again that the only variations we shall ever need are of the form

$$\eta(z)\lambda(u(z)) \tag{1.2.44}$$

with η a Lipschitz function and $\lambda(u)$ a smooth section of TN. Equation (1.2.42) then means

$$\lambda(u) \in T_uM \subset T_uN \qquad \text{whenever } u \in M. \tag{1.2.45}$$

[2] w.r.t. one-dimensional Lebesgue measure.

[3] One can also require that this holds for almost all $z \in \Omega$, and not only for $z \in \partial\Omega$, with $u(z) \in M$. Such a condition arises, when M is impenetrable for the surface $u(\Omega)$, and $u(\Omega)$ may then touch M. The continuity proof in Section 2.3 below works under this condition, too, with a slight modification: cf. [J11] for details.

1.3. Harmonic maps, conformal maps and holomorphic quadratic differentials

We again let Σ be a surface with a local conformal parameter $z = x + iy$.

Lemma 1.3.1: *For a map $f: \Sigma \to N$,*

$$\text{Area}(f(\Sigma)) \leqslant E(f) \tag{1.3.1}$$

(in case these quantities are defined), with equality if and only if f is (almost everywhere) $\pm$ conformal. Here, $\text{Area}(f(\Sigma))$ is counted with appropriate multiplicity.

Proof:

$$\text{Area}(f(\Sigma)) = \int_\Sigma \left\{ g_{ij}\left(\frac{\partial f^i}{\partial x}\frac{\partial f^j}{\partial x}\right) g_{kl}\left(\frac{\partial f^k}{\partial y}\frac{\partial f^l}{\partial y}\right) - \left[g_{ij}\left(\frac{\partial f^i}{\partial x}\frac{\partial f^j}{\partial y}\right)\right]^2 \right\}^{1/2} dx\,dy$$

and the result follows from the relation between the arithmetic and the geometric mean.

q.e.d.

We want to express all quantities on Σ in complex notation, i.e.

$$z = x + iy$$

$$\partial/\partial z = \tfrac{1}{2}(\partial/\partial x - i(\partial/\partial y)) \qquad \partial/\partial \bar{z} = \tfrac{1}{2}(\partial/\partial x) + i(\partial/\partial y))$$

$$dz = dx + i dy \qquad d\bar{z} = dx - i dy.$$

First of all, for $u: \Sigma \to N$

$$E(u) = \int_\Sigma g_{jk}(u_z^j u_{\bar{z}}^k + u_{\bar{z}}^j u_z^k) dz\, d\bar{z} \tag{1.3.2}$$

and u is harmonic, if

$$u_{z\bar{z}}^j + \Gamma_{kl}^j u_z^k u_{\bar{z}}^l = 0 \qquad (j = 1, \ldots, d). \tag{1.3.3}$$

We now recall some results from Section 1.2.

Lemma 1.3.2: *If $k: \Sigma_0 \to \Sigma$ is conformal of degree 1, then*

$$E(u \circ k) = E(u).$$

If u is harmonic and k conformal, then $u \circ k$ is harmonic. In particular, conformal maps between surfaces are harmonic.

If the image is a surface Σ' with local conformal metric $\rho^2\,du\,d\bar{u}$, then (1.3.3) becomes

$$\tau(u) := u_{z\bar{z}} + \frac{2\rho_u}{\rho} u_z u_{\bar{z}} = 0. \tag{1.3.4}$$

As already mentioned, harmonicity of u is independent of the domain metric but depends on the image metric unless u is conformal or anticonformal. (Note that this latter distinction is only meaningful for oriented surfaces; usually, 'conformal' will mean $\pm$ conformal, for simplicity.)

In the present situation, Lemma 1.2.2 becomes

Lemma 1.3.3: *If u: $\Sigma \to \Sigma'$ is harmonic, then*

$$\phi := \rho^2 u_z u_{\bar{z}}\,dz^2 \tag{1.3.5}$$

is a holomorphic quadratic differential.

$$\phi \equiv 0 \Leftrightarrow u \text{ is } \pm \text{ conformal.}$$

In the present case, the computation is much simpler than in the proof of Lemma 1.2.2, namely:

$$\begin{aligned}\phi_{\bar{z}} &= \{2\rho\rho_u u_{\bar{z}} u_z \bar{u}_z + 2\rho\rho_{\bar{u}} \bar{u}_{\bar{z}} u_z \bar{u}_z + \rho^2 u_{z\bar{z}} \bar{u}_z + \rho^2 u_z \bar{u}_{z\bar{z}}\}\,dz^2 \\ &= \rho^2(\bar{u}_z \tau(u) + u_z \bar{\tau}(u)) = 0 \qquad \text{(cf. (1.3.4)).}\end{aligned}$$

We also observe from this computation that if ϕ is holomorphic, then $\tau(u) = 0$, i.e. u is harmonic, with the possible exception of points where $|\bar{u}_z| = |u_z|$, i.e. where the Jacobian $|u_z|^2 - |\bar{u}_z|^2$ vanishes.

In particular:

Lemma 1.3.4: *Suppose ϕ (as in (1.3.5)) is holomorphic, and u: $\Sigma \to \Sigma'$ is of class C^2 or a diffeomorphism (of class C^1). Then u is harmonic.*

Occasionally, the holomorphicity of ϕ has been used as the definition of a harmonic map (cf. [GR], [Sh]).

This definition should be abandoned, however, as the existence theory for harmonic maps (in our sense) and maps with holomorphic ϕ is rather different, cf. Section 4.5. Here, we want to sketch examples of maps with holomorphic ϕ that are not harmonic:

Let Z_i, $i = 1, 2$, be circular cylinders with height b_i, circumference a_i, and boundary curves γ_{i1}, γ_{i2}. Let S^2 be the standard 2-sphere, and p_1 and p_2 be north and sourth poles, respectively. First, we take a differentiable map Ψ_1: $Z_1 \to S^2$ mapping γ_{11} onto p_1, γ_{12} onto p_2, and the interior of Z_1 diffeomorphically onto $S^2 \backslash \{p_1, p_2\}$. By the result of section 4.4 below, Ψ_1 is homotopic to a harmonic map u_1: $Z_1 \to S^2$ with the same boundary values. Let

v_2: $Z_2 \to S^2$ map γ_{21} onto p_1, γ_{22} onto p_2, and Z_2 onto a geodesic arc from p_1 to p_2. Then v_2 is harmonic, if independent of the angle and if the height parameter on Z_2 is proportional to the arc length on this geodesic (this is easily seen, since for a map which is independent of the angular coordinate, (1.2.10) reduces to the equation for geodesics.)

We now choose $a_1 = a_2$ and obtain a torus T by identifying γ_{11} with γ_{21} and γ_{12} with γ_{22}.

We let ϕ_1 be the holomorphic quadratic differential corresponding to u_1. As u_1 maps both boundary curves of Z_1 onto points, $\partial u_1/\partial\varphi = 0$ on ∂Z_1, where φ is the angular coordinate. If r is the height parameter,

$$\phi_1 = \rho^2\left[\frac{\partial u_1}{\partial r}\frac{\partial \bar{u}_1}{\partial r} - \frac{\partial u_1}{\partial \varphi}\frac{\partial \bar{u}_1}{\partial \varphi} + \mathrm{i}\left(\frac{\partial u_1}{\partial r}\frac{\partial \bar{u}_1}{\partial \varphi} + \frac{\partial u_1}{\partial \varphi}\frac{\partial \bar{u}_1}{\partial r}\right)\right].$$

Thus ϕ_1 is real on ∂Z_1, hence

$$\phi_1 \equiv c \in \mathbb{R}$$

as Z_1 is a cylinder (by reflection, ϕ_1 can be made into a holomorphic quadratic differential on a torus which then is constant by Liouville's theorem).

As u_1 cannot be conformal, $c \neq 0$.

On the other hand, for the holomorphic quadratic differential ϕ_2 corresponding to v_2, one computes

$$\phi_2 \equiv (2\pi/b_2)^2.$$

Hence for a suitable choice of b_2,

$$\phi_1 \equiv \phi_2.$$

Thus, the map from T to S^2 patched together from u_1 on Z_1 and v_2 on Z_2 has a constant, and thus in particular holomorphic, quadratic differential. On the other hand, it is only Lipschitz, but not C^1, and therefore cannot be harmonic (cf. Section 2.5). We can also construct an example which has positive Jacobian almost everywhere. For this, on Z_2, we choose a map u_2 of the same type as u_1, and put $b_1 = b_2$. When identifying now the boundaries of Z_1 and Z_2, we can compose either u_1 or u_2 with an arbitrary rotation. This way, we can again construct a map with holomorphic quadratic differential which is Lipschitz but not C^1. (Cf. [J7] for the preceding examples.)

If we pull the image metric of Σ' back under u, we get

$$\rho^2\,\mathrm{d}u\,\mathrm{d}\bar{u} = \rho^2 u_z\bar{u}_z\,\mathrm{d}z^2 + \rho^2(u_z\bar{u}_{\bar{z}} + \bar{u}_z u_{\bar{z}})\,\mathrm{d}z\,\mathrm{d}\bar{z} + \rho^2 u_{\bar{z}}\bar{u}_{\bar{z}}\mathrm{d}\bar{z}^2.$$

Thus, the holomorphic quadratic differential $\rho^2 u_z\bar{u}_z\,\mathrm{d}z^2$ is precisely the (2, 0) part of the pullback of the image metric under u, the (0, 2) part its complex conjugate, and the (1, 1) part is given by the energy density (up to a conformal factor).

1.4. Some applications of holomorphic quadratic differentials. Surfaces in $\mathbb{R}^3$. The Gauss map

In this section, we shall deduce some consequences of:

Lemma 1.4.1: *Let ϕ be a holomorphic quadratic differential on S^2. Then $\phi \equiv 0$. The same holds, if ϕ is a holomorphic quadratic differential on D which is real on ∂D.*

Proof: ϕ gives rise to a holomorphic function φ on $\mathbb{C}$ (via stereographic projection), and $\varphi(z) \to 0$ as $z \to \infty$ because of the transformation properties of quadratic differentials. Hence $\varphi \equiv 0$ by Liouville's theorem. If ϕ is defined on D and real on ∂D, it can be reflected across ∂D to obtain a holomorphic quadratic differential on S^2.

q.e.d.

Corollary 1.4.1: *Suppose $u: S^2 \to N$ is a solution of* (1.2.7). *Then u is conformal. The same is true for $u: D \to N$ provided $\varphi(z)\,\mathrm{d}z^2$ as in Lemma 1.2.2 is real on ∂D. In particular, if u is harmonic, then it represents a (parametric) minimal surface in N. If $u: S^2 \to \mathbb{R}^3$ solves*

$$\Delta u = 2H(u)(D_1 u \wedge D_2 u) \tag{1.4.1}$$

it represents a surface with mean curvature H in $\mathbb{R}^3$.

Proof: This follows from Lemmas 1.2.2 and 1.4.1.

q.e.d.

Similarly, we have the following result of Lemaire [L1] and Wente [W1]:

Corollary 1.4.2: *Let $u: D \to N$ (D = unit disk) be a solution of* (1.2.7) *with*

$$u|_{\partial D} \equiv \text{const.}$$

Then u is constant.

Proof: In polar coordinates (r, θ) on D, the quadratic differential of Lemma 1.2.2 becomes

$$\varphi\,\mathrm{d}z^2 = g_{jk}(u)\left(u_r^j u_r^k - \frac{1}{r^2} u_\theta^j u_\theta^k - \frac{2\mathrm{i}}{r} u_r^j u_\theta^k\right)\mathrm{d}z^2.$$

Because of $u|_{\partial D} \equiv$ const., $u_\theta \equiv 0$ on ∂D. Hence $\varphi\,\mathrm{d}z^2$ is real on ∂D, and Lemma 1.4.1 applies.

q.e.d.

The following result is slightly more subtle as we do not require continuity at ∞. For simplicity we formulate it only for harmonic maps, as we shall need that case below:

Corollary 1.4.3: *Let $H := \{z = x + iy: y > 0\}$, N a Riemannian manifold, $u: H \to N$ harmonic, with*

$$E(u) < \infty.$$

Suppose $g_{jk} u_x^j u_y^k = 0$ on ∂H. Then u is conformal. In particular, if $u|_{\partial H} \equiv \text{const.}$, then also $u \equiv \text{const.}$

Proof: Let (g_{jk}) be the metric tensor of N w.r.t. coordinates u^j, $(j = 1, \ldots, \dim N)$.

$$\varphi := g_{jk}(u_x^j u_x^k - u_y^j u_y^k - 2i u_x^j u_y^k)$$

is holomorphic, and

$$\operatorname{Im} \varphi \equiv 0 \qquad \text{on } \partial H.$$

Thus, we can reflect φ across ∂H to obtain a holomorphic function $\bar{\varphi}$ on $\mathbb{C}$. Because of $E(u) < \infty$, φ and therefore also $\bar{\varphi}$ is in L^1, hence $\varphi \equiv 0$. This means that u is conformal. If $u|_{\partial H} \equiv \text{const.}$, in particular $u_y \equiv 0$ on ∂H. Putting

$$u(x, -y) = u(x, y)$$

we obtain a conformal map of class $C^{1,1}$ on $\mathbb{C}$ which then is also harmonic. Since the extended map is constant on the line ∂H it has to be constant everywhere.

q.e.d.

Let us now recall the basic formulae for surfaces $x(u, v)$ in $\mathbb{R}^3$: The first fundamental form is the metric of the surface induced by the ambient Euclidean metric of $\mathbb{R}^3$:

$$\begin{aligned} \mathrm{I} := dx^2 &= x_u^2\, du^2 + 2x_u x_v\, du\, dv + x_v^2\, dv^2 \\ &=: g_{11}\, du^2 + 2g_{12}\, du\, dv + g_{22}\, dv^2 \\ &=: E\, du^2 + 2F\, du\, dv + G\, dv^2 \end{aligned} \tag{1.4.2}$$

(in the old notation of Gauss).

The normal vector is given by

$$n = \frac{x_u \wedge x_v}{|x_u \wedge x_v|}$$

and the second fundamental form is

$$\mathrm{II} := -dx\, dn =: h_{11}\, du^2 + 2h_{12}\, du\, dv + h_{22}\, dv^2 =: L\, du^2 + 2H\, du\, dv + N\, dv^2. \tag{1.4.3}$$

These forms are related by the Gauss equation

$$g_{l2}(\Gamma^l_{11,2} - \Gamma^l_{12,1} + \Gamma^m_{11}\Gamma^l_{m2} - \Gamma^m_{12}\Gamma^l_{m1}) = LN - M^2 \tag{1.4.4}$$

and the (Mainardi–)Codazzi equations

$$\begin{aligned} L_v - M_u &= \Gamma^1_{12}L + (\Gamma^2_{12} - \Gamma^1_{11})M - \Gamma^2_{11}N \\ M_v - N_u &= \Gamma^1_{22}L + (\Gamma^2_{22} - \Gamma^1_{21})M - \Gamma^2_{21}N \end{aligned} \tag{1.4.5}$$

(of course, as before $\Gamma^i_{jk} = \frac{1}{2}g^{il}(g_{jl,k} + g_{kl,j} - g_{jk,l})$).

The Gauss equation expresses the fact that the Gauss curvature

$$K := \frac{LN - M^2}{EG - F^2}$$

can be expressed in terms of the g_{ij} and their derivatives, i.e. intrinsic quantities of the surface. Hence, K is independent of the embedding of the surface $x(u, v)$ into $\mathbb{R}^3$.

We also need the mean curvature

$$H := \frac{GL - 2FM + EN}{2(EG - F^2)}.$$

K and H are determinant and trace, respectively, of the second fundamental form w.r.t. the first fundamental form.

Finally, the Weingarten equations express the derivatives of n through the derivatives of x:

$$\begin{aligned} n_u &= \frac{1}{W^2}\{(FM - GL)x_u + (FL - EM)x_v\} \\ n_v &= \frac{1}{W^2}\{(FN - GM)x_u + (FM - EN)x_v\} \end{aligned} \tag{1.4.6}$$

with $W^2 := EG - F^2 = |x_u \wedge x_v|^2$.

Since n is always of length 1, $n = n(u, v)$ defines a map into S^2, the spherical image or Gauss map of $x(u, v)$. In the (u, v) parameters, the metric on S^2 then becomes

$$\mathrm{d}n^2 = n_u^2\,\mathrm{d}u^2 + 2n_u n_v\,\mathrm{d}u\,\mathrm{d}v + n_v^2\,\mathrm{d}v^2. \tag{1.4.7}$$

Later on (Section 3.2), we shall see that we can always introduce isothermal parameters, i.e. parameters (u, v) for which

$$E = G \qquad F = 0. \tag{1.4.8}$$

Thus the first fundamental form becomes diagonal:

$$\mathrm{d}x^2 = E(\mathrm{d}u^2 + \mathrm{d}v^2) \tag{1.4.9}$$

and

$$K = \frac{LN - M^2}{E^2} = -\frac{1}{2E}\Delta \log E \tag{1.4.10}$$

and

$$H = \frac{L+N}{2E}. \tag{1.4.11}$$

The Codazzi equations (1.4.5) become

$$\begin{aligned} L_v - M_u &= \frac{E_v}{2E}(L+N) = E_v H \\ M_v - N_u &= -\frac{E_u}{2E}(L+N) = -E_u H \end{aligned} \tag{1.4.12}$$

or, using $EH = \frac{1}{2}(L+N)$,

$$\begin{aligned} \left(\frac{L-N}{2}\right)_u + M_v &= EH_u \\ \left(\frac{L-N}{2}\right)_v - M_u &= -EH_v. \end{aligned} \tag{1.4.13}$$

Finally, the Weingarten equations (1.4.6) become

$$\begin{aligned} n_u &= -\frac{L}{E}x_u - \frac{M}{E}x_v \\ n_v &= -\frac{M}{E}x_u - \frac{N}{E}x_v. \end{aligned} \tag{1.4.14}$$

The following is a special case of a result of Ruh and Vilms [RV].

Lemma 1.4.2: *If $H \equiv$ const., the Gauss map is harmonic (as a map from the surface $x(u,v)$ into S^2). It is $\pm$ conformal, if $H \equiv 0$ (i.e. x is a minimal surface), or $L = N$, $M \equiv 0$ (in which case L/E and N/E are constant and x is (part of) a sphere).*

Proof: Since the map $(u,v) \to x(u,v)$ is conformal, if (u,v) are isothermal parameters, we can work in the (u,v) plane (cf. Lemma 1.3.2). The quadratic differential associated with the Gauss map is (by (1.4.14))

$$\begin{aligned} & n_u^2 - n_v^2 - 2in_u n_v \\ &= L^2 - N^2 - 2iM(L+N) \qquad \text{as } x_u^2 = x_v^2 = E, x_u \cdot x_v = 0 \\ &= 2EH(L-N-2iM). \end{aligned} \tag{1.4.15}$$

Thus, the Gauss map is $\pm$ conformal, if $H \equiv 0$ or if $L \equiv N$, $M \equiv 0$; in the latter case, from (1.4.14), using $n_{uv} = n_{vu}$ with $k := L/E = N/E$,

$$k_v x_u - k_u x_v = 0.$$

Since x_u and x_v are independent, $k_u \equiv k_v \equiv 0$, i.e. $k \equiv$ const.

Moreover, the Codazzi equations (1.4.13) imply

$$(L - N - 2iM)_{\bar{w}} = 2EH_w \quad (w = u + iv) \tag{1.4.16}$$

so that $L - N - 2iM$ is holomorphic, if H is constant. Since the Gauss map is smooth, Lemma 1.3.4 implies that, for constant H, it is harmonic.

q.e.d.

Corollary 1.4.4: *Let S be a closed surface of genus 0, immersed with constant mean curvature into $\mathbb{R}^3$. Then S is a standard sphere.*

Proof: By Theorem 3.1.1, we can introduce global isothermal coordinates (u, v) on S. By Lemma 1.4.2, the Gauss map is harmonic (note that because of the regularity properties of solutions of (1.2.13), (cf. Section 2.5) in isothermal coordinates our surface is smooth). By Corollary 1.4.1 it is conformal. The Weingarten equations then imply $M \equiv 0$, $L/E \equiv N/E$; hence they are constant, as $(L + N)/E = 2H \equiv \text{const.}$

q.e.d.

Alternatively, for the proof of Corollary 1.4.4, we can observe that, by (1.4.16), $L - N - 2iM$ is holomorphic. Since it transforms as a quadratic differential, it vanishes by Lemma 1.4.1. This latter one is the original proof of Hopf [Ho] who invented this kind of argument.

We remark that the result is no longer true for surfaces of genus 1, as shown by Wente [W3, W4]; cf. also [Ab, Wa, Sp]. Combining Corollary 1.4.1 and Corollary 1.4.4, we get:

Corollary 1.4.5: *Let $x: S^2 \to \mathbb{R}^3$ be a solution of*

$$\Delta x = 2Hx_u \wedge x_v \qquad (H \equiv \text{const. and } u, v \text{ isothermal}).$$

Then $x(S^2)$ is a standard sphere of curvature H.

As observed by Nitsche [Ni2], the argument of Hopf also carries over to show:

Corollary 1.4.6: *Let $B(0, 1)$ be the unit ball in $\mathbb{R}^3$, Σ a disk-type surface in $B(0, 1)$, meeting $\partial B(0, 1)$ orthogonally along its boundary $\partial\Sigma$. If Σ has constant mean curvature H, then it is a spherical cap of curvature H. If $H = 0$, Σ is an equatorial disk in $B(0, 1)$.*

Proof: By Theorem 3.2.1, we can find isothermal polar coordinates (r, θ), $0 \leqslant r \leqslant 1$, $\theta \in [0, 2\pi]$, on Σ, i.e. $\Sigma = x(r, \theta)$. The boundary condition means that for $r = 1$

$$x \cdot x = 1 \qquad x \cdot n = 0$$

hence

$$x \cdot n_\theta + x_\theta \cdot n = 0$$

hence

$$x \cdot n_\theta = 0 \qquad \text{(as always } x_\theta \cdot n = 0)$$

and

$$x \cdot x_\theta = 0.$$

As x_r, x_θ, n are orthogonal, it follows that x_r is proportional to x. Therefore,

$$M = -x_r \cdot n_\theta = 0.$$

Thus, the holomorphic quadratic differential $(L - N - 2iM)$ (holomorphic, because of $H \equiv \text{const}$, cf. (1.4.16)) is real for $r = 1$. It can therefore be reflected across $r = 1$ to yield a holomorphic quadratic differential on S^2, and the claim again follows from Lemma 1.4.1. If $H \equiv 0$, then $L + N \equiv 0$, and since $L = N$, $M = 0$, the second fundamental form vanishes identically, so that Σ is planar. q.e.d

More generally, we can also consider surfaces $x(u, v)$ in a three-dimensional manifold N. We denote the metric of N by $\langle \cdot, \cdot \rangle$, the curvature tensor by R, and the covariant derivative by ∇. If our surface and N are orientable, we can choose a unit normal vector n to $x(u, v)$ in a consistent way. We can then express the second fundamental form of $x(u, v)$ as before, namely

$$L := -\left\langle \nabla_{\partial/\partial u} n, \frac{\partial x}{\partial u} \right\rangle$$

$$M := -\left\langle \nabla_{\partial/\partial u} n, \frac{\partial x}{\partial v} \right\rangle = -\left\langle \nabla_{\partial/\partial v} n, \frac{\partial x}{\partial u} \right\rangle$$

$$N := -\left\langle \nabla_{\partial/\partial v} n, \frac{\partial x}{\partial v} \right\rangle.$$

The Codazzi equations in this case take the form

$$L_v - M_u = \Gamma^1_{12} L + (\Gamma^2_{12} - \Gamma^1_{11}) M - \Gamma^2_{11} N + \left\langle R\left(\frac{\partial x}{\partial u}, \frac{\partial x}{\partial v}\right) n, \frac{\partial x}{\partial u} \right\rangle \tag{1.4.17}$$

$$M_v - N_u = \Gamma^1_{22} L + (\Gamma^2_{22} - \Gamma^1_{12}) M - \Gamma^2_{12} N + \left\langle R\left(\frac{\partial x}{\partial u}, \frac{\partial x}{\partial v}\right) n, \frac{\partial x}{\partial v} \right\rangle. \tag{1.4.18}$$

We now note that for a space N of *constant* curvature K,

$$R(X, Y)Z = K(\langle Y, Z \rangle X - \langle X, Z \rangle Y) \tag{1.4.19}$$

(for tangent vectors X, Y, Z). Since $\langle \partial x/\partial u, n \rangle = 0 = \langle \partial x/\partial v, n \rangle$, this implies that in this case the curvature term in the Codazzi equations (1.4.17), (1.4.18) disappears.

Therefore, Hopf's argument carries over to show:

Corollary 1.4.7: *Let S be a closed surface of genus 0, immersed with constant mean curvature into a three-dimensional manifold N of constant sectional curvature. Then S is embedded as a standard sphere. In particular, a closed minimal surface of genus 0 in S^3 is an equatorial 2-sphere.*

This latter special case is due to Calabi [Ca] and Almgren [A1]; the proof uses the same argument. Again, this is no longer true for minimal surfaces of higher genus in S^3; cf. the examples of Lawson [La] and Karcher, Pinkall and Sterling [KPS].

Another example where one can apply Lemma 1.4.1 arises from surfaces of constant Gauss curvature in 3-space. Suppose S is a surface with local coordinates u, v. If the Gauss curvature K of S is positive, then the second fundamental form is positive definite and can hence be diagonalized by introducing new paramenters x, y instead of u, v (i.e. $\mathrm{II} = \lambda(\mathrm{d}x^2 + \mathrm{d}y^2)$, where λ is a positive function). x, y are called isothermal conjugated parameters, and Darboux [Db, Section 725] discovered that u and v as functions of x and y satisfy a system of elliptic equations depending only on the first fundamental form of S, i.e. only on intrinsic geometric quantities, not depending on the immersion of S into 3-space. (This fact was used extensively by Lewy [Lw] and Heinz [H2] for solving the Weyl embedding problem.)

The system discovered by Darboux is the following:

$$\Delta u + \left(\Gamma^1_{11} + \frac{1}{2}\frac{\partial}{\partial u}(\log K)\right)(u_x^2 + u_y^2) + \left(2\Gamma^1_{12} + \frac{1}{2}\frac{\partial}{\partial v}(\log K)\right)(u_x v_x + u_y v_y) + \Gamma^1_{22}(v_x^2 + v_y^2) = 0 \tag{1.4.20}$$

$$\Delta v + \Gamma^2_{11}(u_x^2 + u_y^2) + \left(2\Gamma^2_{12} + \frac{1}{2}\frac{\partial}{\partial u}(\log K)\right)(u_x v_x + u_y v_y) + \left(\Gamma^2_{22} + \frac{1}{2}\frac{\partial}{\partial v}(\log K)\right)(v_x^2 + v_y^2) = 0.$$

We see in particular that, if the Gauss curvature K of S is a positive constant, then the transformation $(x, y) \rightarrow (u, v)$ is harmonic. This implies Liebmann's theorem (cf. [Bl, p. 195]):

Corollary 1.4.8: *The only immersion of a closed surface S of constant positive curvature K into 3-space is given by a standard sphere of radius $1/\sqrt{K}$.*

Proof: By the Gauss–Bonnet theorem, S is topologically a sphere. Since the second fundamental form of a given immersion of S is positive definite, we can use the uniformization theorem to obtain parameters x, y on the sphere S^2

which diagonalize this form. Since $K \equiv$ const., we thus obtain a harmonic map

$$h: S^2 \to S.$$

By Corollary 1.4.1, h is (anti)conformal and therefore also diagonalizes the first fundamental form. Hence the first and the second fundamental form are proportional everywhere, which means that the given immersion satisfies $L \equiv N$, $M \equiv 0$. Since also $K = (LN - M^2)/E^2$ is constant, then the second fundamental form is constant, and we get a standard sphere.

q.e.d.

We finally note that the Weierstrass representation for minimal surfaces can be carried over to our setting in the following way. Let $u: \Sigma \to N$, Σ a Riemann surface, be harmonic, and let $\varphi \, dz^2$ be the associated holomorphic quadratic differential as before. There exists a finite holomorphic branched covering $\bar{\Sigma} \to \Sigma$, $w \to z(w)$ with the property that, on $\bar{\Sigma}$, $\varphi(z(w)) z_w z_w \, dw^2$, the pull-back of $\varphi \, dz^2$, has a square root. This means that there exists a holomorphic 1-form $\psi(w) \, dw$ with

$$\psi(w)^2 \, dw^2 = -\tfrac{1}{4} \varphi(z(w)) z_w z_w \, dw^2.$$

Moreover, there exists a possibly infinite covering $\tilde{\bar{\Sigma}} \to \bar{\Sigma}$, $\zeta \to w(\zeta)$ with the property that

$$\psi(w(\zeta)) w_\zeta \, d\zeta =: \alpha(\zeta) \, d\zeta$$

has no real periods.

We choose some $\zeta_0 \in \tilde{\bar{\Sigma}}$. Then

$$v(\zeta) := \operatorname{Re} \int_{\zeta_0}^{\zeta} \alpha(\omega) \, d\omega$$

is well defined, i.e. independant of the path of integration. Then the map

$$\begin{aligned} &\tilde{\bar{\Sigma}} \to N \times \mathbb{R} \\ &\zeta \to (u(z(w(\zeta))), v(\zeta)) \end{aligned}$$

is harmonic and conformal; for the latter note

$$v_\zeta v_\zeta \, d\zeta^2 = -\tfrac{1}{4} \varphi \, dz^2$$

and thus the holomorphic quadratic differential associated with (u, v) is

$$\varphi \, dz^2 + 4(-\tfrac{1}{4} \varphi \, dz^2) = 0.$$

We summarize the preceding discussion as:

Corollary 1.4.9: *Let $u: \Sigma \to N$ be harmonic, Σ a Riemann surface, N a Riemannian manifold. Then there exists a possibly infinite branched holomorphic covering*

$\tilde{\Sigma} \to \Sigma$, $\zeta \to z(\zeta)$ *and a mapping* $v: \tilde{\Sigma} \to \mathbb{R}$ *with the property that*

$$\tilde{\Sigma} \to N \times \mathbb{R} \qquad \zeta \to (u(z(\zeta)), v(\zeta))$$

is a (possibly) branched minimal surface.

Remark: Such a construction has been employed by Schoen [S] in a regularity proof; cf. Theorem 2.3.4 below.

2 REGULARITY AND UNIQUENESS RESULTS

In this chapter, we assemble the regularity and uniqueness results needed in later chapters. We shall give complete proofs only for those results that are special to two dimensions or not covered in [J6] where the general regularity theory for harmonic maps is worked out[1].

2.1. Harmonic coordinates

In this section, we state the fundamental results of Jost and Karcher [JK] on harmonic coordinates; for proofs, we refer to [JK] or [J6]. Let N be a complete Riemannian manifold of dimension d; we assume that N is of bounded geometry, i.e. that there exists a bound

$$\Lambda^2 := \sup |K| < \infty \qquad (K := \text{sectional curvature}) \tag{2.1.1}$$

$$i(N) := \inf_{p \in N} i(p) > 0 \qquad (i(p) := \text{injectivity radius of } p). \tag{2.1.2}$$

An injective map

$$H: B(p, R) \to \mathbb{R}^d \qquad H = (h^1, \ldots, h^d)$$

$(B(p,R) := \{q \in N : d(p,q) \leqslant R\}, d(\cdot,\cdot) :=$ distance function of $N)$ is said to yield harmonic coordinates if the components h^j $(j = 1, \ldots, d)$ are harmonic.

In this section, $g = (g_{ij})_{i,j=1,\ldots,d}$ will always denote the metric tensor w.r.t. harmonic coordinates.

Theorem 2.1.1: *There exists some $R_0 > 0$, depending solely on $\Lambda^2, i(N)$, and*

[1] We also refer to the survey articles [Hi3] and [S] on the analytic aspects of harmonic maps. References to the original papers can be found in [J6], [Hi3], and [S].

$d = \dim N$, *with the property that for each* $p \in N$, *on* $B(p, R_0)$ *there exist harmonic coordinates.*

Theorem 2.1.2: *There exists* $R_0 > 0$ *depending only on* $i(N)$, Λ^2, d *with the property that for all* $p \in N$, $R \leqslant R_0$, *on* $B(p, R)$ *there exist harmonic coordinates the metric tensor* g *of which satisfies*

$$|dg(x)| \leqslant \frac{c\Lambda^2 R^2}{d(x, \partial B(p, R))} \tag{2.1.3}$$

for all $x \in B(p, R)$ *with* $c = c(d, \Lambda R_0)$, *and on each ball* $B(p, (1-\delta)R)$ *for any* $\alpha \in (0, 1)$

$$|dg|_{C^\alpha} \leqslant \frac{c\Lambda^2 R^2}{\delta^2} \tag{2.1.4}$$

with $c = c(\alpha, d, \Lambda R_0)$.

In particular, the absolute value and the Höldernorms of the corresponding Christofel symbols are bounded in terms of d *and* ΛR_0.

Theorem 2.1.3: *If the curvature tensor to* N *is of class*

$$C^k, C^{k,\beta} (k \in N, \beta \in (0, 1), C^\infty, \text{ or } C^\omega$$

then

$$g \in C^{k+1,\alpha} \text{ (for every } \alpha \in (0, 1)), C^{k+2,\beta}, C^\infty, \text{ or } C^\omega$$

respectively.

Let now $B(p, R) \subset N$ be a ball contained in the image of a harmonic coordinate chart $U(0, 1) := \{x \in \mathbb{R}^d : |x| < 1\}$ with metric tensor $g_{ij}(x)$.

We define a new metric on $U(0, 1)$ via

$$g_{ij,\rho}(x) := g_{ij}(\rho x) \qquad \text{for } \rho \in \mathbb{R}, \rho \geqslant 1.$$

By Theorem 2.1.2, $g_{ij,\rho}$ converges in $C^{1,\alpha}$ to the Euclidian metric δ_{ij} in $U(0, 1)$ as $\rho \to \infty$.

This rescaling process, of course, is equivalent to replacing $B(p, R)$ by $B(p, R/\rho)$ and multiplying the metric by the factor ρ.

Corollary 2.1.1 (Rescaling principle): *Let* $\lambda(g)$ *be any continuous expression defined in* $U(0, 1)$ *involving a metric* $g = (g_{ij})$ *and its first derivatives. If* $\lambda((\delta_{ij})) > 0$ *(where* (δ_{ij}) *is the Euclidian metric) then also* $\lambda((g_{ij,\rho})) > 0$ *if* ρ *is large enough.*

2.2. Uniqueness of harmonic maps

Let us first give the uniqueness and stability theorem of Jäger and Kaul [JäK]:

Theorem 2.2.1: *Suppose that* $u_i : \bar{\Omega} \to N$ $(i = 1, 2)$ *are harmonic maps of class*

$C^0(\bar{\Omega}, N) \cap C^2(\Omega, N)$, Ω is a bounded domain in some Riemannian manifold, and $u_i(\bar{\Omega}) \subset B(p, \rho)$, where $B(p, \rho)$ is a geodesic ball in N, disjoint to the cut locus of p and with radius $\rho < \pi/2\kappa$ where κ^2 is an upper bound for the sectional curvature of $B(p, \rho)$. Then the function θ,

$$\theta(x) := \frac{q_\kappa(d(u_1(x), u_2(x)))}{\cos(\kappa d(p, u_1(x)))\cos(\kappa d(p, u_2)))}$$

where

$$q_\kappa(t) := \begin{cases} \dfrac{1}{\kappa^2}[1 - \cos(\kappa t)] & \text{if } \kappa > 0 \\ \dfrac{t^2}{2} & \text{if } \kappa = 0 \end{cases}$$

satisfies the maximum principle

$$\sup_\Omega \theta \leqslant \sup_{\partial\Omega} \theta. \tag{2.2.1}$$

In particular, if $u_1 | \partial\Omega = u_2 | \partial\Omega$, then

$$u_1 \equiv u_2.$$

As we shall need a generalization below, we will give the variant of the proof displayed in [J6], but will refer to [J6, Section 2.5] for the necessary Jacobi field estimates.

We shall also need the chain rule for a harmonic $u \in C^2(\Omega, N)$ and $h \in C^2(N, \mathbb{R})$:

$$\Delta(h \circ u) = D^2 h(u_{e^\alpha}, u_{e^\alpha}) \tag{2.2.2}$$

where e^α is an orthonormal frame on Ω and $D^2 h$ is the second fundamental form of h.

Proof: We assume that θ has a positive maximum at some interior point $x_0 \in \Omega$. Then, θ is positive in a neighbourhood of x_0, and $\log\theta > -\infty$ in this neighbourhood.

We define

$$\Psi(x) := Q_\kappa(u_1(x), u_2(x)) = \begin{cases} \dfrac{1}{\kappa^2}[1 - \cos(\kappa d(u_1(x), u_2(x)))] & \text{if } \kappa > 0 \\ \dfrac{1}{2} d^2(u_1(x), u_2(x)) & \text{if } \kappa = 0 \end{cases}$$

$$\phi_i(x) = \cos(\kappa d(p, u_i(x))) \qquad i = 1, 2.$$

Then $\theta = \Psi/(\phi_1 \cdot \phi_2)$, and consequently

$$\operatorname{grad} \log \theta = \frac{\operatorname{grad} \Psi}{\Psi} - \frac{\operatorname{grad} \phi_1}{\phi_1} - \frac{\operatorname{grad} \phi_2}{\phi_2} \tag{2.2.3}$$

and

$$\Delta \log \theta = \frac{\Delta \Psi}{\Psi} - \frac{|\operatorname{grad} \Psi|^2}{\Psi^2} - \frac{\Delta \phi_1}{\phi_1} + \frac{|\operatorname{grad} \phi_1|^2}{\phi_1^2} - \frac{\Delta \phi_2}{\phi_2} + \frac{|\operatorname{grad} \phi_2|^2}{\phi_2^2}. \tag{2.2.4}$$

Since $x \to u(x) = (u_1(x), u_2(x)) \in B(p, \rho) \times B(p, \rho)$ is also harmonic, we can make use of the chain rule (2.2.2) in order to apply Lemma 2.5.1 of [J6]. This yields

$$\Delta \Psi \geqslant \frac{|\operatorname{grad} \Psi|^2}{2\Psi} - \kappa^2 \Psi(|du_1|^2 + |du_2|^2) \tag{2.2.5}$$

since

$$|\operatorname{grad} \Psi|^2 = \sum_\alpha \langle (\operatorname{grad} Q_\kappa) \circ u, du(e_\alpha) \rangle^2$$

where e_α is an orthonormal frame on Ω.

Similarly, from (2.5.2) of [J6], since

$$\phi_i(x) = 1 - \kappa^2 Q_\kappa(p, u_i(x))$$

we obtain

$$\Delta \phi_i(x) \leqslant -\kappa^2 \phi_i |du_i|^2. \tag{2.2.6}$$

Finally, by (2.2.3)

$$\begin{aligned} &-\frac{1}{2}\frac{|\operatorname{grad} \Psi|^2}{\Psi^2} + \frac{|\operatorname{grad} \phi_1|^2}{\phi_1^2} + \frac{|\operatorname{grad} \phi_2|^2}{\phi_2^2} \\ &\geqslant -\left\langle \operatorname{grad} \log \theta, \tfrac{1}{2} \operatorname{grad} \log \theta + \frac{\operatorname{grad} \phi_1}{\phi_1} + \frac{\operatorname{grad} \phi_2}{\phi_2} \right\rangle. \end{aligned} \tag{2.2.7}$$

Putting

$$k(x) := \tfrac{1}{2} \operatorname{grad} \log \theta + \frac{\operatorname{grad} \phi_1}{\phi_1} + \frac{\operatorname{grad} \phi_2}{\phi_2}$$

and plugging (2.2.5), (2.2.6), and (2.2.7) into (2.2.4), we obtain

$$\Delta \log \theta + \langle \operatorname{grad} \log \theta, k(x) \rangle \geqslant 0.$$

Therefore, the assumption that θ has a positive maximum in the interior contradicts Hopf's maximum principle.

q.e.d.

The following extension was noted in [GJ].

Theorem 2.2.2: *Suppose Ω is a bounded domain in a Riemannian manifold, $\partial\Omega = D_1 \cup D_2$, where D_1 is of class C^2. Let $B(p,\rho)$ be a ball in a Riemannian manifold N, disjoint from the cut locus of p, with $\rho < \pi/2\kappa$, where κ^2 is an upper bound for the sectional curvature of $B(p,\rho)$. Then there is some $\delta > 0$ with the following property:*

If $\Gamma \subset N$ is a submanifold with $p \in \Gamma$, with $|\nabla^2\Gamma| \leqslant \delta$ in $B(p,\rho)$, where $\nabla^2\Gamma$ denotes the second fundamental form of Γ, if $u_i \in C^0(\bar{\Omega}, N) \cap C^2(\Omega \cup D_1, N)$ $(i = 1, 2)$ are harmonic maps with $u_i(\bar{\Omega}) \subset B(p,\rho)$,

$$u_1 = u_2 \qquad \text{on } D_2$$

and $u_i(D_1) \subset \Gamma$, and u_i is stationary among such maps(i.e. u_i solves a mixed boundary value problem requiring $u_i(D_1) \subset \Gamma, (\partial u_i/\partial n) \perp \Gamma$ on D_1, where $\partial/\partial n$ denotes the derivative in the direction of the exterior normal of Ω at D_1) then

$$u_1 \equiv u_2.$$

More generally, a similar stability result as (2.2.1) (*for a suitably modified function $\tilde{\theta}$*) *holds.*

Proof: Let us first treat the case $\delta = 0$, i.e. where Γ is totally geodesic. Using the notation of the proof of Theorem 2.2.1, we have to show that θ cannot assume its maximum on D_1 as by the preceding proof it cannot assume its maximum in Ω either. However, on D_1, $\operatorname{grad}\theta$ is tangential to Γ, because Γ is totally geodesic and $p \in \Gamma$, and hence the argument of the preceding proof also rules out a maximum on D_1. For the general case, we slightly modify the preceding construction.

Given $\epsilon > 0$, we can find $\delta > 0$ with the property that if $|\nabla^2\Gamma| < \delta, p \in \Gamma$, we can modify the distance function $d(\cdot,\cdot)$ to obtain a function $\bar{d}(\cdot,\cdot)$ with the following properties:

(i) $\nabla_x \bar{d}(x,y)$ is tangential to Γ if $x, y \in \Gamma$
(ii) $|d - \bar{d}|_{C^2} \leqslant \epsilon$
(iii) $\bar{d}(x,y) = 0 \qquad$ iff $x = y$.

We put

$$\bar{\Psi}(x) := \begin{cases} \dfrac{1}{\kappa^2}[1 - \cos(\kappa \bar{d}(u_1(x), u_2(x)))] & \text{if } \kappa > 0 \\ \dfrac{1}{2}\bar{d}^2(u_1(x), u_2(x)) & \text{if } \kappa = 0 \end{cases}$$

$$\left.\begin{aligned} \bar{\varphi}_i(x) &:= \cos(\bar{\kappa}\bar{d}(p, u_i(x))) \\ \hat{\varphi}_i(x) &:= \cos(\bar{\kappa} d(p, u_i(x))) \end{aligned}\right\} \quad \text{with } \kappa < \bar{\kappa} < \pi/2\rho \qquad \begin{aligned} &(i = 1, 2) \\ &(i = 1, 2) \end{aligned}$$

$$\bar{\theta} := \frac{\bar{\Psi}}{\bar{\varphi}_1 \cdot \bar{\varphi}_2} \qquad \hat{\theta} := \frac{\Psi}{\hat{\varphi}_1 \cdot \hat{\varphi}_2} \qquad \text{with } \Psi \text{ as above.}$$

We shall again exclude that $\bar{\theta}$ assumes a maximum on $\Omega \cup D_1$ to prove the result.

By property (ii) and (2.2.2)

$$|\Delta(\bar{\Psi} - \Psi)| \leqslant \epsilon(|du_1|^2 + |du_2|^2) \tag{2.2.8}$$

and similarly for $\bar{\varphi}_1$ and $\bar{\varphi}_2$.

Likewise

$$||\text{grad}\,\Psi|^2 - |\text{grad}\,\bar{\Psi}|^2| \leqslant \epsilon(|du_1|^2 + |du_2|^2) \tag{2.2.9}$$

and again similarly for $\bar{\varphi}_1$ and $\bar{\varphi}_2$.

On the other hand now

$$-\Delta\hat{\varphi}_i(x) \geqslant \bar{\kappa}^2\hat{\varphi}_i|du_i|^2 \tag{2.2.10}$$

with $\bar{\kappa} > \kappa$.

From the proof of Theorem 2.2.1, we therefore conclude

$$\Delta \log \hat{\theta} + \langle \text{grad} \log \hat{\theta}, \hat{k}(x) \rangle \geqslant (\bar{\kappa}^2 - \kappa^2)(|du_1|^2 + |du_2|^2)$$

(for a suitable $\hat{k}(x)$ as above), and this allows us to compensate the error terms arising from (2.2.8) and (2.2.9) if we do the computation with $\bar{\theta}$ instead of $\hat{\theta}$, to obtain, if ϵ and hence δ is small enough,

$$\Delta \log \bar{\theta} + \langle \text{grad} \log \bar{\theta}, \bar{k}(x) \rangle \geqslant 0$$

for some $\bar{k}(x)$.

Since $\text{grad}\,\bar{\theta}$ is always trangential to D_1 because of property (i), $\bar{\theta}$ cannot assume its maximum on $\Omega \cup D_1$. Since $\bar{\theta}$ vanishes on D_2, property (iii) implies $u_1 \equiv u_2$.

q.e.d.

We shall also need the global uniqueness theorem of Al'ber and Hartman (see [Alb1], [Alb2], [Ht], [SY3]; cf. also [J6, Section 3.8] for a proof).

Theorem 2.2.3: *Let $u: M \to N$ be a harmonic map between compact Riemannian manifolds (without boundary). Suppose N has negative sectional curvature. Then u is the unique harmonic map in its homotopy class unless $u(M)$ is a point or a closed geodesic.*

If the sectional curvature of N is non-positive, then for any two homotopic harmonic $u_0, u_1: M \to N$, there exists a family $u_t: M \to N$ of harmonic maps, with the property that the curves $u_t(x)$, for fixed $x \in M, t \in [0, 1]$ varying, constitute a family of parallel geodesics, parametrised proportionally to arc length. In particular, all maps u_t have the same energy.

2.3. Continuity of weak solutions

In order to show the essential idea of the subsequent arguments, we shall first consider the trivial case of a weak minimal surface $h(z)$ in $\mathbb{R}^d$, $z = x + iy \in D$.

The argument will depend on Grüter's observation [Gr1] that the standard monotonicity formula for minimal surfaces can be pulled back under a (weakly) conformal map.

Thus we let $z_0 \in D, B(z_0, r) \subset D, z_1 \in B(z_0, r)$, and put

$$a := h(z_1).$$

We assume that for almost all $z \in \partial B(z_0, r)$

$$|h(z) - a| > \rho' \tag{2.3.1}$$

or in other words that the minimal surface $h(B(z_0, r))$ has no boundary in the ball $B(a, \rho')$.

We let $0 \leqslant \rho \leqslant \rho'$ and choose the test vector

$$\varphi(z) = \psi\left(\frac{|h-a|}{\rho}\right) \cdot (h - a) \qquad (h = h(z))$$

where $\psi(t) \equiv 1$ for $t \leqslant \frac{1}{2}, \psi(t) \equiv 0$ for $t \geqslant 1, \psi'(t) \leqslant 0$. (In the end, we shall choose a sequence ψ_n of such functions tending to the characteristic function $\chi_{(-\infty,1)}$.)

Since h is a weak minimal surface,

$$\int (h_x \varphi_x + h_y \varphi_y) \mathrm{d}x \, \mathrm{d}y = 0 \tag{2.3.2}$$

(note that supp $\varphi \subset B(z_0, r)$ by (2.3.1)). We put

$$I(\rho) := \frac{1}{2} \int_{B(z_0, r)} |\nabla h|^2 \psi\left(\frac{|h-a|}{\rho}\right)$$

(for $\psi = \chi_{(-\infty,1)}$, this is nothing but the area of the minimal surface $h(B(z_0, r))$ inside the ball $B(a, \rho)$.) Then, with $\gamma(t) = \psi(t/\rho)$, thus $t\gamma'(t) = -\rho \mathrm{d}\psi(t/\rho)/\mathrm{d}\rho$, we have

$$I'(\rho) = -\frac{1}{2\rho} \int_{B(z_0, r)} |\nabla h|^2 |h - a| \gamma'(|h - a|).$$

We compute

$$h_x \varphi_x + h_y \varphi_y = \gamma(|h-a|)|\nabla h|^2 + \gamma'(|h-a|) \frac{1}{|h-a|} \{[(h-a)h_x]^2 + [(h-a)h_y]^2\}.$$

On the other hand, since h is weakly conformal,

$$h_x h_x = h_y h_y \qquad h_x h_y = 0 \qquad \text{almost everywhere}$$

and consequently

$$[(h-a)h_x]^2 + [(h-a)h_y]^2 \leqslant \tfrac{1}{2} |\nabla h|^2 |h-a|^2. \tag{2.3.3}$$

Therefore, we can deduce altogether from (2.3.2) that

$$2I(\rho) - \rho I'(\rho) \leqslant 0.$$

(The factor 2 here comes from the factor $\frac{1}{2}$ in (2.3.3), and here the conformality thus enters in an essential way.) Integrating, we obtain for $0 < \rho_1 \leqslant \rho_2 \leqslant \rho'$

$$\frac{I(\rho_1)}{\rho_1^2} \leqslant \frac{I(\rho_2)}{\rho_2^2}$$

and with $\psi = \chi_{(-\infty,1)}$ this is the fundamental *monotonicity formula for minimal surfaces.*

Letting ρ_1 tend to zero, the left-hand side tends to 2π times the density of the minimal surface $h(B(z_0, r))$ at a.

(This density is at least 1 for every $a \in h(B(z_0, r))$. This follows first if a is a smooth point of the minimal surface, e.g. by the asymptotic expansion of Corollary 2.6.1 below. By standard results about Sobolev functions, h is continuous almost everywhere, hence smooth almost everywhere (cf. e.g. Theorem 2.5.1) and the monotonicity formula then easily implies that the density of the minimal surface $h(B(z_0, r))$ at a is an upper semicontinuous function of $a \in h(B(z_0, r))$. Hence the density at $a \in h(B(z_0, r))$ is at least 1).

Therefore, if we choose $\rho_2 = \rho'$, this gives an upper bound for ρ'. In particular, if r tends to zero, $I(\rho')$ tends to zero, and consequently also ρ'. Thus, we can control

$$\inf_{z \in \partial B(z_0, r)} |h(z) - h(z_1)| \leqslant \gamma' \to 0 \qquad \text{as } r \to 0. \tag{2.3.4}$$

On the other hand, by Lemma 1.1.1, given $r_0 < 1$, we find $r \in (r_0, \sqrt{r_0})$ with

$$|h(z) - h(z')| \leqslant \frac{2\pi^{1/2}}{(\log 1/r)^{1/2}} \left(\int_{B(z_0, r^{1/2})} |\nabla h|^2 \right)^{1/2} \qquad \text{for } z, z' \in \partial B(z_0, r)$$

and this also goes to zero as $r \to 0$.

Hence for such r, and $z_1, z_1' \in B(z_0, r)$, we choose z, z' for which the infimum is attained in (2.3.4), and continuity then follows from the triangle inequality.

q.e.d.

Let us also discuss the idea for the proof of continuity of a minimal surface in $\mathbb{R}^d$ with a free boundary M, again for the easiest case, namely where M is an (affine) linear subspace L. For $h \in \mathbb{R}^d$, we then define $\xi(h)$ to be the nearest point projection of h onto L, and we define

$$\tilde{h} := 2\xi(h) - h$$

as the reflection of h across L.

In the following, P_{hq} denotes parallel transport from h to q and, for $q \in L$, $\tau(q)$ and $v(q)$ denote the projections onto L and the direction orthogonal to L, respectively.

We then use the test vector

$$\varphi(h)=\psi\left(\frac{|h-a|}{\rho}\right)(h-a)$$
$$+\psi\left(\frac{|\tilde{h}-a|}{\rho}\right)P_{\xi(h)h}\{\tau(\xi(h))P_{\tilde{h}\xi(h)}(\tilde{h}-a)-\nu(\xi(h))P_{\tilde{h}\xi(h)}(\tilde{h}-a)\}$$

(a,ψ as above), and define

$$\tilde{I}(\rho):=\frac{1}{2}\int_{B(z_0,r)}|\nabla h|^2\left(\psi\left(\frac{|h-a|}{\rho}\right)+\psi\left(\frac{|\tilde{h}-a|}{\rho}\right)\right).$$

For $\psi=\chi_{(-\infty,1)}$, $\tilde{I}(\rho)$ yields the area of the minimal surface $h(B(z_0,r))$ inside the ball $B(a,\rho)$ plus the area inside the ball reflected across L. (For technical reasons, in the following it will be more convenient to reflect the balls in the monotonicity formula rather than to reflect the minimal surface itself across L.) Since in this special case, the map $h\to\tilde{h}$ is a linear isometry, one computes as above

$$2\tilde{I}(\rho)-\rho\tilde{I}'(\rho)\leqslant 0$$

and obtains after integration the *monotonicity formula for a minimal surface with a* (linear) *free boundary*

$$\frac{\tilde{I}(\rho_1)}{\rho_1^2}\leqslant\frac{\tilde{I}(\rho_2)}{\rho_2^2}.$$

One then deduces continuity by similar arguments as before.

In the following, we shall study minimal surfaces in Riemannian manifolds and/or with free boundaries that are curved manifolds themselves, and if we then choose suitable analogues of the above test vector φ, we get additional terms that need to be estimated with the help of standard Jacobi field estimates. This will then lead us to suitable approximate versions of the above monotonicity formula, and these formulae with then allow us to derive continuity as above.

Before we proceed, let us first state the necessary Jacobi field estimates. They are taken from [BK, Section 6.3] and can also be found in [J6, Section 2.2].

Denote the curvature tensor of N by R, its sectional curvatures by K, and the metric as usual by $\langle\cdot,\cdot\rangle$. If $c(s)$ is a geodesic in N, we put for abbreviation

$$|c'|^2:=\left\langle\frac{dc}{ds},\frac{dc}{ds}\right\rangle.$$

If $c(s,t)=c_t(s)$ is a family of geodesics parametrized by t, then

$$J_t(s):=\frac{\partial}{\partial t}c(s,t)$$

is a Jacobi field, i.e. it satisfies

$$\nabla_{\partial/\partial s}\nabla_{\partial/\partial s}J_t(s) + R\left(\frac{\partial c}{\partial s}, J_t\right)\frac{\partial c}{\partial s} = 0$$

which easily follows from $\nabla_{\partial/\partial s}(\partial c/\partial s) = 0$ (the equation for geodesics) and the definition of the curvature tensor R. Let us assume the following curvature bounds on N:

$$-\omega^2 \leqslant K \leqslant \kappa^2$$

$$|K| \leqslant \Lambda^2 \qquad (= \max(\omega^2, \kappa^2)).$$

The required Jacobi field estimates are then

Estimate (J1) Let $c(s)$ be a geodesic with $|c'| \equiv r$, and $J(s)$ a Jacobi field along c with $J(0) = 0$. Assume $r \leqslant \pi/2\kappa$. Then

$$|J(s) - sJ'(s)| \leqslant |J(1)| \cdot \tfrac{1}{2}\Lambda^2 s^2 r^2$$

(and, of course, a corresponding estimate holds for Jacobi fields with $J(1) = 0$; one has merely to replace s by $(1 - s)$).

Estimate (J2) Let J be a Jacobi field along a geodesic $c(|c'| \equiv r)$, with $J(0)$ and $J'(0)$ linearly dependent; P_s = parallel transport along c. Then

$$|J(s) - P_s(J(0) + sJ'(0))| \leqslant |J(0)|(\cosh(\Lambda rs) - 1) + |J|'(0)\left(\frac{1}{\Lambda}\sinh(\Lambda rs) - rs\right).$$

Moreover, we shall need an estimate on the path dependence of parallel transport which will be taken from [BK, Section 6.2]:

Estimate (P) Let $\gamma_t(s), s, t \in [0, 1]$ be a smooth family of curves from p to q in N. Assume

$$\int_0^1 \int_0^1 \left|\frac{\partial}{\partial t}\gamma_t(s) \wedge \frac{\partial}{\partial s}\gamma_t(s)\right| \mathrm{d}s\,\mathrm{d}t \leqslant F.$$

Let $s \to v_i(s)$ be parallel vector fields along γ_i, i.e.

$$\nabla_{\partial/\partial s} v_i = 0 \qquad (i = 0, 1)$$

Suppose $v_0(0) = v_1(0) \in T_p N$. Then

$$|v_0(1) - v_1(1)| \leqslant \Lambda^2 F |v_0(0)|$$

(note that $|v_0(s)| = \text{const.} = |v_1(s)|$).

Theorem 2.3.1: *Let N be a d-dimensional Riemannian manifold of bounded geometry, i.e.*

$$i(N) := \inf_{p \in N} i(p) > 0$$

($i(p)$ is the injectivity radius of p) and

$$\Lambda^2 := \max |K| < \infty$$

(K denotes the sectional curvature of N). Let $h: D \to N$ be a weak minimal surface (cf. Definition 1.2.4). Then h is continuous in the interior of D.

Proof: Let

$$\rho_0 < \tfrac{1}{2}\min\left(\frac{\pi}{2\Lambda}, i(N)\right).$$

Let $z_0 \in D, r < d(z_0, \partial D)$. We assume that $z_1 \in B(z_0, r)$ and that for almost all $z \in D$ with $|z - z_0| = r$.

$$d(h(z), h(z_1)) > \rho' \tag{2.3.5}$$

with $0 < \rho' \leqslant \rho_0$.

We want to control ρ'. We let $a := h(z_1)$ and $0 < \rho \leqslant \rho'$, and choose $\gamma \in C^1(\mathbb{R})$ with

$$\begin{aligned} \gamma(t) &\equiv 1 && \text{for } t \leqslant \rho/2 \\ \gamma(t) &\equiv 0 && \text{for } t \geqslant \rho \\ \gamma'(t) &\leqslant 0 && \text{for } t \in \mathbb{R} \end{aligned}$$

(eventually, we shall let γ increase to the characteristic function $\chi_{(-\infty,\rho)}$. We want to use the test vector

$$\phi(h) := \gamma(d(h, a))(-\exp_h^{-1} a) \in T_h N. \tag{2.3.6}$$

This is the appropriate generalization of the Euclidean test vector. Thus, $\phi(h(z))$ is a vector field along $h(z)$. One also checks (for example using the estimate (2.3.9) below)

$$\int \langle \mathrm{d}\phi, \mathrm{d}\phi \rangle \leqslant c \int \langle \mathrm{d}h, \mathrm{d}h \rangle < \infty.$$

Finally, because of (2.3.5), $\phi(h(z))$ has compact support in $B(z_0, r)$. Therefore, $\phi(h(z))$ is an admissible variation, and since h is stationary

$$\int \langle \mathrm{d}h, \mathrm{d}\phi \rangle = 0. \tag{2.3.7}$$

Now

$$\mathrm{d}\phi = \nabla_{\partial/\partial x} \varphi \, \mathrm{d}x + \nabla_{\partial/\partial y} \varphi \, \mathrm{d}y$$

(here $\nabla_{\partial/\partial x}$ denotes the covariant derivatives of $\phi(h(x, y))$ in the direction h_x), and, for example,

$$\begin{aligned} \nabla_{\partial/\partial x} \phi = {} & \gamma(d(h, a)) \nabla_{\partial/\partial x}(-\exp_h^{-1} a) \\ & + \gamma'(d(h, a)) \langle (-\exp_h^{-1} a), h_x \rangle (-\exp_h^{-1} a) d(h, a)^{-1} \end{aligned} \tag{2.3.8}$$

We estimate $\nabla_{\partial/\partial t}\exp_{h(t)}^{-1} a$ as follows. Let $h(t)$ be a curve in $B(a,\rho)$ and

$$c(s,t) := \exp_{h(t)}(s\exp_{h(t)}^{-1} a).$$

Then

$$\frac{\partial}{\partial s}c(s,t)|_{s=0} = \exp_{h(t)}^{-1} a$$

and

$$\begin{aligned}\nabla_{\partial/\partial t}\exp_h^{-1} a &= \nabla_{\partial/\partial t}\frac{\partial}{\partial s}c(s,t)|_{s=0,t=0}\\ &= \nabla_{\partial/\partial s}\frac{\partial}{\partial t}c(s,t)|_{s=0,t=0}.\end{aligned}$$

For fixed $t, J_t(s) := (\partial/\partial t)c(s,t)$ is a Jacobi field along the geodesic $c(\cdot,t)$ with $J_t(0) = \dot{h}(t)$ and $J_t(1) = 0 \in T_aN$. Therefore

$$\nabla_{\partial/\partial t}\exp_h^{-1} a = J_t'(0)$$

($\cdot$denotes a derivative w.r.t. t,$'$ a derivative w.r.t. s). Estimate (J1) then implies

$$|\nabla_{\partial/\partial t}\exp_h^{-1} a + \dot{h}(t)| \leqslant \tfrac{1}{2}\Lambda^2 d^2(h,a)|\dot{h}(t)|. \tag{2.3.9}$$

This provides the desired estimate for $\nabla_{\partial/\partial t}\exp_h^{-1} a$.

We can now basically proceed as before. The conformality relations satisfied by h(i.e. $\langle h_x, h_x\rangle = \langle h_y, h_y\rangle, \langle h_x, h_y\rangle = 0$) imply (noting $|\exp_h^{-1} a| = d(h,a)$, $|\mathrm{d}h|^2 = |h_x|^2 + |h_y|^2 = 2|h_x|^2$)

$$\langle \exp_h^{-1} a, h_x\rangle^2 + \langle \exp_h^{-1} a, h_y\rangle^2 \leqslant \tfrac{1}{2}d^2(h,a)\cdot|\mathrm{d}h|^2. \tag{2.3.10}$$

We put

$$\gamma(t) = \psi(t/\rho)$$

(i.e. $\psi(t) \equiv 1$ for $t \leqslant \frac{1}{2}, \psi(t) \equiv 0$ for $t \geqslant 1, \psi'(t) \leqslant 0$) and

$$I(\rho) := \frac{1}{2}\int_{B(z_0,r)} |\mathrm{d}h|^2 \psi\left(\frac{d(h,a)}{\rho}\right).$$

Thus

$$t\gamma'(t) = -\rho\frac{\mathrm{d}}{\mathrm{d}\rho}\psi\left(\frac{t}{\rho}\right)$$

$$I'(\rho) = -\frac{1}{2\rho}\int_{B(z_0,r)} |\mathrm{d}h|^2 d(h,a)\gamma'(d(h,a))$$

because of (2.3.5) (note $\rho \leqslant \rho'$). Now

$$\begin{aligned}\langle \mathrm{d}h, \mathrm{d}\varphi\rangle = \gamma(&\langle h_{x'} - \nabla_{\partial/\partial x}\exp_h^{-1} a\rangle + \langle h_{y'} - \nabla_{\partial/\partial y}\exp_h^{-1} a\rangle)\\ &+ \gamma'(\langle -\exp_h^{-1} a, h_x\rangle^2 + \langle -\exp_h^{-1} a, h_y\rangle^2)\frac{1}{d(h,a)}.\end{aligned}$$

From (2.3.9) and (2.3.10) we then obtain from $\int\langle dh, d\varphi\rangle = 0$ that

$$2I(\rho) - \rho I'(\rho) \leqslant c\rho I(\rho) \tag{2.3.11}$$

where c depends only on Λ. With

$$\sigma(\rho) := (1 + c\rho)e^{2c\rho}$$

this implies in a standard way for $0 < \rho_1 \leqslant \rho_2 \leqslant \rho'$ that

$$\frac{\sigma(\rho_1)}{\rho_1^2} I(\rho_1) \leqslant \frac{\sigma(\rho_2)}{\rho_2^2} I(\rho_2). \tag{2.3.12}$$

We let ψ approach the characteristic function $\chi_{(-\infty,1)}$, let ρ_1 tend to zero (which lets the left-hand side tend to 2π times the density of the minimal surface $h(D)$ at $a = h(w_1)$) and choose $\rho_2 = \rho'$ and obtain

$$(\rho')^2 \leqslant \frac{\sigma(\rho')}{2\pi} \int_{B(z_0,r_0)} |dh|^2.$$

Given $\rho' \in (0, \rho_0)$, we can find r_0 so that

$$\int_{B(z_0,r)} |dh|^2 < \frac{4\pi}{\sigma(\rho_0)} (\rho')^2$$

and hence (2.3.5) is not possible, i.e. if $0 < r \leqslant r_0$, $z_1 \in B(z_0, r)$ then

$$\inf_{z \in \partial B(z_0,r)} d(h(z), h(z_1)) \leqslant \rho'. \tag{2.3.13}$$

Also, by the intermediate value theorem, we can find r with $\frac{1}{2}r_0 \leqslant r \leqslant r_0$

$$d(h(z), h(z')) \leqslant 2\pi(\log 2)^{-1/2} \left(\int_{B(z_0,r)} |dh|^2 \right)^{1/2} \tag{2.3.14}$$

for all z, $z' \in \partial B(z_0, r)$.

We can furthermore choose r_0 so small that

$$2\pi(\log 2)^{-1/2} \left(\int_{B(z_0,r_0)} |dh|^2 \right)^{1/2} \leqslant \rho'. \tag{2.3.15}$$

Thus, if $z_1, z_1' \in B(z_0, r)$, we can find $z, z' \in \partial B(z_0, r)$ for which the infimum is attained in (2.3.13), and using (2.3.14) and (2.3.15)

$$d(h(z_1), h(z_1')) \leqslant c_1 \rho'$$

$c_1 = c_1(\Lambda, d)$, and continuity follows.

Theorem 2.3.2: *Suppose N is a d-dimensional Riemannian manifold of bounded geometry, M a C^2-submanifold with bounded second fundamental form and suppose there exists $\rho_1 > 0$ so that the nearest point projection $\xi: \{q \in N: d(q, M) \leqslant \rho_1\} \to M$ is uniquely defined and differentiable. Suppose $h \in H^{1,2}(D, N)$ is a weak minimal surface with free boundary M. Then h is continous on the closure of D.*

Proof: Let Λ^2 be a bound for the absolute value of the sectional curvature of N, and μ a bound for the norm of the second fundamental form of M. For $q \in M$ we have the orthogonal decomposition

$$T_q N = \tau(q) + \nu(q)$$

into the subspace $\tau(q)$ tangential to M and the subspace $\nu(q)$ normal to M. We let τ and ν also denote the orthogonal projection of $T_q N$ onto the respective subspace.

We choose $\rho_0 < \frac{1}{2}\min(\pi/2\Lambda, i(N))$ so small that the nearest point projection

$$\xi: \{q \in M: d(q, M) \leqslant \rho_0\} \to M$$

is uniquely defined and smooth, and in particular that ρ_0 is smaller than the focal radius of M. Moreover, we assume

$$\tfrac{3}{2}[(\mu/\Lambda)\sinh(\Lambda\rho_0) + \cosh(\Lambda\rho_0)]d < 1.$$

If $d(p, q) \leqslant 2\rho_0$, we denote by P_{pq} parallel transport along the (unique) shortest geodesic from p to q. As already indicated, $\xi(q)$ is the nearest point projection of q onto M. Then

$$q = \exp_{\xi(q)}(d(q, \xi(q))\nu(\xi(q))).$$

We also define

$$\tilde{q} = \tilde{q}(q) := \exp_{\xi(q)}(-d(q, \xi(q))\nu(\xi(q)))$$

as the reflection of q across M. Continuity in the interior of D follows from Theorem 2.3.1, but the present argument actually can handle continuity in the interior and at the boundary simultaneously. Note that D is always the closed unit disk. Let $z_0 \in D$. We assume that $z_1 \in B(z_0, r)$ and that for almost all $z \in D$ with $|z - z_0| = r$

$$d(h(z), h(z_1)) > \rho'$$

and (2.3.16)

$$d(\tilde{h}(z), h(z_1)) > \rho'$$

with

$$0 < \rho' \leqslant \rho_0.$$

As before, we want to control ρ', and as before, we let

$$a := h(z_1)$$

and $0 < \rho \leqslant \rho'$, and choose $\gamma \in C^1(\mathbb{R})$ with

$$\begin{aligned} \gamma(t) &\equiv 1 && \text{for } t \leqslant \rho/2 \\ \gamma(t) &\equiv 0 && \text{for } t \geqslant \rho \\ \gamma'(t) &\leqslant 0 && \text{for } t \in \mathbb{R}. \end{aligned}$$

We want to use the test vector

$$\begin{aligned} \phi(h) := \gamma(d(h, a))(-\exp_h^{-1} a) + \gamma(d(\tilde{h}, a))P_{\xi(h)h}\{\tau(\xi(h))P_{\tilde{h}\xi(h)}(-\exp_{\tilde{h}}^{-1} a) \\ - \nu(\xi(h))P_{\tilde{h}\xi(h)}(-\exp_{\tilde{h}}^{-1} a)\} \in T_h N. \end{aligned}$$

$\phi(h(z))$ is a vector field along $h(z)$. Noting that if $h \in M$, $h = \tilde{h} = \xi(h)$, we see

$$\phi(h) \in \tau(h) \qquad \text{for } h \in M.$$

One also checks (for example, using the estimates (1)–(5) below)

$$\int \langle \mathrm{d}\phi, \mathrm{d}\phi \rangle \leqslant c \int \langle \mathrm{d}h, \mathrm{d}h \rangle < \infty.$$

Finally, because of (2.3.16), $\phi(h(z))$ has compact support in $B(z_0, r)$. Therefore $\phi(h(z)) \in T_{h(z)}N$, the corresponding vector field along h, is an admissible variation, and since h is stationary

$$\int \langle \mathrm{d}h, \mathrm{d}\phi \rangle = 0.$$

Now for example,

$$\begin{aligned}
\nabla_{\partial/\partial x}\phi = {} & \gamma(\mathrm{d}(h,a))\nabla_{\partial/\partial x}(-\exp_h^{-1}a) \\
& + \gamma'(d(h,a))((-\exp_h^{-1}a)\cdot h_x)(-\exp_h^{-1}a)d(h,a)^{-1} \\
& + \gamma(d(\tilde{h},a))\nabla_{\partial/\partial x}(P_{\xi(h)h})(\tau(\xi(h))P_{\tilde{h}\xi(h)}(-\exp_{\tilde{h}}^{-1}a) \\
& - v(\xi(h))P_{\tilde{h}\xi(h)}(-\exp_{\tilde{h}}^{-1}a) \\
& + \gamma(d(\tilde{h},a))P_{\xi(h)h}((\tau - v)\nabla_{\partial/\partial x}(P_{\tilde{h}\xi(h)})(-\exp_{\tilde{h}}^{-1}a)) \\
& + \gamma(d(\tilde{h},a))P_{\xi(h)h}((D\tau(\xi(h))D\xi(h)\cdot h_x \\
& - Dv(\xi(h))D\xi(h)\cdot h_x)P_{\tilde{h}\xi(h)}(-\exp_{\tilde{h}}^{-1}a) \\
& + \gamma(d(\tilde{h},a))P_{\xi(h)h}((\tau - v)P_{\tilde{h}\xi(h)}\nabla_{\partial/\partial x}(-\exp_{\tilde{h}}^{-1}a)) \\
& + \gamma'(d(\tilde{h},a))(-\exp_{\tilde{h}}^{-1}a\cdot D\tilde{h}\cdot h_x) \\
& \times P_{\xi(h)h}((\tau - v)P_{\tilde{h}\xi h)}(-\exp_{\tilde{h}}^{-1}a))d(\tilde{h},a)^{-1}.
\end{aligned}$$

In order to control all these terms, we need several estimates.

Estimate (1) We estimate $\nabla_{\partial/\partial t}\exp_{h(t)}^{-1}a$ as before (cf. (2.3.9)), namely

$$|\nabla_{\partial/\partial t}\exp_h^{-1}a + \dot{h}(t)| \leqslant \tfrac{1}{2}\Lambda^2 d^2(h,a)|\dot{h}(t)|.$$

Estimate (2) The derivatives of τ and v are controlled by the second fundamental form of M:

$$|D\tau| \leqslant \mu \qquad |Dv| \leqslant \mu.$$

Estimate (3) We estimate $D\xi$. Let $M(\sigma) := \{q \in N, d(q, M) = \sigma\}$, $0 < \sigma \leqslant \rho_0$. Let $h \in M(\sigma_0)$, $0 < \sigma_0 \leqslant \rho_0$, $h \in T_h M(\sigma_0) \subset T_h N$, $h(t)$ a curve in $M(\sigma_0)$ with $h(0) = h$, $\dot{h}(0) = v$.

Let ξ_σ be the nearest point projection from $M(\sigma_0)$ onto $M(\sigma)$, $0 < \sigma \leqslant \sigma_0$, which is also well defined, and let $v_\sigma(t)$ be the unit normal vector of $M(\sigma)$ at $\xi_\sigma(h(t))$ pointing into the direction of $M(\sigma_0)$, i.e.

$$h(t) = \exp_{\xi_\sigma(h(t))}(\sigma_0 - \sigma)v_\sigma(t).$$

We look at the family of geodesics

$$c_\sigma(s,t) := \exp_{\xi_\sigma(h(t))} s(\sigma_0 - \sigma) v_\sigma(t).$$

$J_\sigma(s) := \partial/\partial t c_\sigma(s,t)|_{t=0}$ is a Jacobi field with

$$\begin{aligned} J_\sigma(0) &= D\xi_\sigma v \\ J'_\sigma(0) &= (\sigma_0 - \sigma) D v_\sigma D\xi_\sigma \cdot v \\ J_\sigma(1) &= v. \end{aligned}$$

Now $\xi_\sigma: M(\sigma_0) \to M(\sigma)$, $0 < \sigma \leqslant \sigma_0$, is a diffeomorphism; hence $D\xi_\sigma: T_h M(\sigma_0) \to T_{\xi_\sigma(h)} M(\sigma)$ is surjective. Hence we can find a basis $v_\sigma^1, \ldots, v_\sigma^{d-1}$ of unit vectors of $T_h M(\sigma_0)$ for which the $D\xi_\sigma v_\sigma^i$ $(i = 1, \ldots, d-1)$ are mutually orthogonal principal curvature directions, i.e. $D\xi_\sigma v_\sigma^i$ and $Dv_\sigma \cdot D\xi_\sigma v_\sigma^i$ are linearly dependent. Our Jacobi field estimate [J2] implies

$$\begin{aligned} &|D\xi_\sigma \cdot v_\sigma^i + (\sigma_0 - \sigma) D v_\sigma D\xi_\sigma \cdot v_\sigma^i - P_{h(t)\xi_\sigma(h(t))} v_\sigma^i| \\ &\leqslant |D\xi_\sigma v_\sigma^i| (\cosh(\Lambda\sigma_0) - 1) \\ &\quad + \frac{\langle D v_\sigma D\xi_\sigma v_\sigma^i, D\xi_\sigma v_\sigma^i \rangle}{|D\xi_\sigma v_\sigma^i|} \left(\frac{1}{\Lambda} \sinh(\Lambda\sigma_0) - \sigma_0 \right). \end{aligned} \tag{2.3.17}$$

By choosing $\sigma_0 > 0$ small enough, and σ small compared to σ_0 (actually, we want to let σ tend to zero), we infer from the preceding inequality that $Dv_\sigma D\xi_\sigma v_\sigma^i$ remains bounded as $\sigma \to 0$. Therefore, as $\sigma \to 0$, the vectors $Dv_\sigma D\xi_\sigma v_\sigma^i$ converge to mutually orthogonal vectors in $T_{\xi(h)} N$. Also, of course, the v_σ^i converge to the unit vectors $v^i \in T_h M(\sigma_0)$, and $v^1, \ldots, v^{d-1}$ again constitute a basis of $T_h M(\sigma_0)$.

If $|D\xi_\sigma v_\sigma^i|$ stays bounded away from zero, $D\xi_\sigma v_\sigma^i$ and hence also $Dv_\sigma D\xi_\sigma v_\sigma^i$ converges to a vector tangent to M, whereas when $|D\xi_\sigma v_\sigma^1| \to 0$, $Dv_\sigma D\xi_\sigma v_\sigma^i$ has to converge to a vector normal to M, as $\sigma \to 0$. (Actually, if then $\sigma_0 \to 0$, this normal vector becomes unbounded at the rate $1/\sigma_0$, as one sees from (2.3.17).) From these observations and (2.3.17), we deduce, for $i = 1, \ldots, d-1$, that

$$|D\xi \cdot v^i - \tau(P_{h\xi(h)} v^i)| \leqslant c(\mu, \Lambda, \sigma_0) |v^i| \tag{2.3.18}$$

where μ is a bound for the second fundamental form of M, and $c = c(\mu, \Lambda, \sigma_0)$ is a constant depending on μ, Λ, and σ_0. Also

$$c(\mu, \Lambda, \sigma_0) \to 0 \qquad \text{as } \sigma_0 \to 0$$

Finally, if v_σ^d is a unit vector normal to $M(\sigma)$, then $D\xi \cdot v_\sigma^d = 0$, and (2.3.18) thus trivially also holds for v^d.

Moreover, (2.3.18) implies that, for small enough σ_0, $v^1, \ldots, v^d$ is close to being an orthonormal basis of $T_h M(\sigma_0)$, and then simple linear algebra and the triangle inequality imply that for an arbitrary $v \in T_h N$, $h \in M(\sigma_0)$, σ_0 sufficiently small,

$$|D\xi \cdot v - \tau(P_{h\xi(h)} v)| \leqslant c_1(\mu, \Lambda, \sigma_0, d) |v|. \tag{2.3.19}$$

(Namely, let v^k, $k = 1, \ldots, l$, be those vectors for which $D\xi \cdot v^k =: e^k$ is non-zero. We can choose e^k as an orthonormal basis of $T_{\xi(h)} M$ (of course, then v^k need no longer be a unit vector). We can then choose the remaining vectors v^j, $j=l+1,\dots,d$, as mutually orthonormal and orthogonal to the v^k, $k=1,\dots,l$, i.e.

$$\langle v^i, v^j \rangle = \delta_{ij} \qquad i=1,\dots,d \quad j=l+1,\dots,d.$$

Then (2.3.18) holds for this basis, and putting

$$e^j := P_{h\xi(h)} v^j \qquad \text{for } j=l+1,\dots,d$$

the vectors e^i form an orthonormal basis of $T_{\xi(h)}M$, and we obtain from (2.3.18)

$$|e^i - P_{h\xi(h)} v^i| \leqslant c(\mu, \Lambda, \sigma_0)|v^i|$$

hence also

$$|e^i - P_{h\xi(h)} v^i| \leqslant \frac{c}{1-c}|e^i|.$$

For simplicity, we shall omit the symbol $P_{h\xi(h)}$ for the rest of the argument. In particular, for small enough σ_0, the v^i are linearly independent. Thus, we can determine a matrix B with

$$Bv^i = e^i \qquad i=1,\dots,d$$

and we have to estimate $|Bv - v|$ for arbitrary v. By the last estimate, however,

$$|e^i - B^{-1}e^i| \leqslant \frac{c}{1-c}$$

hence

$$\|I - B^{-1}\| \leqslant \frac{dc}{1-c}$$

hence $B^{-1} \to I$ as $\sigma_0 \to 0$, hence $B \to I$ as $\sigma_0 \to 0$ and thus

$$|Bv - v| \leqslant c_1(\mu, \Lambda, \sigma_0, d)|v|$$

which is (2.3.19).)

Estimate (4) Equation (2.3.19) already gives the necessary control of $D\xi$. In the following, however, we shall need one more consequence of (2.3.17). Let (for $i=1,\dots,d-1$)

$$\omega_{\sigma 0}(v^i) := \lim_{\sigma \to 0} (\sigma_0 - \sigma) Dv_\sigma D\xi_\sigma \cdot v^i_\sigma \in T_{\xi(h)}N.$$

If $\lim_{\sigma\to 0}|D\xi_\sigma v^i_\sigma| > 0$ then $\omega_{\sigma 0}(v^i) = 0$, whereas if $\lim_{\sigma \to 0}|D\xi_\sigma v^i_\sigma| = 0$, (2.3.17) yields

$$|\omega_{\sigma 0}(v^i) - P_{h\xi(h)} v^i| \leqslant |\omega_{\sigma 0}(v^i)| \left(\frac{1}{\Lambda\sigma_0} \sinh(\Lambda\sigma_0) - 1 \right).$$

If v is normal to $M(\sigma_0)$, we put

$$\omega_{\sigma 0}(v) := P_{h\xi(h)} v.$$

Therefore, in any case, for $v \in T_h N$ with $h \in M(\sigma_0)$

$$|\omega_{\sigma_0}(v) - vP_{h\xi(h)}v| \leqslant c_2(\mu, \Lambda, \sigma_0, d). \tag{2.3.20}$$

Estimate (5) We control $D\tilde{h}$.

We consider $\tilde{h}$ as a map

$$\tilde{h}: M(\sigma_0) \to M(\sigma_0).$$

We look at the following family of geodesics, letting $c(t) \in M(\sigma_0)$, $h(0) = h$, $\dot{h}(0) = v \in T_h M(\sigma_0)$:

$$c(s,t) := \exp_{h(t)}(s \exp_{h(t)}^{-1} \xi(h(t)))$$

and the associated Jacobi fields

$$J_t(s) = \frac{\partial}{\partial t} c(s,t)$$

$$c(0,t) = h(t) \qquad c(1,t) = \xi(h(t)) \qquad c(2,t) = \tilde{h}(t)$$

$$J_0(1) = D\xi \cdot v \qquad J_0'(1) = Dv \cdot D\xi v \qquad J_0(2) = D\tilde{h} \cdot v.$$

If $D\xi \cdot v \neq 0$ and is a principle curvature vector of M, we obtain as in estimate (3)

$$|D\xi \cdot v - P_{\tilde{h}(t)\xi(h(t))} D\tilde{h} \cdot v| \leqslant c(\mu, \Lambda, \sigma_0) |D\xi \cdot v|$$

and hence, using (2.3.18) (and various times the triangle inequality),

$$|D\tilde{h} \cdot v - P_{h(t)\tilde{h}(t)} D\tilde{h} \cdot v| \leqslant \frac{2c}{1-c} |v|.$$

Likewise, if $v \in T_h M(\sigma_0)$ satisfies $D\xi \cdot v = 0$, we obtain with the help of (2.3.20)

$$|D\tilde{h} \cdot v + P_{h(t)\tilde{h}(t)} v| \leqslant c_3(\mu, \Lambda, \sigma_0, d) |v|.$$

Finally, if v is orthogonal to $T_h M(\sigma_0)$, then

$$D\tilde{h} \cdot v = -P_{\tilde{h}(t)h(t)} v.$$

Altogether, we obtain, for any $v \in T_h N$

$$|P_{\tilde{h}(t)\xi(h(t))} D\tilde{h} \cdot v - (\tau - v) P_{h(t)\xi(h(t))} v| \leqslant c_4(\mu, \Lambda, \sigma_0, d) |v|. \tag{2.3.21}$$

Estimate (6) We estimate $\nabla_{\partial/\partial t} P_{h(t)\xi(h(t))}$.

Let $v \in T_h N$, $h(t)$ be a curve in N with $h(0) = h$. Let $v_1(t) \in T_{h(t)} N$ be the vector obtained by transporting v parallely along $h(\sigma)$ from $h(0)$ to $h(t)$, and let $v_2(t)$ be the vector obtained by transporting v first parallely along the geodesic from $h(0)$ to $\xi(h(0))$, then along $\xi(h(\sigma))$ from $\xi(h(0))$ to $\xi(h(t))$ and finally along the geodesic from $\xi(h(t))$ to $h(t)$. Then Jacobi field estimate (P) yields

$$|v_1(t) - v_2(t)| \leqslant \rho\Lambda \int_0^t |\dot{h}(\sigma)| \, d\sigma \cdot |v|.$$

Therefore

$$|\nabla_{\partial/\partial t} P_{h(t)\xi(h(t))}| \, |_{t=0} \leqslant \rho\Lambda |\dot{h}(0)|. \tag{2.3.22}$$

Similarly

$$|\nabla_{\partial(\partial_t P_{h(t)\tilde{h}(t)}}| |_{t=0} \leqslant 2\rho\Lambda|\dot{h}(0)|. \tag{2.3.23}$$

We have now estimated all terms occurring in $d\varphi$. As before, we have inequality (2.3.10), and we put again

$$\gamma(t)=\psi(t/\rho).$$

We then define

$$\tilde{I}(\rho):=\frac{1}{2}\int_{B(z_0,r)}|dh|^2\left(\psi\left(\frac{d(h,a)}{\rho}\right)+\psi\left(\frac{d(\tilde{h},a)}{\rho}\right)\right)dz.$$

Thus (with $r\gamma'(t)=-\rho(d/d\rho)\psi(t/\rho)$)

$$\tilde{I}'(\rho)=-\frac{1}{2\rho}\int_{B(z_0,r)}|dh|^2(d(h,a)\gamma'(d(h,a))+d(\tilde{h},a)\gamma'(d(\tilde{h},a)))\,dz.$$

If we then evaluate $\int\langle dh, d\phi\rangle=0$ and use all the preceding estimates, we obtain this time

$$2\tilde{I}(\rho)-\rho\tilde{I}'(\rho)\leqslant k_1\rho\tilde{I}(\rho)+k_2\rho^2\tilde{I}'(\rho)$$

where k_1 and k_2 depend only on μ and Λ. With

$$k:=\max(k_1,k_2)$$
$$\sigma(\rho):=(1+k\rho)e^{2k\rho}$$

integrating this differential inequality again implies in a standard way for $0<\rho_1\leqslant\rho_2\leqslant\rho'$

$$\frac{\sigma(\rho_1)}{\rho_1^2}\tilde{I}(\rho_1)\leqslant\frac{\sigma(\rho_2)}{\rho_2^2}\tilde{I}(\rho_2). \tag{2.3.24}$$

We let ψ approach the characteristic function $\chi_{(-\infty,1)}$, let ρ_1 tend to zero (which lets the left-hand side tend to 2π times the density of the minimal surface $h(D)$ at $a=h(z_1)$ as observed earlier in this section) and choose $\rho_2=\rho'$ and obtain

$$(\rho')^2\leqslant\frac{\sigma(\rho')}{4\pi}\int_{B(z_0,r)}|dh|^2.$$

Given $\rho'\in(0,\rho_0)$, we can find r_0 so that

$$\int_{B(z_0,r)}|dh|^2<\frac{4\pi}{\sigma(\rho_0)}(\rho')^2$$

and hence (2.3.16) is not possible, i.e. if $0<r\leqslant r_0$, $z_1\in B(z_0,r)$ then

$$\inf_{z\in\partial B(z_0,r)}\{d(h(z),h(z_1)),d(\tilde{h}(z),h(z_1))\}\leqslant\rho'. \tag{2.3.25}$$

Also, by the intermediate value theorem, we can find r with $\frac{1}{2}r_0 \leqslant r \leqslant r_0$ and

$$d(h(z_2), h(z_3)) \leqslant 2\pi(\log 2)^{-1/2} \left(\int_{B(z_0,r)} |dh|^2 \right)^{1/2} \tag{2.3.26}$$

for all z_2, $z_3 \in \partial B(z_0, r)$. We can furthermore choose r_0 so small that

$$2\pi(\log 2)^{-1/2} \left(\int_{B(z_0,r)} |dh|^2 \right)^{1/2} \leqslant \rho'. \tag{2.3.27}$$

Finally, we can also choose r in (2.3.26) in such a way that for the corner points $z_4, z_5 \in \partial B(z_0, r) \cap D$

$$h(z_4), h(z_5) \in M$$

that is

$$h(z_4) = \tilde{h}(z_4) \qquad h(z_5) = \tilde{h}(z_5).$$

Then (2.3.26) and the estimate (5) imply

$$\begin{aligned} d(\tilde{h}(z_2), \tilde{h}(z_3)) &\leqslant k_3 d(h(z_2), h(z_3)) \\ &\leqslant k_4 \left(\int_{B(z_0,r)} |dh|^2 \right)^{1/2} \end{aligned} \tag{2.3.28}$$

and also

$$d(\tilde{h}(z_2), h(z_3)) \leqslant k_4 \left(\int_{B(z_0,r)} |dh|^2 \right)^{1/2} \tag{2.3.29}$$

($z_2, z_3 \in \partial B(z_0, r)$, k_3 and k_4 depending only on μ, Λ and d).

Thus, if $z_1, z_1' \in B(z_0, r)$, we can find z, $z' \in \partial B(z_0, r)$ for which the infimum is attained in (2.3.25), and using (2.3.26), (2.3.28) or (2.3.29) and the triangle inequality and (2.3.27)

$$d(h(z_1), h(z_1')) \leqslant k_6 \rho'$$

$k_6 = k_6(\mu, \Lambda, d)$ and continuity follows.

q.e..d.

Remarks: Continuity of minimal surfaces with free boundaries, without imposing an area-minizing condition, was first obtained by Grüter, Hildebrandt and Nitsche [GHN1] and Dziuk [Dz1]. Extensions of these results were found by Dziuk [Dz2] and Ye [Ye1]. These results assume codim $M = 1$ and stronger conditions than in Theorem 2.3.2 for the manifold N or the free boundary M.

Theorem 2.3.2 was obtained in [J10] for codim $M = 1$ and in [J11] for arbitrary codimension.

Since the codimension of M in Theorem 2.3.2 is arbitrary, it also applies to the situation of the Plateau problem without requiring monotonicity of the boundary values.

Let us also point out that one can even prove continuity if the free boundary is only piecewise C^2 provided the various pieces (of possibly varying dimensions) satisfy similar assumptions as above, and the angle between two pieces is always bounded away from zero; see [J11].

Concerning weak H-surfaces, we have Grüter's theorem [Gr1]:

Theorem 2.3.3: *Let N again be a manifold of bounded geometry, $h \in H^{1,2}(D,N)$ be a weak H-surface in N, i.e. a weakly conformal solution of* (1.2.25), *and assume that H is (measurable and) bounded. Then h is continuous in the interior of D.*

Proof: We shall give the argument only for $N = \mathbb{R}^d$, because a general Riemannian manifold can then be treated with the constructions used in the proof of Theorem 2.3.1; for simplicity, we also only treat the case $H \in C^\alpha$, $0 < \alpha < 1$, in detail. We use the same notations as in the beginning of this section where we showed continuity of a weak minimal surface in $\mathbb{R}^d$. In particular, we use again the test vector

$$\varphi(h) = \psi\left(\frac{|h-a|}{\rho}\right)(h-a)$$

where ψ will eventually increase to the characteristic function $\chi_{(-\infty,1)}$. Since h is a weak H-surface

$$\begin{aligned}\int_D (h_x\varphi_x + h_y\varphi_y) &= -\int 2H(h_x, h_y)\varphi \\ &= -\int 2H(h_x, h_y)\psi\left(\frac{|h-a|}{\rho}\right)(h-a).\end{aligned}$$

With

$$I(\rho) := \tfrac{1}{2}\int_{B(z_0,r)} |\mathrm{d}h|^2 \psi\left(\frac{|h-a|}{\rho}\right)$$

we compute

$$2I(\rho) - \rho I'(\rho) \leqslant -\int 2H(h)(h_x, h_y)\psi\left(\frac{|h-a|}{\rho}\right)(h-a).$$

Therefore

$$\frac{\mathrm{d}}{\mathrm{d}\rho}\left(\frac{I(\rho)}{\rho^2}\right) \geqslant -\Gamma\frac{I(\rho)}{\rho^2}$$

with

$$\Gamma := 2\|H\|_{L^\infty} \qquad (<\infty \text{ by assumption}).$$

This yields for $0<\rho_1\leqslant\rho_2\leqslant\rho'$

$$e^{\Gamma\rho_1}\frac{I(\rho_1)}{\rho^2}\leqslant e^{\Gamma\rho_2}\frac{I(\rho_2)}{\rho_2^2} \tag{2.3.30}$$

the monotonicity formula for H-surfaces.

If $\rho_1\to 0$, the limit of the left-hand side is again bounded from below by 2π, if $a=h(z_1)$, $z_1\in D$ for almost all $z_1\in D$. (This is a consequence of the weak conformality by a direct measure theoretic argument, using e.g. the fact that an $H^{1,2}(D,\mathbb{R}^d)$ map is almost everywhere approximately differentiable in the sense of Federer [F, Section 3.2.16] cf. [Gr1]; if $H\in C^\alpha$, this also follows from Corollary 2.6.1 below.) One can therefore use (2.3.30) to obtain continuity for weak H-surfaces in the same way as was done above for weak minimal surfaces.

q.e.d.

Remark: One can also show continuity of weak H-surfaces with free boundaries; cf. [GHN2].

If u is not necessarily weakly conformal, but only stationary in the sense of Definition 1.2.5, u is still continuous by the following result of Schoen [S].

Theorem 2.3.4: *Let $u\in H^{1,2}(D,N)$, N a Riemannian manifold of bounded geometry, $H\in L^\infty(N)$. Let u be a weak solution of the Euler–Lagrange equations for a two-dimensional geometric variational problem, i.e. satisfy* (1.2.23), (1.2.24), *or, more generally,* (1.2.25). *Suppose u is also stationary in the sense of Definition 1.2.5. Then u is continuous in the interior of D.*

Proof: (We shall use a variant of Schoen's argument, due to Grüter [Gr3].) By Lemma 1.2.4,

$$\varphi(z):=g_{jk}(u_x^j u_x^k-u_y^j u_y^k-2\mathrm{i}u_x^j u_x^k)$$

($z=x+\mathrm{i}y\in D$, (g_{jk}) the metric tensor of N in local coordinates) is holomorphic. We now look at maps

$$h=(u,v)\colon D\to N\times\mathbb{C}.$$

We want to achieve that v is harmonic and h (weakly) conformal. h is (weakly) conformal, if

$$\frac{\partial}{\partial z}v\cdot\frac{\partial}{\partial z}\bar{v}=-\tfrac{1}{2}\varphi,\qquad (v=v^1+\mathrm{i}v^2,\bar{v}=v^1-\mathrm{i}v^2) \tag{2.3.31}$$

Since φ is holomorphic, there exists a holomorpic Ψ with

$$\Psi'=\frac{\partial}{\partial z}\Psi=-\tfrac{1}{2}\varphi.$$

Then

$$v:=\bar{z}+\Psi(z)$$

satisfies (2.3.31), as well as $\Delta v = 0$. Hence h is weakly conformal, and by Theorem 2.3.3 h and therefore also u is continuous.

q.e.d.

Remark. Recently, Helein [Hl 2, 3] showed that if N has enough symmetries, e.g. $N = s^n$, Theorem 2.3.4 holds even without the assumption that n is stationary in the sense of Definition 1.2.5.

2.4. Removability of isolated singularities

In this section, we want to prove the following result of Sacks and Uhlenbeck [SkU1]:

Theorem 2.4.1: *Suppose Ω is open in $\mathbb{R}^2$, $x_0 \in \Omega$, $u \in C^2(\Omega \backslash \{x_0\})$, $\nabla u \in L^2_{\text{loc}}(\Omega)$, and u solves* (1.2.7), *i.e. the Euler–Lagrange equations of a two-dimensional conformally invariant variational integral $I(u)$, in $\Omega \backslash \{x_0\}$. Suppose g_{ij}, $b_{ij} \in C^{1,\alpha} \cap L^\infty$, and $g_{ij}\xi^i\xi^j \geqslant \lambda |\xi|^2$ with $\lambda > 0$ for all $\xi \in \mathbb{R}^d$. Then the singularity at x_0 is removable, i.e. u may be extended to all of Ω as a C^2-solution of* (1.2.7).

In order to make this result applicable to the case where the target is a manifold M, we embed this manifold differentiably into some $\mathbb{R}^d$. This will enable us in particular to exploit the linear structure of $\mathbb{R}^d$ in order to construct suitable comparison functions below.

We shall display Grüter's presentation [Gr2] of the proof. If we do not need the special structure of (1.2.7), we write

$$\Delta u^i = \Gamma^i(u, \nabla u) \tag{2.4.1}$$

with

$$|\Gamma^i(u, \nabla u)| \leqslant \Gamma |\nabla u|^2 \qquad \Gamma = \text{const.} < \infty. \tag{2.4.2}$$

The proof of Theorem 2.4.1 depends on several lemmata. W.l.o.g.

$$x = 0$$

and with $B_r := B(0, r)$

$$B_2 \subset\subset \Omega.$$

We first estimate higher integral norms in terms of the Dirichlet integral of u:

Lemma 2.4.1: *There exists $\epsilon = \epsilon(\Gamma) > 0$, with the property that for all solutions $u \in C^2(B_1, \mathbb{R}^d)$ of* (2.4.1) *satisfying $\int_{B_1} |\nabla u|^2 < \epsilon$ the following holds: for any $p \geqslant 1$ and $0 < \rho < 1$ there exists $K_1 = K_1(p, \rho, \epsilon) > 0$ with*

$$\| \nabla u \|_{H^{1,p}(B_\rho)} \leqslant K_1 \| \nabla u \|_{L^2(B_1)}. \tag{2.4.3}$$

Proof: Let $0 < \rho_1 < \rho_2 \leqslant 1$. Let $\eta \in C_0^\infty(B_{\rho_2})$ be a cut-off function with $\eta = 1$ on B_{ρ_1}, $|\nabla \eta| \leqslant 2/(\rho_2 - \rho_1)$.

The differential equation (2.4.1) implies for $x \in B_1$

$$|\Delta(\eta u)(x)| \leqslant \|\Delta\eta\|_{L^\infty(B_1)}|u(x)| + 2\|\nabla\eta\|_{L^\infty(B_1)}|\nabla u(x)| + \Gamma|\nabla(\eta u)(x)||\nabla u(x)| + \Gamma\|\nabla\eta\|_{L^\infty(B_1)}|\nabla u(x)||u(x)|.$$

We put $c_1 := 2/(\rho_2 - \rho_1)$ and obtain for $r \geqslant 1$

$$\|\Delta(\eta u)\|_{L^r(B_{\rho_2})} \leqslant \Gamma \| |\nabla(\eta u)||\nabla u| \|_{L^r(B_{\rho_2})} + c_1\Gamma\| |u||\nabla u| \|_{L^r(B_{\rho_2})} + c_1\|u\|_{H^{1,r}(B_{\rho_2})}. \tag{2.4.4}$$

As ηu has compact support in B_{ρ_2}, the Calderon–Zygmund inequality implies for every $r > 1$ the existence of a number $\gamma(r) > 0$ with

$$\|\eta u\|_{H^{2,r}(B_{\rho_2})} \leqslant \gamma(r)\|\Delta(\eta u)\|_{L^r(B_{\rho_2})}.$$

Together with (2.4.4) this yields

$$\|\eta u\|_{H^{2,r}(B_{\rho_2})} \leqslant \gamma(r)\Gamma\| |\nabla(\eta u)||\nabla u| \|_{L^r(B_{\rho_2})} + \gamma(r)c_1\Gamma\| |u||\nabla u| \|_{L^r(B_{\rho_2})} + \gamma(r)c_1\|u\|_{H^{1,r}(B_{\rho_2})}. \tag{2.4.5}$$

We observe that if $c \equiv$ const., then $v := u - c$ solves

$$\Delta v^i = \Gamma^i(v - c, \nabla v)$$

i.e. an equation of the same type as (2.4.1). Therefore, our considerations so far, in particular (2.4.5), also hold for $u + c$, $c \equiv$ const. As a consequence of this observation, we may assume $\int_{B_1} u = 0$.

We first use (2.4.5) for $r = 4/3$, $\rho_1 = (\rho + 1)/2$, $\rho_2 = 1$. Since $2r/(2 - r) = 4$, the Poincaré and Sobolev inequalities imply

$$\| |\nabla(\eta u)||\nabla u| \|_{L^r(B_1)} \leqslant \|\nabla u\|_{L^2(B_1)}\|\nabla(\eta u)\|_{L^4(B_1)} \leqslant c_2\|\nabla u\|_{L^2(B_1)}\|\eta u\|_{H^{2,r}(B_1)}$$

$$\|u\|_{H^{1,r}(B_1)} \leqslant c_3\|\nabla u\|_{L^2(B_1)}$$

$$\| |u||\nabla u| \|_{L^r(B_1)} \leqslant \|\nabla u\|_{L^2(B_1)}\|u\|_{L^4(B_1)} \leqslant c_4\|\nabla u\|_{L^2(B_1)}$$

with suitable constants c_2, c_3, c_4.

Using these inequalitis in (2.4.5), we obtain

$$(1 - \gamma(4/3)c_2\Gamma\sqrt{\epsilon})\|\eta u\|_{H^{2,4/3}(B_1)} \leqslant (\gamma(4/3)c_1c_4\Gamma\sqrt{\epsilon} + \gamma(4/3)c_1c_3)\|\nabla u\|_{L^2(B_1)}. \tag{2.4.6}$$

For small enough $\epsilon > 0$, $1 - \gamma(4/3)c_2\Gamma\sqrt{\epsilon} \geqslant \frac{1}{2}$, and since also

$$\|\nabla u\|_{L^4(B_{\rho_1})} \leqslant c_5\|\eta u\|_{H^{2,4/3}(B_1)}$$

where c_5 behaves like $(1 - \rho)^{-1}$, we obtain from (2.4.6) (noting $\rho_1 = (\rho + 1)/2$)

$$\|\nabla u\|_{L^4(B_{(\rho+1)/2})} \leqslant c_6\|\nabla u\|_{L^2(B_1)} \tag{2.4.7}$$

where c_6 depends on ρ, Γ, and ϵ.

In order to prove the lemma, we now estimate the $H^{1,p}$-norm of ∇u on B_ρ by the L^4-norm of ∇u on B_{ρ_1} ($\rho_1 = (\rho+1)/2$). As observed above, we may add a constant to u without changing the validity of (2.4.5). We consider $\bar{u} = u - \fint_{B_{\rho_2}} u$ where $\fint_A u$ denotes the mean value of u on the set A.

We now apply (2.4.5) with $\rho_1 = (3\rho+1)/4$, $\rho_2 = (\rho+1)/2$, $r = 2$. Equation (2.4.5) implies with Hölder's inequality (note $\nabla\bar{u} = \nabla u$)

$$\|\eta\bar{u}\|_{H^{2,2}(B_{\rho_2})} \leqslant c_7 \|\nabla(\eta\bar{u})\|_{L^4(B_{\rho_2})} \|\nabla u\|_{L^4(B_{\rho_2})} + c_8 \|\bar{u}\|_{L^4(B_{\rho_2})} \|\nabla u\|_{L^4(B_{\rho_2})} + c_9 \|\bar{u}\|_{H^{1,2}(B_{\rho_2})} \tag{2.4.8}$$

where the constants depend on Γ and ρ. By Poincaré's inequality

$$\|\bar{u}\|_{L^4(B_{\rho_2})} \leqslant c_{10} \|\nabla u\|_{L^4(B_{\rho_2})}$$

and by Hölder's inequality then

$$\|\bar{u}\|_{H^{1,2}(B_{\rho_2})} \leqslant c_{11} \|\nabla u\|_{L^4(B_{\rho_2})}$$

with constants depending on ρ.

Since $\eta \equiv 1$ on B_{ρ_1}, (2.4.8) therefore implies

$$\begin{aligned}\|\nabla u\|_{H^{1,2}(B_{\rho_1})} &\leqslant \|\bar{u}\|_{H^{1,2}(B_{\rho_1})}\\ &\leqslant c_{12}(1 + \|\nabla u\|_{L^4(B_{\rho_2})}) \|\nabla u\|_{L^4(B_{\rho_2})}\end{aligned}$$

c_{12} depending again on Γ and ρ.

We next consider

$$\bar{\bar{u}} := u - \fint_{B_{(3\rho+1)/4}} u$$

The Poincaré and Sobolev inequalities then give ($\nabla\bar{\bar{u}} = \nabla u$):

$$\begin{aligned}\|\bar{\bar{u}}\|_{H^{1,2p}(B_{(3\rho+1)/4})} &\leqslant c_{13} \|\nabla u\|_{H^{1,2}(B_{(3\rho+1)/4})}\\ &\leqslant c_{14}(1 + \|\nabla u\|_{L^4(B_{(\rho+1)/2})}) \|\nabla u\|_{L^4(B_{(\rho+1)/2})}.\end{aligned} \tag{2.4.9}$$

We then use (2.4.5) with $r = p$, $\rho_1 = \rho$, $\rho_2 = (3\rho+1)/4$ and obtain

$$\begin{aligned}\|\bar{u}\|_{H^{2,p}(B_{\rho_1})} &\leqslant \|\eta\bar{\bar{u}}\|_{H^{2,p}(B_{(\rho_2)})}\\ &\leqslant c_{15}(\|\nabla(\eta\bar{\bar{u}})\|_{L^{2p}(B_{\rho_2})} \|\nabla u\|_{L^{2p}(B_{\rho_2})} + \|\bar{\bar{u}}\|_{L^{2p}(B_{\rho_2})} \|\nabla u\|_{L^{2p}(B_{\rho_2})} + \|\bar{\bar{u}}\|_{H^{1,p}(B_{\rho_2})})\end{aligned}$$

with c_{15} depending on ρ and Γ; hence

$$\|\bar{u}\|_{H^{2,p}(B_\rho)} \leqslant c_{16}(1 + \|\nabla u\|^3_{L^4(B_{(\rho+1)/2})}) \|\nabla u\|_{L^4(B_{(\rho+1)/2})} \tag{2.4.10}$$

from (2.4.9).

Combining (2.4.10) with (2.4.7) (note $\rho_1 = (\rho+1)/2$ in (2.4.7)), the claim follows.

q.e.d

We now use Lemma 2.4.1 to obtain a pointwise gradient estimate away from the singularity:

Lemma 2.4.2: *Let* ϵ, K_1 *be the constants of Lemma 2.4.1 with* $\rho = \frac{1}{2}$, $\rho = 4$.

Let $u \in C^2(B_2 \setminus \{0\})$ be a solution of (2.4.1) with

$$\int_{B_2} |\nabla u|^2 < \epsilon.$$

There exists a constant K_2 which can be computed in terms of K_1 with the property that for any $x_0 \in B_1 \setminus \{0\}$

$$|\nabla u(x_0)| \leqslant K_2 \left(\fint_{B_{3|x_0|/2}} |\nabla u|^2 \right)^{1/2}. \tag{2.4.11}$$

Proof: We define for $x_0 \in B_1 \setminus \{0\}$

$$\tilde{u}(x) := u(x_0 + x(|x_0|/2)) \qquad x \in B_1.$$

Then $\tilde{u} \in C^2(B_1)$ is a solution of (2.4.1) on B_1. The conformal invariance of the Dirichlet integral implies

$$\int_{B_1} |\nabla \tilde{u}|^2 = \int_{B(x_0, |x_0|/2)} |\nabla u|^2 \leqslant \int_{B_2} |\nabla u|^2 < \varepsilon$$

so that the assumptions of Lemma 2.4.1 are satisfied. Together with Sobolev's inequality we conclude

$$|\nabla \tilde{u}(0)| \leqslant c_{17} \| \nabla \tilde{u} \|_{H^{1/4}(B_{1/2})} \leqslant c_{17} K_1 \| \nabla \tilde{u} \|_{L^2(B_1)}.$$

This yields (2.4.11).

q.e.d.

We recall that since u is a solution of (1.2.7) on $B_2 \setminus \{0\}$,

$$\varphi(z) := g_{jk}(u_x^j u_x^k - u_y^j u_y^k - 2i u_x^j u_y^k)$$

is holomorphic on $B_2 \setminus \{0\}$.

We next show that $z\varphi(z)$ is holomorphic on all of B_2, if u has finite Dirichlet integral:

Lemma 2.4.3: *Let $u \in C^2(B_2 \setminus \{0\})$ be a solution of* (1.2.7) *with*

$$\int_{B_2} |\nabla u|^2 < \infty.$$

Then $\psi(z) := z\varphi(z)$ is holomorphic on B_2.

Proof: Since ψ is holomorphic in $B_2 \setminus \{0\}$, we only need to consider the situation near 0. We may therefore assume

$$\int_{B_2} |\nabla u|^2 < \epsilon$$

with ϵ as in Lemma 2.4.2. We now show $\lim_{z \to 0} z\psi(z) = 0$. Then $\psi(z)$ has a removable singularity at 0 and is holomorphic. For $z \neq 0$

$$|\varphi(z)| \leqslant \mu |\nabla u(z)|^2$$

with $\mu := 2\sup g_{jk}$. From (2.4.11) then

$$|z\psi(z)| = |z^2\varphi(z)| \leqslant \mu|z|^2|\nabla u(z)|^2$$
$$\leqslant c_{18}\int_{B_{3|z|/2}} |\nabla u|^2$$

and this tends to 0 as z tends to 0 by the absolute continuity of the integral.

q.e.d.

We now write things in polar coordinates $z = re^{i\theta}$.
We first observe

$$\frac{1}{r^2} g_{jk}(u)(u_\theta^j u_\theta^k + r^2 u_r^j u_r^k) = g_{jk}(u)\nabla u^j \cdot \nabla u^k. \tag{2.4.12}$$

Also

$$\mathrm{Re}(z\psi(z)) = g_{jk}(u)(u_\theta^j u_\theta^k - r^2 u_r^j u_r^k). \tag{2.4.13}$$

By Lemma 2.4.3, we may apply Cauchy's integral theorem to obtain

$$\int_{|z|=r} \psi(z)\,\mathrm{d}z = 0.$$

Hence

$$0 = \mathrm{Im}\left(\int_{|z|=r} \psi(z)\,\mathrm{d}z\right) = \mathrm{Im}\left(\mathrm{i}\int_0^{2\pi} \varphi(re^{i\theta})(re^{i\theta})^2\,\mathrm{d}\theta\right) = \int_0^{2\pi} \mathrm{Re}(z\psi)(z))\,\mathrm{d}\theta.$$

Recalling (2.4.13), we get

$$\int_0^{2\pi} g_{jk}(u)u_\theta^j u_\theta^k\,\mathrm{d}\theta = r^2\int_0^{2\pi} g_{jk}(u)u_r^j u_r^k\,\mathrm{d}\theta. \tag{2.4.14}$$

Proof of Theorem 2.4.1: We may assume w.l.o.g.

$$x_0 = 0 \qquad \Omega = B_2,$$
$$\int_{B_2} |\nabla u|^2 < \epsilon$$

where ϵ depends on Γ and on λ, μ satisfying $\lambda|\xi|^2 \leqslant g_{ij}\xi^i\xi^j \leqslant \mu|\xi|^2$; the precise value of ϵ will be determined below.

The principal idea now is to choose a rotationally symmetric, piecewise harmonic comparison function q. To this end, we let $\mu_m (m = 0, 1, 2, \ldots)$ be the mean value of u on the circle with radius 2^{-m}, i.e.

$$\mu_m := \frac{1}{2\pi}\int_0^{2\pi} u(2^{-m}e^{i\theta})\,\mathrm{d}\theta.$$

We also put

$$\alpha_m := \frac{1}{\log 2}(\mu_{m-1} - \mu_m)$$

$$\beta_m := m\mu_{m-1} + (1-m)\mu_m.$$

We then define

$$q(|x|) := \alpha_m \log|x| + \beta_m \qquad \text{for } 2^{-m} \leqslant |x| < 2^{-m+1}.$$

q is continuous on $\bar{B}_1 \setminus \{0\}$, harmonic on the annulus

$$A_m := \{x : 2^{-m} < |x| < 2^{-m+1}\}$$

with boundary values μ_m and μ_{m-1}, respectively.

Furthermore, for $x \in A_m$, we can control the difference between $u(x)$ and $q(|x|)$ as follows:

$$\begin{aligned} |q(|x|) - u(x)| &\leqslant |q(|x|) - \mu_{m-1}| + |\mu_{m-1} - u(x)| \\ &\leqslant |\mu_m - \mu_{m-1}| + |\mu_{m-1} - u(x)| \\ &\leqslant 2 \max_{x_1, x_2 \in \bar{A}_m} |u(x_1) - u(x_2)| \\ &\leqslant 2\pi 2^{-m+1} \max_{x_1 \in \bar{A}_m} |\nabla u(x_1)| \end{aligned}$$

and by Lemma 2.4.2, if ϵ is chosen to satisfy the assumption of Lemma 2.4.2, then

$$|q(|x|) - u(x)| \leqslant c_{19}\sqrt{\epsilon}.$$

We have thus obtained the estimate

$$\sup_{0<|x|\leqslant 1} |q(|x|) - u(x)| \leqslant c_{19}\sqrt{\epsilon}. \tag{2.4.15}$$

We now write

$$\int_{A_m} |\nabla(q-u)|^2 = -\int_{A_m} (q-u)\cdot\Delta(q-u) + \int_{\partial A_m} (q-u)\cdot\frac{\partial}{\partial n}(q-u)$$

and

$$\begin{aligned} &\int_{\partial A_m} (q-u)\cdot\frac{\partial}{\partial n}(q-u) \\ &= 2^{1-m}\int_0^{2\pi} (q(2^{1-m}) - u(2^{1-m}e^{i\theta}))\cdot(q'(2^{1-m}) - u_r(2^{1-m}e^{i\theta}))\,d\theta \\ &\quad - 2^{-m}\int_0^{2\pi} (q(2^{-m}) - u(2^{-m}e^{i\theta}))\cdot(q'(2^{-m}) - u_r(2^{-m}e^{i\theta}))\,d\theta. \end{aligned}$$

where q' is the derivative of q for $2^{-m} \leqslant |x| \leqslant 2^{-m+1}$. We obtain

$$\int_{B_1 \setminus B_{2^{-m}}} |\nabla(q-u)|^2 = -\int_0^{2\pi} (q(1)-u(e^{i\theta}))\cdot u_r(e^{i\theta})\, d\theta + 2^{-m}\int_0^{2\pi} (q(2^{-m})-u(2^{-m}e^{i\theta}))\cdot u_r(2^{-m}e^{i\theta})\, d\theta + \int_{B_1 \setminus B_{2^{-m}}} (q-u)\Delta u$$

because the boundary integrals containing q' vanish by choice of q, because the remaining boundary integrals cancel each other except those for $|x|=1$ and $|x|=2^{-m}$ when we write

$$\int_{B_1 \setminus B_{2^{-m}}} = \sum_{\mu=1}^{m} \int_{A_m}$$

and because q is harmonic on A_m.

Now, by (2.4.11) and (2.4.15),

$$\left| 2^{-m}\int_0^{2\pi} (q(2^{-m})-u(2^{-m}e^{i\theta}))\cdot u_r(2^{-m}e^{i\theta})\, d\theta \right| \leqslant c_{20}\sqrt{\epsilon}\left(\int_{B_{2^{1-m}}} |\nabla u|^2\right)^{1/2} \to 0 \qquad \text{as } m \to \infty.$$

Moreover, from (2.4.1) and (2.4.15)

$$\left| \int_{B_1 \setminus B_{2^{-m}}} (q-u)\Delta u \right| \leqslant c_{21}\sqrt{\epsilon}\int_{B_1} |\nabla u|^2. \tag{2.4.16}$$

Consequently, we obtain for $m \to \infty$

$$\begin{aligned} \int_{B_1} |\nabla(q-u)|^2 &= \int_0^{2\pi} (u(e^{i\theta})-q(1))\cdot u_r(e^{i\theta})\, d\theta + \int_{B_1} (q-u)\Delta u \\ &\leqslant \left(\int_0^{2\pi} |u(e^{i\theta})-q(1)|^2\, d\theta\right)^{1/2} \left(\int_0^{2\pi} |u_r(e^{i\theta})|^2\, d\theta\right)^{1/2} \\ &\quad + \delta\int_{B_1} |\nabla u|^2 \qquad \text{by (2.4.16)} \end{aligned}$$

for any $\delta > 0$, provided $\epsilon \leqslant (\delta/c_{21})^2$,

$$\leqslant c_{22}\left(\int_0^{2\pi} |u_\theta(e^{i\theta})|^2\, d\theta\right)^{1/2} \left(\int_0^{2\pi} |u_r(e^{i\theta})|^2\, d\theta\right)^{1/2} + \delta\int_{B_1} |\nabla u|^2$$

by Poincaré's inequality. Thus

$$\int_{B_1} |\nabla(q-u)|^2 \leqslant c_{22} \int_0^{2\pi} |\nabla u(e^{i\theta})|^2 \, d\theta + \delta \int_{B_1} |\nabla u|^2. \tag{2.4.17}$$

Now since q is a function of $|x|$

$$\begin{aligned}
\int_{B_1} |\nabla(q-u)|^2 &\geqslant \int_0^1 \int_0^{2\pi} \frac{1}{r} u^i_\theta u^i_\theta \, d\theta \, dr \\
&\geqslant \frac{1}{\mu} \int_0^1 \int_0^{2\pi} \frac{1}{r} g_{jk}(u) u^j_\theta u^k_\theta \, d\theta \, dr \\
&= \frac{1}{2\mu} \int_0^1 \int_0^{2\pi} \left(r g_{jk}(u) u^j_r u^k_r + \frac{1}{r} g_{jk}(u) u^j_\theta u^k_\theta\right) d\theta \, dr \qquad \text{by (2.4.14)} \\
&= \frac{1}{2\mu} \int_{B_1} g_{jk}(u) \nabla u^j \nabla u^k \qquad \text{by (2.4.12)} \\
&\geqslant \frac{\lambda}{2\mu} \int_{B_1} |\nabla u|^2.
\end{aligned}$$

We now choose $\delta < \lambda/2\mu$ in (2.4.17) and obtain with $\gamma = (\lambda/2\mu - \delta)/c_{22}$

$$\gamma \int_{B_1} |\nabla u|^2 \leqslant \int_0^{2\pi} |\nabla u(e^{i\theta})|^2 \, d\theta.$$

More generally, the previous reasoning is valid for $u(\rho x)$, $0 < \rho \leqslant 1$, and we obtain

$$\begin{aligned}
\gamma \int_{B_\rho} |\nabla u|^2 &\leqslant \rho \int_0^{2\pi} |\nabla u(\rho e^{i\theta})|^2 \, d\theta \\
&= \rho \frac{d}{d\rho}\left(\int_{B_\rho} |\nabla u|^2\right).
\end{aligned}$$

This differential inequality implies for $\rho \leqslant 1$

$$\int_{B_\rho} |\nabla u|^2 \leqslant \rho^\gamma \int_{B_1} |\nabla u|^2.$$

Lemma 2.4.2 then yields for $|y| \leqslant \frac{1}{2}$

$$|\nabla u(y)| \leqslant c_{23} |y|^{\gamma/2 - 1} \left(\int_{B_1} |\nabla u|^2\right)^{1/2}$$

and consequently for $|x| \leqslant \frac{1}{4}$, $r \leqslant \frac{1}{4}$

$$\int_{B(x,r)} |\nabla u|^2 \leqslant c_{24} \epsilon \int_{B(x,r)} |x|^{\gamma - 2} = m\rho^\gamma$$

with $m = (2\pi\epsilon c_{24})/\gamma$.

Morrey's Dirichlet growth theorem (cf. [M3; Theorem 3.5.2]) then implies that u is Hölder continuous on $B_{1/4}$. Since u is a weak solution of (1.2.7), Theorem 2.5.1 below then implies $u \in C^2(B_{1/4})$. This concludes the proof of Theorem 2.4.1. q.e.d.

2.5. Higher regularity

In this section, we state some results on the differentiability of continuous weak solutions—mostly without proofs as these can be readily found in [J6].

Theorem 2.5.1 [JK]: *Let $u \in H^{1,2}(D,N) \cap C^0(\mathring{D},N)$ be a weak solution of (1.2.10), where N is a Riemannian manifold of bounded geometry. Then*

$$u \in C^{2,\alpha}(\mathring{D},N) \qquad \text{for every } \alpha \in (0,1),$$

and one also has corresponding a priori *estimates for the $C^{2,\alpha}$-norm of u, depending on $i(N)$, $\sup|K|$ (K = sectional curvature of N), $d = \dim N$, and the modulus of continuity of u. Likewise, for a continuous weak solution of (1.2.7) with $(b_{jk}) \in C^{1,\alpha}$, also $u \in C^{2,\alpha}(\mathring{D},N)$ with corresponding estimates. (If $(b_{jk}) \in C^{0,1}$, then $u \in C^{1,\alpha}(\mathring{D},N)$.) Similarly, if under these assumptions,*

$$u = g \qquad \text{on } \partial D$$

with $g \in C^{1,\alpha}(\partial D,N)$, then u is of class $C^{2,\alpha}$ on the closure of D. Likewise, if we only assume $g \in C^{1,\alpha}(\partial D,N)$, then $u \in C^{1,\alpha}(\bar{D},N)$. If N, g, (b_{jk}) are of class C^∞, then so is u.

(The proof of Theorem 2.5.1 is based on Theorem 2.1.2.)

We shall also need the interior gradient bound of [JK]:

Theorem 2.5.2: *Let $u: D \to N$, N as above, be a harmonic map, $B(x_0,r) \subset D$. Then for all $R \leqslant R_0$*

$$|du(x_0)| \leqslant c_0 \max_{x \in B(x_0,r)} \frac{d(u(x),u(x_0))}{R} \tag{2.5.1}$$

where c_0 depends on $i(N)$, $\sup|K|$, $d = \dim N$, and the modulus of continuity of u (if $u(B(x_0,R_0)) \subset B(p,\rho)$ for some $\rho \in N$ with $\rho < \min(\pi/2\kappa, i(p))$ where $K \leqslant \kappa^2$, then no further knowledge of the properties of u is needed; if $d = \dim N = 2$, then it suffices that $B(p,\rho)$ is topologically a disk and $\rho < \pi/2\kappa$.) If

$$u = g \qquad \text{on } \partial D$$

with $g \in C^{1,\alpha}(\partial D,N)$, then one gets a gradient bound on $\bar{D}$.

As a consequence of Theorem 2.5.2, we have (cf. [J6; Section 4.8]):

Corollary 2.5.1: *Let $u: D \to N$ be harmonic, N as above. Suppose*

$$u = g \qquad \text{on } \partial D \text{ with} \qquad g \in C^\alpha \qquad (\alpha \in (0,1))$$

Then $u \in C^\alpha(\bar{D})$ with the same α, and

$$d(u(x_1), u(x_2)) \leqslant c_1 d(x_1, x_2)^\alpha \tag{2.5.2}$$

for all $x_1, x_2 \in \bar{D}$, where c_1 depends on $|g|_{C^\alpha}$ and the above geometric quantities and the modulus of continuity of u (which is automatically controlled under assumptions as in Theorem 2.5.2).

Proof: If $B(x_0, 2R) \subset D$ and x_1, $x_2 \in B(x_0, R)$, then (2.5.1) implies

$$\frac{d(u(x_1), u(x_2))}{|x_1 - x_2|^\alpha} \leqslant \frac{c_0}{R^\alpha} \max_{x \in B(x_0, 2R)} d(u(x), u(x_0)). \tag{2.5.3}$$

Elementary applications of the triangle inequality then show that it suffices to consider the case

$$x_1 \in \partial B(x_0, 2R) \qquad x_1 \in D \qquad x_2 \in B(x_0, 2R).$$

Since we assume that we control the modulus of continuity of u (cf. for this point [HKW2]; the result is also reproduced in [J6; Theorem 4.7.1]), we can assume that

$$u(B(x_0, 2R)) \subset B(p, s/4)$$

for some $p \in N$, $s < \min(i(N), \pi/2\kappa)$ ($K \leqslant \kappa^2$ on N). If $u(x_1) \neq u(x_2)$, we connect $u(x_2)$ to $u(x_1)$ by the (unique) geodesic arc in $B(p, s/4)$ and continue this arc beyond $u(x_1)$ until a distance $s/8$, reaching a point $q \in B(p, 3s/8)$. In particular, $u(B(x_0, 2R)) \subset B(q, s)$ so that

$$d^2(u(x), q)$$

is subharmonic on $B(x_0, 2R)$ (cf. [J6, Lemmas 1.7.1 and 2.3.2]). Let

$$h: B(x_0, 2R) \to \mathbb{R}$$

satisfy

$$\begin{aligned} \Delta h &= 0 && \text{in } B(x_0, 2R) \\ h(x) &= d^2(u(x), q) && \text{on } \partial B(x_0, 2R). \end{aligned} \tag{2.5.4}$$

By the maximum principle

$$d^2(u(x), q) \leqslant h(x) \qquad \text{in } B(x_0, 2R). \tag{2.5.5}$$

Then

$$\begin{aligned} d(u(x_1), u(x_2)) &= d(u(x_2), q) - d(u(x_1), q) \qquad \text{by choice of } q \\ &\leqslant (4/s)(d^2(u(x_2), q) - d^2(u(x_1), q)) \\ &\leqslant (4/s)(h(x_2) - h(x_1)) \end{aligned}$$

by (2.5.5) and (2.5.4).

Thus, the Hölder continuity of u is reduced to the Hölder continuity of the harmonic function h with Hölder continuous boundary values. The latter is a result from potential theory.

q.e.d.

Remark: The preceding proof refines the argument of [HKW1].

Theorem 2.5.3: *Let N be a complete Riemannian manifold, M a complete submanifold of class C^2. Let $u \in H^{1,2}(D, N)$ be a continuous weakly harmonic map with free boundary M. Then*

$$u \in C^{1,\alpha}(\bar{D}, N) \qquad (0 < \alpha < 1).$$

Proof: Making use of conformal transformations, we can assume that the domain is an upper half-plane $H = \{x + iy \in \mathbb{C}: y > 0\}$, and it suffices to show regularity at $0 \in \partial H$.

Since u is continuous, we can find a neighbourhood Ω of 0 in $\mathbb{C}$ for which $u(\Omega \cap \bar{H})$ is contained in a normal neighbourhood U of M, i.e. for which each point in $u(\Omega \cap \bar{H})$ has a unique projection onto M along a shortest geodesic. For a point u in this neighbourhood U, we let $\xi(u)$ be this projection.

To each such u, we associate a point $\tilde{u}$, and we let $\bar{U}$ be the union of U and $\tilde{U} := \{\tilde{u}: u \in U\}$, where u and $\tilde{u}$ are identified for $u \in M$. W.l.o.g. we can assume that U is convex, and we use the distance function $d(\cdot,\cdot)$ of N to make $\bar{U}$ into a metric space by defining

$$d(\tilde{u}_1, \tilde{u}_2) := d(u_1, u_2)$$

and

$$d(u_1, \tilde{u}_2) := \inf_{v \in M} (d(u_1, v) + d(u_2, v)) \qquad \text{for } u_1, u_2 \in U.$$

This then also induces a reflection of the metric tensor of U across M. Since M is C^2, the reflection of the metric tensor is Lipschitz across M. (To see this, one takes a moving orthonormal frame $e_1, \ldots, e_d$ on U for which $e_l, \ldots, e_d$ are normal to M for $u \in M$. We can then extend $e_1, \ldots, e_d$ to $\bar{U}$; for $e_1, \ldots, e_{l-1}$ nothing happens, as they are tangential to M, whereas $e_l, \ldots, e_d$ are reflected in a Lipschitz continuous way.) Of course, $\bar{U}$ is not a manifold, as it is branched along M, but this difficulty disappears when we pull everything back via the reflected map

$$\bar{u}: \Omega \to N$$

defined by

$$\bar{u}(\bar{z}) = u(z) \qquad \text{for } z \in \Omega \qquad (z = x + iy, \bar{z} = x - iy).$$

Thus we have a reflection T in Ω, a reflection S in $\bar{u}(\Omega)$, and a reflection $\bar{S}$ in $\bar{u}^{-1}TN$ (the pull-back of the tangent bundle of N on $\bar{U}$ under $\bar{u}$). If φ is a section of $\bar{u}^{-1}TN$, defined on Ω, we have

$$2\varphi = \rho + \psi$$

where $\rho = \varphi - \bar{S}\circ\varphi\circ T$, $\psi = \varphi + \bar{S}\circ\varphi\circ T$. Furthermore $\psi(x)$ is tangential to M if $u(z)\in M$.

Since u has M as free boundary for $z_0\in\partial\Omega$ and sufficiently regular φ

$$2\int_{B(z_0,R)} \langle du, d\varphi \rangle = \int_{B(z_0,R)} \langle du, d\rho \rangle$$

$$= \int_{B(z_0,R)} \langle du(z), d\varphi(z) \rangle - \langle du(z), d_z\bar{S}\varphi(Tz) > dz.$$

(the notation d_z here means that the derivative is to be taken w.r.t. z)

$$= \int_{B(z_0,R)} (\langle du(z), d\varphi(z) - \langle d_{Tz}u(Tz), d_{Tz}\bar{S}\varphi(z) \rangle)\, dz$$

since the change of variables $z \to Tz$ is an involutive isometry

$$= \int_{B(z_0,R)} (\langle du(z), d\varphi(z) \rangle - \langle d_{Tz}Su(z), d_{Tz}\bar{S}\varphi(z) \rangle)\, dz$$

$$= \int_{B(z_0,R)} (\langle du(z), d\varphi(z) \rangle - \langle du(z), d\varphi(z) \rangle)\, dz$$

since by construction (for this point we had to reflect the metric in the image) all reflections are isometries with determinant (-1). Hence

$$2\int_{B(z_0,R)} \langle du, d\varphi \rangle = 0.$$

By approximation, this continues to hold for all sections $\varphi\in L^\infty \cap H_0^{1/2}$ $(\Omega, \bar{u}^{-1}TN)$, i.e. which are bounded and satisfy

$$\int_\Omega \langle d\varphi, d\varphi \rangle < \infty. \tag{2.5.6}$$

(Note for this point that $\bar{u}$ is continuous.) Equation (2.5.6) here makes sense, since $E(\bar{u}) < \infty$, and in local coordinates

$$d\varphi = \frac{\partial \bar{u}^i}{\partial z^\alpha}\frac{\partial}{\partial \bar{u}^i}\, dz^\alpha + \varphi^k \tilde{\Gamma}^i_{jk} \frac{\partial \bar{u}^i}{\partial z^\alpha}\frac{\partial}{\partial \bar{u}^j}\, dz^\alpha$$

where $z = z^1 + iz^2 \in\Omega$, and the $\tilde{\Gamma}^i_{jk}$ are the Christoffel symbols of the reflected metric tensor $\tilde{g}_{jk}$; since the latter is Lipschitz, $\tilde{\Gamma}^i_{jk}(\bar{u})$ is in L^∞.

For such φ, we thus have

$$\int_\Omega \langle d\bar{u}, d\varphi \rangle = 0. \tag{2.5.7}$$

In local coordinates, (2.5.7) reads as

$$0 = \int_\Omega \left(\tilde{g}_{ij} \frac{\partial \bar{u}^i}{\partial z^\alpha}\frac{\partial \psi^j}{\partial z^\alpha} - \tilde{\Gamma}^i_{jk}(\bar{u}) \frac{\partial \bar{u}^i}{\partial z^\alpha}\frac{\partial \bar{u}^k}{\partial z^\alpha} \psi^i \right) \tag{2.5.8}$$

for all compactly supported $\psi \in H^{1,2} \cap L^\infty$ (cf. Lemma 1.2.3). Since $\bar{u}$ is continuous, this implies $\bar{u} \in C^{1,\alpha}(\Omega)$ (for every $\alpha \in (0,1)$) by the basic regularity theorem of Ladyženskaya and Ural'ceva (cf. [LU]) which shows $\bar{u} \in C^1$ and linear elliptic theory which then shows $\bar{u} \in C^{1,\alpha}$.

q.e.d.

Remark: Theorem 2.5.3 is due to Baldes [Ba]; the preceding proof is taken from [GJ].

Corollary 2.5.2: *Let M, N be in Theorem 2.5.3. Let*

$$\begin{aligned} D^+ &:= \{x + iy \in \mathbb{C} : y \geq 0,\ x^2 + y^2 \leq 1\} \\ \partial_+ D^+ &:= \{y \geq 0, x^2 + y^2 = 1\} \\ \partial_0 D^+ &:= \{y = 0, x^2 < 1\}. \end{aligned}$$

Suppose $u \in H^{1,2}(D^+, N) \cap C^0(D^+, N)$, u maps $\partial_+ D^+$ in a $C^{1,\alpha}$ way onto $\gamma \subset N$ with endpoints on M, and γ meets M orthogonally. Finally suppose

$$\int_{D^+} \langle du, d(\eta\varphi) \rangle = 0$$

whenever $\varphi \in L^\infty \cap H^{1,2}(D^+, u^{-1}TN)$, $\varphi(v) \in T_v M$ whenever $v \in M$, and η is a Lipschitz function with compact support in $\mathring{D}^+ \cup \partial_0 D^+$. (In other words, u solves a mixed boundary value problem, with a Dirichlet condition on $\partial_+ D^+$, and a free boundary condition on $\partial_0 D^+$.) Then

$$u \in C^{1,\alpha}(D^+, N).$$

Proof: Since γ hits M orthogonally, we can use the construction of the preceding proof to reflect γ across M in such a way that the tangent vector to γ is Lipschitz continuous across M. The argument of the preceding proof can then be applied.

q.e.d.

We also have the following result of Baldes [Ba]:

Theorem 2.5.4: *Under the assumptions of Theorem 2.5.3,*

$$u \in H^{2,2}(D, N).$$

The same holds for the reflected map $\bar{u}$, constructed in the proof of Theorem 2.5.3, and $\bar{u}$ satisfies almost everywhere

$$\Delta \bar{u}^i + \tilde{\Gamma}^i_{jk}(\bar{u}) \frac{\partial \bar{u}^j}{\partial z^\alpha} \frac{\partial \bar{u}^k}{\partial z^\alpha} = 0 \tag{2.5.9}$$

where the $\tilde{\Gamma}^i_{jk}$ are the reflected Christoffel symbols constructed in the proof of Theorem 2.5.3.

Proof: This follows from (2.5.8), $\bar{u} \in C^1$, and $\tilde{\Gamma}^i_{jk}(\bar{u}) \in L^\infty$ by linear elliptic theory. q.e.d.

We shall also need a gradient bound at a free boundary similar to Theorem 2.5.2, namely:

Theorem 2.5.5: *Suppose the assumptions of Theorem 2.5.3 hold, and $B(x_0, R) \subset D$. Then*

$$|d\bar{u}(x_0)| \leqslant c_0 \max_{x \in B(x_0, R)} \frac{d(\bar{u}(x), u(x_0))}{R} \tag{2.5.10}$$

where $\bar{u}$ is the reflection constructed in the proof of Theorem 2.5.3, and c_0 depends on the modulus of continuity of u and on the geometry of M and N.

Proof: The proof is similar to the interior estimate, cf. [J6]. The essential idea of the proof is due to Heinz [H1]. We put $R_0 := R/2$ and

$$\mu := \max_{x \in B(x_0, R_0)} (R_0 - d(x, x_0)) \cdot |d\bar{u}(x)|.$$

There exists $x_1 \in B(x_0, R_0)$ with

$$\mu = (R_0 - d(x_1, x_0)) \cdot |d\bar{u}(x_1)| \tag{2.5.11}$$

$$d := R_0 - d(x_1, x_0)$$

and

$$|d\bar{u}(x_0)| \leqslant \mu / R_0. \tag{2.5.12}$$

Furthermore,

$$\delta = \delta(\theta) := \max_{x \in B(x_1, d\theta)} d(\bar{u}(x), \bar{u}(x_1)).$$

Since a control of the modulus of continuity of u and hence also of $\bar{u}$ is assumed, δ can be made as small as desired by choosing θ sufficiently small, in other words

$$\theta \leqslant \theta_0 \Rightarrow \delta \leqslant \delta_0.$$

Then

$$\mu / d = |d\bar{u}(x_1)| \leqslant \frac{1}{\pi d^2 \theta^2} \int_{\partial B(x_1, d\theta)} |\bar{u}(x) - \bar{u}(x_1)| \, |dx| + \frac{1}{2\pi} \int_{B(x_1, d\theta)} \frac{|\Delta \bar{u}(x)|}{d(x, x_1)} dx. \tag{2.5.13}$$

Here, we have used an easy consequence of Green's formula. We have also used $\bar{u} \in H^{2,2}$, cf. Theorem 2.5.4. Finally, the absolute values are taken w.r.t. to our local coordinates:

$$|\bar{u}(x) - \bar{u}(x_1)| \leqslant c_1 \cdot \delta \tag{2.5.14}$$

$$|\Delta u(x)| \leqslant c_2 \cdot |du(x)|^2 \leqslant c_2 \cdot \frac{\mu^2}{d^2(1-\theta)^2}, \tag{2.5.15}$$

where c_1 and c_2 depend on the geometry of M and N (in particular bounds for $|\tilde{\Gamma}^i_{jk}|$ are involved).

Equations (2.5.13)–(2.5.15) imply

$$\frac{\mu}{d} \leqslant \frac{2c_1\delta}{d\theta} + c_2 \frac{\mu^2\theta}{d(1-\theta)^2}$$

or, if $\theta \leqslant \theta_0 \leqslant \frac{1}{2}$,

$$\mu \leqslant \frac{a}{2}\frac{\delta(\theta_0)}{\theta} + \frac{b}{2}\theta\mu^2 \qquad \text{for all } \theta \leqslant \theta_0. \tag{2.5.16}$$

As we have noted above, we can choose $\theta_0 > 0$ so small (depending only on ω, κ, and M) that

$$ab\delta(\theta) \leqslant ab\delta(\theta_0) < 1 \qquad \text{for all } \theta \leqslant \theta_0. \tag{2.5.17}$$

Now

$$\mu < \frac{a\delta(\theta_0)}{\theta_0} \tag{2.5.18}$$

because otherwise we could choose θ in (2.5.16) as

$$\theta = \frac{a\delta(\theta_0)}{\mu} \leqslant \theta_0$$

and obtain

$$\mu \leqslant (\tfrac{1}{2} + \tfrac{1}{2}ab\delta(\theta_0))\mu$$

contradicting (2.5.17).

Equation (2.5.18) then implies (2.5.10) by definition of μ.

q.e.d.

Corollary 2.5.3: *Suppose that the situation of Corollary 2.5.2 holds, with the exception that we only require*

$$u|_{\partial_+ D^+} \in C^\alpha \qquad (\alpha \in (0,1)) \tag{2.5.19}$$

and that the image $u(\partial_+ D^+)$ need not hit M orthogonally. Then

$$u \in C^\alpha(\bar{D}^+, N)$$

with the same α as in (2.5.19).

Proof: The argument is the same as for Corollary 2.5.1, using Theorem 2.5.5 instead of Theorem 2.5.2. We also need to use the construction of the proof of Theorem 2.2.2; we note that locally the second fundamental form of M can be made arbitrarily small by rescaling (here, again the assumption of continuity enters in a crucial way). We omit the details.

q.e.d.

Remark: All results of this section actually hold for domains of arbitrary dimensions, not necessarily two. The proofs can be carried over as well to this more general situation.

A consequence of the regularity and uniqueness results is:

Theorem 2.5.6: *Let Σ be a closed surface, $(N_t)_{t\in\mathbb{R}}$ a C^∞-family of negatively curved Riemannian manifolds, $\alpha\in[\Sigma, N_t]$ (independent of t) a homotopy class of maps that does not contain constant maps or maps onto a closed curve. Let $u_t\colon \Sigma\to N_t$ be the harmonic map in the class α. Then the map $t\to u_t$ is of class C^∞.*

Proof: Let $(g_{ij,t})$ be C^∞-local Riemannian metrics on the family (N_t), with Christoffel symbols $\Gamma^i_{jk,t}$. Locally, u_t is a solution of

$$\Delta u^i_t = -\Gamma^i_{jk,t}(u_t)\left(\frac{\partial u^j_t}{\partial x}\frac{\partial u^k_t}{\partial x}+\frac{\partial u^j_t}{\partial y}\frac{\partial u^k_t}{\partial y}\right) \qquad (z=x+iy). \qquad (2.5.20)$$

As a consequence of Theorems 2.2.3 and 2.5.1, $(u_t)_{t\in\mathbb{R}}$ is a continuous family in C^∞.

Moreover,

$$h_t := \partial u_t/\partial t$$

(this is well defined by uniqueness (Theorem 2.2.3)) satisfies

$$\Delta h^i_t = -\frac{\partial\Gamma^i_{jk,t}}{\partial t}(u_t)\left(\frac{\partial u^j_t}{\partial x}\frac{\partial u^k_t}{\partial x}+\frac{\partial u^j_t}{\partial y}\frac{\partial u^k_t}{\partial y}\right)-\frac{\partial\Gamma^i_{jk,t}}{\partial u^l}h^l_t\left(\frac{\partial u^j_t}{\partial x}\frac{\partial u^k_t}{\partial x}+\frac{\partial u^j_t}{\partial y}\frac{\partial u^k_t}{\partial y}\right)$$
$$-2\Gamma^i_{jk,t}(u_t)\left(\frac{\partial h^j_t}{\partial x}\frac{\partial u^k_t}{\partial x}+\frac{\partial h^j_t}{\partial y}\frac{\partial u^k_t}{\partial y}\right). \qquad (2.5.21)$$

Equation (2.5.21) is a linear elliptic system for h_t, and the coefficients of the right-hand side are of class C^∞. Elliptic regularity theory implies that h_t is also of class C^∞. Iterating, one deduces that the derivatives of u_t w.r.t. t of any order are of class C^∞.

q.e.d.

Remark: Theorem 2.5.6 was obtained in [EL2] by a more abstract argument based on the implicit function theorem in Banach manifolds.

2.6. The Hartman–Winter Lemma and some of its consequences. Asymptotic expansions at branch points

In this section, we shall first prove the following version of a lemma discovered by Hartman and Winter [HtW; pp. 455–458]. We shall apply the complex notation also to real-valued (vector) functions. For $u\colon D\to\mathbb{R}^d$, we have

$$u_z\colon D\to\mathbb{C}^d$$

with $u_z=\frac{1}{2}(u_x-iu_y)$, $u_{\bar z}=\frac{1}{2}(u_x+iu_y)$, with component-wise differentiation.

Lemma 2.6.1: *Suppose* $u \in C^{1,1}(D, \mathbb{R}^d)$ *satisfies almost everywhere*

$$|u_{z\bar{z}}| \leqslant K(|u_z| + |u|) \tag{2.6.1}$$

where K *is a fixed positive constant. If*

$$u(z) = o(|z|^n) \tag{2.6.2}$$

for some $n \in \mathbb{N}$, *then*

$$\lim_{z \to 0} u_z \cdot z^{-n} \tag{2.6.3}$$

exists. If (2.6.2) *holds for all* $n \in \mathbb{N}$, *then*

$$u \equiv 0.$$

Proof: First we note that since $u \in C^{1,1}$, $u_{z\bar{z}}$ exists almost everywhere by Rademacher's Theorem. If B is a closed subdomain of D with Lipschitz boundary ∂B and $g \in C^1(B, \mathbb{C})$, then

$$\oint_{\partial B} g u_z = \int_B (u_z g_{\bar{z}} + u_{z\bar{z}} g)\, dz. \tag{2.6.4}$$

Suppose that

$$u_z = o(|z|^{k-1}) \qquad \text{for some } k \in \mathbb{N}. \tag{2.6.5}$$

In (2.6.4), we take $B = \{z: \epsilon \leqslant |z| \leqslant R,\ |z - z_0| \geqslant \epsilon\}$, $R < \min(1, \pi/4K)$, and $g = z^{-k}(z - z_0)^{-1}$, $z_0 \neq 0$. Since $g_{\bar{z}} = 0$ in B, we obtain by letting $\epsilon \to 0$ from (2.6.4):

$$2\pi u_z(z_0) z_0^{-k} = \oint_{|z|=R} u_z z^{-k}(z - z_0)^{-1} - \int_{B(0,R)} u_{z\bar{z}} z^{-k}(z - z_0)^{-1}\, dz \tag{2.6.6}$$

and hence from (2.6.1)

$$\begin{aligned} 2\pi |u_z(z_0) z_0^{-k}| \leqslant & \oint_{|z|=R} |u_z z^{-k}(z - z_0)^{-1}|\,|dz| \\ & + K \int_{B(0,R)} (|u_z| + |u|)|z|^{-k}|z - z_0|^{-1}\, dz. \end{aligned} \tag{2.6.7}$$

We now want to estimate

$$I_k := \int_{B(0,R)} |u_z|\,|z|^{-k}|z - z_0|^{-1}\, dz$$

uniformly as $z_0 \to 0$. If that can be achieved, then the second integral in (2.6.6) converges to some limit as $z_0 \to 0$, and hence

$$\lim_{z \to 0} u_z \cdot z^{-k}$$

exists.

Moreover, if $k<n$, then this limit vanishes because of (2.6.2), and hence (2.6.5) holds for $k+1$. On the other hand, the first assertion of the lemma is trivial if $n=0$, and $n\geqslant 1$ implies (2.6.5) for $k=1$. Thus, by induction, (2.6.5) holds for $k=n$, which implies the first assertion of the lemma.

In order to control I_k, we multiply (2.6.7) by $|z_0-z_1|^{-1}$, use

$$\int_{B(0,R)} |z-z_0|^{-1}\,dz<2R \qquad \text{if } |z_0|<R \tag{2.6.8}$$

and

$$(z-z_0)^{-1}(z_0-z_1)^{-1}=(z-z_1)^{-1}((z-z_0)^{-1}+(z_0-z_1)^{-1})$$

and obtain, integrating w.r.t. z_0, $|z_0|<R$,

$$\begin{aligned}2\pi\int_{B(0,R)} &|u_z||z|^{-k}|z-z_0|^{-1}\,dz\\ &\leqslant 4R\oint_{|z|=R}|u_z z^{-k}(z-z_1)^{-1}||dz|\\ &\quad +4KR\int_{B(0,R)}(|u_z|+|u|)|z|^{-k}|z-z_0|^{-1}\,dz.\end{aligned} \tag{2.6.9}$$

By choice of $R\,(R<\pi/2K)$, and since $k<n$, i.e. $|uz^{-k}|=o(|z|)$ by induction, (2.6.9) controls I_k, putting $z_1=z_0$. In particular, by induction, (2.6.9) holds for $k=n-1$, and letting $z_1\to 0$, we obtain

$$\begin{aligned}(2\pi-4KR)&\int_{B(0,R)}|u_z z^{-n}|\,dz\\ &\leqslant 4R\oint_{|z|=R}|u_z z^{-n}||dz|+4KR\int_{B(0,R)}|u||z|^{-n}\,dz.\end{aligned} \tag{2.6.10}$$

Now $|u(z)|\leqslant\int_0^1|zu_z(tz)|\,dt$, which leads to

$$\begin{aligned}\int_{B(0,R)}|uz^{-n}|\,dz&\leqslant\int_0^1\int_{B(0,R)}|z^{-n+1}u_z(tz)|\,dz\,dt\\ &=\int_0^1 t^{n-3}\int_{B(0,tR)}|z^{-n+1}u_z(tz)|\,dz\,dt.\end{aligned}$$

Hence for $n\geqslant 3$

$$\int_{B(0,R)}|uz^{-n}|\,dz\leqslant\int_{B(0,R)}|u_z z^{-n+1}|\,dz.$$

Since $|z|\leqslant R\leqslant 1$ by choice of R, we see that the second integral in (2.6.10) can be absorbed into the left-hand side, yielding

$$(2\pi-8KR)\int_{B(0,R)}|u_z z^{-n}|\,dz\leqslant 2R\int_{|z|=R}|u_z z^{-n}||dz|. \tag{2.6.11}$$

Since $(2\pi - 8KR) > 0$ by choice of R, it is not difficult to see that (2.6.11) for all $n \in \mathbb{N}$ implies $u_z \equiv 0$ and hence $u \equiv 0$ which completes the proof. (Namely, otherwise, there would exist z_0 with $|z_0| < R$ and $u_z(z_0) = c \neq 0$, and the left-hand side of (2.6.11) would grow like $c|z_0|^{-n}$ and the right-hand side like $c_0 R^{-n}$, c and c_0 being independent of n, and hence (2.6.11) could not hold for all n.)

q.e.d.

Remark: Instead of (2.6.1), it suffices for the proof to assume the integral relation

$$\left|\int_{\partial B} g u_z \, dz\right| \leqslant K \int_B (|g_{\bar{z}}||u_z| + |g|(|u_z| + |u|)) \, dz \, d\bar{z} \tag{2.6.12}$$

for every $B \subset D$ with Lipschitz boundary ∂B and every $g \in C^1(B, \mathbb{C}) \cap \mathrm{Lip}(\bar{B}, \mathbb{C})$.

As a first consequence, we have the following asymptotic expansion of solutions of our problems at points where the derivative vanishes.

Corollary 2.6.1: *Suppose* $u \in C^2(D, \mathbb{R}^d)$ *(or, more generally,* $u \in C^{1,1}(D, \mathbb{R}^d)$*) satisfies*

$$|\Delta u| \leqslant K(|u_x|^2 + |u_y|^2) \tag{2.6.13}$$

with some constant K. (For example, u can be a solution of (1.2.10)*, or, more generally, of* (1.2.7)*—with* $(b_{jk}) \in C^{1,\alpha}$*, as otherwise u will not be in* C^2*).*

Then, for each $z_0 \in \mathring{D}$*, there exist a non-negative integer m, and* $a \in \mathbb{C}^d$*,* $a \neq 0$*, with (near* z_0*)*

$$u_z = a(z - z_0)^m + o(|z - z_0|^m); \tag{2.6.14}$$

$m \geqslant 1$ *if* $u_z(z_0) = 0$*. In particular, the zeros of* u_z *are isolated. If u is conformal, i.e.*

$$4 g_{jk} u_z^j u_z^k = g_{jk}(u_x^j u_x^k - u_y^j u_y^k - 2i u_x^j u_y^k) = 0$$

then

$$a^2 := g_{jk} a_j a_k = 0 \qquad a = (a_1, \ldots, a_d). \tag{2.6.15}$$

Proof: $u - u(z_0)$ satisfies (2.6.13), if u does. (Thus, we can apply Lemma 2.6.1 to $u - u(z_0)$.) All claims then follow directly.

q.e.d.

Similar results hold at the boundary for solutions of free boundary problems. Here, we only treat the special case of a harmonic map with a Plateau-type boundary condition (Cf. [GHN1], [GHN2] for other results.) Together with the asymptotic expansion, we shall also give a new proof of $C^{1,\alpha}$-regularity in this case. (This of course, is a special case of Theorem 2.5.3; the present, strictly two-dimensional proof is more elementary, however)

Theorem 2.6.1: *Let S be a compact surface with smooth boundary* ∂S*,* $u: S \to N$ *a harmonic map with a* C^2*-curve* $\Gamma \subset N$ *as a free boundary, which is continuous on S (including the boundary). Suppose u also satisfies* (1.2.35)*. Then* $u \in C^{1,\alpha}(S, N)$*.*

Moreover, the asymptotic expansion (2.6.14) *also holds for* $z_0 \in \partial S$, *with* (2.6.15) *if* u *is conformal. If* $\Gamma \in C^{2,\alpha}$, *then* $u \in C^{2,\alpha}(S, N)$.

Proof: We let γ be a boundary curve of S, and let $z_0 \in \gamma$. We represent γ as the unit circle in the plane. We can find a neighbourhood of $u(z)$ in which Γ is written in the form

$$\Gamma = \{u^1 = u^2 = \cdots = u^{d-1} = 0, \, |u^d| < 1\}$$

(in suitable local coordinates) and for the metric in these coordinates, we may assume

$$g_{ij} \in C^1 \qquad g_{ij}(u(z_0)) = \delta_{ij} \tag{2.6.16}$$

(because $\Gamma \in C^2$).

By Lemma 1.2.5

$$\varphi := 4 g_{jk} u_z^j u_z^k$$

is holomorphic and smooth up to the boundary, and $z^2 \varphi(z)$ (noting $\gamma = \{|z| = 1\}$) is real on γ.

We now rewrite the equation

$$\varphi = 4 g_{ij} u_z^i u_z^j \tag{2.6.17}$$

as

$$\left(u_z^d + \frac{g_{\alpha d}}{g_{dd}} u_z^\alpha \right)^2 = \left(\frac{g_{\alpha d}}{g_{dd}} u_z^\alpha \right)^2 - \frac{g_{\alpha\beta}}{g_{dd}} u_z^\alpha u_z^\beta + \frac{\varphi}{g_{dd}} \tag{2.6.18}$$

where α and β have to be summed from 1 to $(d-1)$. Using (2.6.16) and the continuity of u, (2.6.18) implies that in some neighbourhood of z_0

$$|\nabla u^d|^2 \leqslant \text{const.} \sum_{\alpha=1}^{d-1} |\nabla u^\alpha|^2 + \beta \tag{2.6.19}$$

where β depends on the magnitude of φ (therefore, we only get regularity, but no *a priori* estimates). On the other hand, since u is harmonic

$$\Delta u^i + \Gamma_{jk}^i (u_x^j u_x^k + u_y^j u_y^k) = 0. \tag{2.6.20}$$

Hence, for the vector $y = (u^1, \ldots, u^{d-1})$, we get from using (2.6.19) in (2.6.20)

$$|\Delta y| \leqslant c |\nabla y|^2 + \beta \qquad \text{near } z_0$$

$$y = 0 \qquad \text{on } \gamma.$$

This implies, for example by referring to [H1], that

$$y \in C^{1,\beta} \qquad \text{near } z_0.$$

Equation (2.6.19) then implies that also ∇u^d is bounded. Moreover, we compute from (2.6.18)

$$\left(\mathrm{i} z u_z^d + \frac{g_{\alpha d}}{g_{dd}} \mathrm{i} z u_z^\alpha \right)^2 = \frac{g_{\alpha\beta}}{g_{dd}} (z u_z^\alpha)(z u_z^\beta) - \left(\frac{g_{\alpha d}}{g_{dd}} z u_z^\alpha \right)^2 - \frac{z^2 \varphi}{g_{dd}}. \tag{2.6.21}$$

On $\gamma = \{|z| = 1\}$, we have

$$zu_z^k = \tfrac{1}{2}(u_r^k - \mathrm{i}u_\varphi^k).$$

Since $u_\varphi^\alpha = 0\,(\alpha = 1, \ldots, d-1)$ on γ, and since $z^2\varphi$ is real on γ, the right-hand side of (2.6.21) is real. We conclude

$$0 \equiv \left(u_r^d + \frac{g_{\alpha d}}{g_{dd}} u_r^\alpha\right)\left(u_\varphi^d + \frac{g_{\alpha d}}{g_{dd}} u_\varphi^\alpha\right) = \left(u_r^d + \frac{g_{\alpha d}}{g_{dd}} u_r^\alpha\right) u_\varphi^d \qquad \text{on } \gamma. \tag{2.6.22}$$

Let $z_0 \in \gamma$ as before. We want to create a new function $\tilde{u}$ defined on a full neighbourhood of z_0 by reflecting u across γ (locally): for $\alpha = 1, \ldots, d-1$, we put

$$\tilde{u}^\alpha(re^{\mathrm{i}\varphi}) = \begin{cases} u^\alpha(re^{\mathrm{i}\varphi}) & r \leqslant 1 \\ -u^\alpha(r^{-1}e^{\mathrm{i}\varphi}) & r \geqslant 1 \end{cases}$$

and

$$\tilde{u}^d(re^{\mathrm{i}\varphi}) = \begin{cases} u^d(re^{\mathrm{i}\varphi}) & r \leqslant 1 \\ 2u^d(e^{\mathrm{i}\varphi}) - u^d(r^{-1}e^{\mathrm{i}\varphi}) & r \geqslant 1. \end{cases}$$

$\tilde{u}$ then is a C^1-function on a neighbourhood $U(z_0)$ of z_0, and $\tilde{u} \in C^2(U(z_0)\backslash\gamma)$. Hence $\tilde{u} \in C^{1,1}(U(z_0))$. Moreover, because of (2.6.20) and since $\tilde{u}$ is of class C^1

$$|\tilde{u}_{z\bar{z}}| \leqslant \alpha |\tilde{u}_z|. \tag{2.6.23}$$

For $h \in C_0^1(U(z_0))$, and $G \subset\subset U(z_0)$, we obtain

$$\frac{1}{2\mathrm{i}} \int_{\partial G} h\tilde{u}_z \, dz = \int_{G - \gamma_1} (\tilde{u}_z h_{\bar{z}} + \tilde{u}_{z\bar{z}} h) \, dz \, d\bar{z}$$

and hence from (2.6.23)

$$\left| \frac{1}{2i} \int_{\partial G} h\tilde{u}_z \right| dz = \int_{G - \gamma_1} (|h_{\bar{z}}| + \alpha |h|) |\tilde{u}_z| \, dz \, d\bar{z}. \tag{2.6.24}$$

Lemma 2.6.1 implies (using either (2.6.23) or (2.6.24), cf. (2.6.12))

$$\tilde{u}_z = a(z - z_0)^m + o(|z - z_0|^m)$$

where $m \in \mathbb{N}, a \in \mathbb{C}^n, a \neq 0$, which is the desired asymptotic expansion. This in turn implies that u_φ^d either vanishes identically or only at isolated points. The first case is trivial. In the second case, we can conclude from (2.6.22)

$$u_r^d = -\frac{g_{\alpha d}}{g_{dd}} u_r^\alpha \qquad \text{on } \gamma_1. \tag{2.6.25}$$

Also, from (2.6.20)

$$\Delta u^d = -\Gamma_{jk}^d (u_x^j u_x^k + u_y^j u_y^k) \tag{2.6.26}$$

and the right-hand side is bounded as u has a bounded gradient. Since also $u^\alpha \in C^{1,\beta}$, it follows from (2.6.25), (2.6.26) that $u^d \in C^{1,\beta}$. Then $C^{2,\beta}$-regularity for $C^{2,\beta}$ bounded curves follows in the same way.

q.e.d.

Remark: The preceding argument is taken from [GJ]. It generalizes the argument of [H6] and [HH].

2.7. Estimates from below for the functional determinant of univalent harmonic mappings

It turns out that for univalent harmonic maps between surfaces, one can estimate the Jacobian from below in terms of the geometric data involved. Such estimates were given in [JK] (cf. also [J4]). The interior estimate is based on the fundamental work of Heinz (see [H5]). In this section, we shall give a considerably simplified version of the estimate for harmonic maps (note, however, that the results of Heinz apply to a more general class of elliptic systems; on the other hand, we point out that the present arguments also apply if we have an additional terms of the form $h(u)(u_x^1 u_y^2 - u_x^2 u_y^1)$ on the right-hand side, or for the system (1.4.20) from the study of which Heinz derived the motivation for his estimates.)

We start with some auxiliary results. The first one is the so-called similarity principle of Bers and Vekua (cf. e.g. [BJS]; a detailed proof along classical potential theoretic lines can also be found in [H1]).

Lemma 2.7.1: *Let $f: B(0,R) \to \mathbb{C}$ satisfy*

$$|f_{\bar z}| \leqslant \mu |f| \qquad \textit{almost everywhere } (\mu = const.). \tag{2.7.1}$$

Then

$$f(z) = e^{s(z)} h(z) \tag{2.7.2}$$

where h is holomorphic, and $s \in C^\alpha$ $(0 < \alpha < 1)$, *and satisfies*

$$|s(z)| \leqslant 4\mu R \qquad \textit{for } |z| < R. \tag{2.7.3}$$

Proof: We put

$$\sigma(z) := \begin{cases} f_{\bar z}(z)/f(z) & \text{if } f(z) \neq 0 \\ 0 & \text{otherwise.} \end{cases}$$

Then

$$|\sigma(z)| \leqslant \mu \qquad \text{almost everywhere.} \tag{2.7.4}$$

We then solve the equation

$$s_{\bar z} = \sigma. \tag{2.7.5}$$

A solution of (2.7.5) is given by

$$s(z) = -\frac{1}{\pi} \int_{B(0,R)} \sigma(\zeta) \left(\frac{1}{\zeta - z} - \frac{1}{\zeta} \right) d\xi \, d\eta + c_0$$

$$(\zeta = \xi + i\eta \in B(0,R), c_0 = \text{const.}) \tag{2.7.6}$$

and by choosing c_0 in such a way that $s(0)=0$, we obtain (2.7.3). Also, $s \in C^{0,\alpha}$ for every $\alpha < 1$.

(In fact, we have for every p with $2 < p < \infty$:

$$|s(z_1) - s(z_2)| \leqslant c_p \|\sigma\|_{L^p} \cdot |z_1 - z_2|^{1-2/p}$$

by Hölder's inequality.) Moreover,

$$h := f\,\mathrm{e}^{-s} \tag{2.7.7}$$

then satisfies

$$h_{\bar{z}} = f_{\bar{z}}\,\mathrm{e}^{-s} - \sigma \cdot f \cdot \mathrm{e}^{-s} = 0$$

almost everywhere and is hence holomorphic.

q.e.d.

Lemma 2.7.2: *Let $g \in C^2(B(0,R), \mathbb{R})$ satisfy*

$$|g_{z\bar{z}}| \leqslant \mu |g_z| \qquad (\mu = const.) \tag{2.7.8}$$

and suppose

$$g_z \neq 0 \qquad \text{in } B(0,R) \tag{2.7.9}$$

Then

$$|g_z(z_1)| \leqslant k_0\,\mathrm{e}^{8\mu R} |g_z(z_2)| \qquad \textit{for } z_1, z_2 \in B(0, R/2) \tag{2.7.10}$$

with a universal constant k_0.

Proof: $f := g_z$ satisfies (2.7.1); hence, by Lemma 2.7.1,

$$f(z) = \mathrm{e}^{s(z)} h(z) \tag{2.7.11}$$

with

$$|s(z)| \leqslant 4\mu R. \tag{2.7.12}$$

Since $h(z) \neq 0$ in $B(0,R)$ because of (2.7.9), we can apply the Harnack inequality to the non-negative harmonic function

$$-\log \frac{h(z)}{\kappa} \qquad \text{with } \kappa := \sup_{B(0,R)} |h(z)|$$

(if $|h|$ is unbounded on $B(0,R)$, we take $R - \epsilon$ instead) to obtain

$$|h(z_1)| \leqslant k_0 |h(z_2)| \qquad \text{for } z_1, z_2 \in B(0, R/2) \tag{2.7.13}$$

and (2.7.10) then follows from (2.7.11), (2.7.12), (2.7.13).

q.e.d.

Lemma 2.7.3: *Let Σ be a surface with a complete Riemannian metric, $u: D \to \Sigma$ harmonic. Assume that the functional determinant $J(u) \neq 0$ in $\mathring{D}$, and*

$$|\nabla u(z)| \leqslant k_1 \qquad \textit{for } z \in B(0, \tfrac{1}{2}(1+\rho)) \tag{2.7.14}$$

for some $\rho < 1$. *Then*

$$|J(u)(z_1)| \geqslant c|J(u)(z_2)| \qquad \textit{for any } z_1, z_2 \in B(0,\rho) \tag{2.7.15}$$

where c *depends on* ρ *and the geometry of* Σ, *and on* k_1.

Proof: Because of (2.7.14), $u(B(0, \frac{1}{2}(1+\rho))$ is contained in a compact subset Σ_0 of Σ. Therefore, we can find some $r > 0$ with the property that for any $p \in \Sigma_0$ and any geodesic arc c through p, there exist geodesic parallel coordinates in $B(p, r)$ based on c. In these coordinates (u^1, u^2), for $u^2 \equiv 0, u^1$ is the arclength parameter of c, whereas the curves $u^1 \equiv$ const. are geodesics normal to c parametrized by arclength u^2; consequently the curves $u^2 \equiv$ const. are parallel curves of c. In these coordinates, we have for the metric tensor

$$\begin{aligned} g_{11}(u^1, 0) &= 1 \\ g_{12}(u^1, u^2) &\equiv 0 \\ g_{22}(u^1, u^2) &\equiv 1. \end{aligned}$$

Therefore, the only non-vanishing Christoffel symbols are

$$\begin{aligned} \Gamma^1_{11} &= \tfrac{1}{2} g^{11} g_{11,1} \\ \Gamma^1_{12} &= \tfrac{1}{2} g^{11} g_{11,2} \\ \Gamma^2_{11} &= -\tfrac{1}{2} g_{11,2}. \end{aligned} \tag{2.7.16}$$

Hence, the equations (1.2.10) for u take the form

$$\begin{aligned} \Delta u^1 &= -\Gamma^1_{11}(u^1_x u^1_x + u^1_y u^1_y) - \Gamma^1_{12}(u^1_x u^2_x + u^1_y u^2_y) \\ \Delta u^2 &= -\Gamma^2_{11}(u^1_x u^1_x + u^1_y u^1_y). \end{aligned} \tag{2.7.17}$$

In particular, we have

$$|\Delta u^1| \leqslant \mu |u^1_z| \tag{2.7.18}$$

where μ depends on k_1 and the magnitude of $|\Gamma^1_{11}|$ and $|\Gamma^1_{12}|$.

For $z \in B(z_1, \rho_1) \subset B(0, \frac{1}{2}(1+\rho))$, (2.7.14) implies

$$d(u(z), u(z_1)) \leqslant \rho_1 k_1.$$

We choose

$$\rho^* := \min(\tfrac{1}{4}(1-\rho), r/2k_1) \qquad (r \text{ as above}). \tag{2.7.19}$$

Then, for $z_1 \in B(0, \rho)$,

$$B(z_1, 2\rho^*) \subset B(0, \tfrac{1}{2}(1+\rho)) \tag{2.7.20}$$

and

$$u(B(z_1, 2\rho^*)) \subset B(u(z_1), r). \tag{2.7.21}$$

By assumption, for geodesic parallel coordinates (u^1, u^2) on $B(u(z_1), r)$, we have

$$0 < \frac{|J(u)(z)|}{c_0 k_1} \leqslant |\nabla u^1(z)| \leqslant c_0 k_1 \tag{2.7.22}$$

(c_0 depends on the geometry of Σ).

We now choose the geodesic through $u(z_1)$, entering in the definition of our coordinates, in such a way that

$$|\nabla u^1(z_1)| \leqslant |J(u)(z_1)|^{1/2} \tag{2.7.23}$$

(note $g_{11}(0,0)=1$).

Because of (2.7.18) and (2.7.20), we may apply Lemma 2.7.2 (on $B(z_1,\rho^*)$ instead of $B(0,R/2)$) to u^1. From (2.7.22) and (2.7.23) we then conclude for $z\in B(z_1,\rho^*)$

$$|J(u)(z_1)| \geqslant c_1 |J(u)(z)| \tag{2.7.24}$$

where c_1 depends on k_1 and the geometry of Σ_0.

Covering $B(0,\rho)$ by a finite number of balls of radius ρ^* satisfying (2.7.19) then shows the claim.

q.e.d.

We also have the following Lemma of Heinz [H5]:

Lemma 2.7.4: *Suppose $u\colon D\to\Sigma$ is a univalent (i.e. injective) harmonic map. Then*

$$J(u)(z)\neq 0 \qquad \text{for all } z\in \mathring{D}. \tag{2.7.25}$$

Proof: Assume

$$J(u)(z_1)=0 \qquad \text{for some } z_1\in D. \tag{2.7.26}$$

As in the preceding proof, we use geodesic parallel coordinates in some $B(u(z_1),r)$. This time, we choose the geodesic arc in such a way that

$$\nabla u^1(z_1)=0. \tag{2.7.27}$$

Using (2.7.18) again, we infer from Lemma 2.7.1 (or from Lemma 2.6.1)

$$u_z^1(z)=(z-z_1)^n+o(|z-z_1|^n) \qquad \text{for some } n\geqslant 1 \tag{2.7.28}$$

in a neighbourhood of z_1 (possibly after a linear change of coordinates). Then

$$u^1(z)=\mathrm{Re}((z-z_1)^{n+1})+o(|z-z_1|^{n+1}) \tag{2.7.29}$$

near z_1.

In polar coordinates (r,θ) centred at z_1,

$$u^1(r\,e^{i\theta})=r^{n+1}\cos((n+1)\theta)+o(r^{n+1})$$

and in particular for $k=0,1,\ldots,2n+1$,

$$u^1(r\,e^{i\pi k/(n+1)})=r^{n+1}(-1)^k+o(r^{n+1}). \tag{2.7.30}$$

For sufficiently small $\epsilon>0$ and $r\leqslant\epsilon$, the sign of the left-hand side of (2.7.30) is therefore $(-1)^k$.

Therefore, if z traverses a closed Jordan curve in $B(z_1,\epsilon)\backslash\{z_1\}$ with index 1 w.r.t. z_1, then $u_1(z)$ changes sign at least $2(n+1)$ times. Since on the other hand,

u is a homeomorphism by assumption, we can choose δ so small that

$$u^{-1}\{(u^1)^2+(u^2)^2\leqslant\delta\}\subset B(z_1,\epsilon).$$

Therefore, the pre-image of $\{(u^1)^2+(u^2)^2\leqslant\delta\}$ is a Jordan curve with the properties stated above. However, as z traverses this curve, u^1 changes sign exactly twice, contradicting $n\geqslant 1$. Therefore, (2.7.27) cannot happen, and thus neither (2.7.26).

q.e.d.

Theorem 2.7.1: *Suppose $u: D\to u(D)\subset\Sigma$ is a bijective harmonic map, where Σ is a surface with a (complete) Riemannian metric.*

Assume

$$u(D)\subset B(p,r) \qquad \text{for some } p\in\Sigma \tag{2.7.31}$$

with

$$r<\pi/2\kappa \tag{2.7.32}$$

where κ^2 is an upper curvature bound. Then for $\rho<1$ and $z\in B(0,\rho)$

$$|J(u)(z)|\geqslant\delta^{-1} \tag{2.7.33}$$

where δ depends on ρ, r, the geometry of Σ, and in addition on some kind of normalization, like the following:

(i) *prescribing $u(0)$;*
(ii) *prescribing $u(z_1), u(z_2), u(z_3)$ for three distinct points $z_1, z_2, z_3\in\partial D$;*
(iii) $|u_{|\partial D}|_{C^\alpha}$;
(iv) *assuming* $\operatorname{meas}(u(B(0,\sigma)))\geqslant\mu>0$ *for some $\sigma\in(0,1)$.*

In the first three cases, δ also depends on $\operatorname{meas} u(D)$, *and in the first two cases in addition on $E(u)$.*

Proof: By Lemma 2.7.4, $J(u)\neq 0$ in $\mathring{D}$. Because of (2.7.31), (2.7.32), Theorem 2.5.2 yields a gradient bound of the type (2.7.14) in $B(0,\frac{1}{2}(1+\rho))$. Lemma 2.7.3 can therefore be applied, and in order to show the claim, we need to find some $z_0\in B(0,\rho)$ with

$$|J(u)(z_0)|\geqslant\tau \tag{2.7.34}$$

for some $\tau>0$.

If $\operatorname{meas}(u(B(0,\sigma)))\geqslant\mu$ for some $\sigma\in(0,1)$, then

$$\int_{B(0,\sigma)}|J(u)(z)|\,\mathrm{d}z\geqslant\mu \tag{2.7.35}$$

and we can find $\tau>0$ with (2.7.34).

For any $\mu<\operatorname{meas} u(D)$, we can find $\epsilon>0$ with

$$\operatorname{meas}\{q\in u(D): d(q,\partial u(D))\geqslant\epsilon\}\geqslant\mu. \tag{2.7.36}$$

If we normalize u by prescribing the image of 0 or of three boundary points, then the Courant–Lebesgue Lemma allows us to estimate the modulus of continuity in terms of $E(u)$, because u is univalent (cf. Lemma 1.1.1 and also the proof of Lemma 3.1.2 below for more details).

Finally, Corollary 2.5.1 allows us to estimate the modulus of continuity of u in terms of the C^α-norm of its boundary values. Since by injectivity, $\partial u(D) = u(\partial D)$, we can in any case compute $\sigma \in (0,1)$ with

$$d(u(z), \partial u(D)) \leqslant \epsilon \qquad \text{if } z \in D \backslash B(0, \sigma).$$

Since u is univalent

$$u(B(0,\sigma)) \supset \{q \in u(D) \colon d(q, \partial u(D)) \geqslant \epsilon\}$$

and hence from (2.7.36)

$$\operatorname{meas}(u(B(0,\sigma)) \geqslant \mu.$$

q.e.d.

Remark: The optimal estimate would involve a dependence of δ on the geometry of Σ only through curvature bounds. This does not follow from the preceding argument, as Γ_{11}^1 from (2.7.16) which enters into the constant μ in (2.7.18) cannot be estimated in terms of curvature bounds alone. On the other hand, such an optimal estimate is obtained in [JK] (cf. also [J4]); the proof is more complicated, however, as it needs an analogue of Lemma 2.7.2 for functions satisfying

$$|g_{z\bar{z}}| \leqslant \mu_1 |g_z| + \mu_2 |g|$$

(cf. [JK] for details).

In any case, the present estimate is explicit, as δ in (2.7.33) can be computed in terms of the data involved.

We now want to give an indirect argument, also due to Heinz [H2], to show that an estimate with the optimal geometric dependence holds.

Lemma 2.7.5: *Let $u \colon D \to \Sigma$ be a harmonic map with $J(u) \not\equiv 0$, which is a C^1-limit of diffeomorphisms $v_n \colon D \to \Sigma$. Then*

$$J(u) \neq 0 \qquad \text{in } \mathring{D}. \tag{2.7.37}$$

Proof: Suppose

$$J(u)(z_1) = 0 \tag{2.7.38}$$

for some $z_i \in \mathring{D}$. W.l.o.g. $z_1 = 0$.

We shall again use geodesic parallel coordinates based on a geodesic arc c through $u(z_1)$, but this time we shall work with u^2 as this coordinate can be better controlled in terms of geometric data. (We want to make use of this geometric control below; for the present proof it is not really necessary).

$\operatorname{grad} u^2(q)$ is the unit tangent vector of the geodesic through q which is orthogonal to c. If $Y, Z \in T_q\Sigma$, and one of them is parallel to grad $u^2(q)$, we have

$$D^2u^2(q)(Y, Z) = \langle \nabla_Y \operatorname{grad} u^2(q), Z \rangle = 0 \tag{2.7.39}$$

(here D^2 denotes the second fundamental form of a map) whereas for $Y \in T_q\Sigma$ orthogonal to grad $u^2(q)$

$$D^2u^2(q)(Y, Y) = \kappa_g(q)|Y|^2 \tag{2.7.40}$$

where $\kappa_g(q)$ is the geodesic curvature of that curve through q that is parallel to c.
We shall now use:

Lemma 2.7.6: *Suppose c is geodesic, $q \in \Sigma$ is not further away from c than $\pi/2\Lambda$ ($\Lambda^2 := \max|K|, K =$ curvature) or a cut point. Then*

$$|D^2u^2(q)(Y, Y)| \leqslant \Lambda|\tan(\Lambda u^2(q))| \cdot |Y|^2 \tag{2.7.41}$$

for $Y \in T_q\Sigma$.

Proof: For the geodesic curvature $\kappa_g(h)$ of the parallel curves of c at distance h, we have the differential equation

$$\kappa_g'(h) = \kappa_g^2(h) + K(h) \tag{2.7.42}$$

and thus

$$\left|\frac{1}{\Lambda} \operatorname{arctg}\left(\frac{1}{\Lambda}\kappa_g\right)'\right| \leqslant 1$$

and since by assumption $\kappa_g(0) = 0$,

$$|\kappa_g(h)| \leqslant \Lambda|\tan \Lambda h| \qquad (h = u^2) \tag{2.7.43}$$

and (2.7.41) now follows from (2.7.39) and (2.7.40).

q.e.d.

Continuation of proof of Lemma 2.7.5: Let

$$|\nabla u(z)| \leqslant c \qquad \text{in a neighbourhood of } z_1$$

Lemma 2.7.6 and the chain rule (2.2.2) then imply

$$|\Delta u^2(z)| \leqslant \Lambda^2 c^2 |u^2(z)|. \tag{2.7.44}$$

If now (2.7.38) holds, we can choose the coordinates in such a way that

$$u_z^2(z_1) = 0.$$

Since also

$$u^2(z_1) = 0$$

Lemma 2.6.1 yields (after a linear change of coordinates)

$$u_z^2 = z^n + o(|z|^n) \tag{2.7.45}$$

for some $n \geqslant 1$. (Note that $J(u) \equiv 0$ by assumption.) Let

$$f^n := u^2(v_n(z)).$$

Since the v_n are diffeomorphisms,

$$f_z^n \neq 0 \tag{2.7.46}$$

W.l.o.g. $z_1 = 0$. We find some $r > 0$ and some $n_0 \in \mathbb{N}$ with

$$|f_z^{n_0}(z) - u_z^2(z)| < |u_z^2(z)|$$

for $|z| = r$, because of (2.7.45) and since $v_n \to u$ in C^1 by assumption.

Then, by elementary properties of winding numbers (cf. Rouche's Theorem),

$$\frac{1}{2\pi}\int_{|z|=r} d\arg f_z^{n_0}(z) = \frac{1}{2\pi}\int_{|z|=r} d\arg u_z^2(z) = n > 0 \tag{2.7.47}$$

and hence, we find some z_0 with $|z_0| < r$ and $f_z^{n_0}(z_0) = 0$, contradicting (2.7.46).

This shows that (2.7.38) is impossible.

q.e.d.

Theorem 2.7.2: *Given numbers $\omega^2, \kappa^2, 0 < r, 0 < \rho < 1, k_1$, there exists a number $\tau(\omega^2, \kappa^2, r, \rho, k_1)$ with the following property:*

If Σ is a surface with a complete Riemannian metric with curvature K satisfying

$$-\omega^2 \leqslant K \leqslant \kappa^2, \tag{2.7.48}$$

if $u: D \to u(D) \subset \Sigma$ is a bijective harmonic map with

$$B(p, r) \subset u(D) \qquad \text{for some } p \in \Sigma \tag{2.7.49}$$

and if the modulus of continuity is controlled by k_1, then

$$|J(u)(z)| \geqslant \tau^{-1} \tag{2.7.50}$$

for all $z \in B(0, \rho)$.

Proof: Otherwise, we find Riemannian metrics on Σ with the stated curvature bounds, and corresponding harmonic diffeomorphisms $u_n: D \to \Sigma$ (using Lemma 2.7.4; note that in the proof of Lemma 2.7.4, one can work with u^2 instead of u^1 as in the proof of Lemma 2.7.5, in order to get the dependence on the geometry straight) satisfying (2.7.49), and points $z_n \in B(0, \rho)$, with

$$|J(u_n)(z_n)| \to 0. \tag{2.7.51}$$

From Theorems 2.1.1 and 2.1.2, we can infer that after selection of a subsequence, the metrics on Σ coverage to a metric with (2.7.48), and from Theorems 2.5.1 and 2.5.2 that the maps u_n coverage locally in $C^{2,\alpha}$ to a harmonic map

$$u: D \to \Sigma \qquad (\Sigma \text{ equipped with the limit metric})$$

with (2.7.49), and finally also $z_n \to z_1 \in B(0, \rho)$. Lemma 2.7.5 becomes applicable to show

$$J(u) \equiv 0 \qquad \text{in } \mathring{D} \tag{2.7.52}$$

as

$$J(u)(z_1)=0 \qquad \text{by (2.7.51).}$$

In the same way as in the proof of Theorem 2.7.1, we find $\sigma>0$ with

$$B(p,r/2)\subset u(B(0,\sigma))$$

because the u_n are equicontinuous by assumption. This, however, is incompatible with (2.7.52). This proves the result.

q.e.d.

We now turn to estimates at the boundary; the following result is taken from [JK]. It is remarkable that here one does not need to assume that u is bijective; it suffices that u maps D to one side of $u(\partial D)$. Also this case is much easier than the interior one.

Theorem 2.7.3: *Suppose $u: D\to\Sigma$ is harmonic, and $u(D)\subset B(p,r)$, where $B(p,r)$ is a disc with radius $r<\pi/2\kappa$. Suppose that $\partial u(D)=u(\partial D)$ and that $g:=u|\partial D:\partial D\to \partial u(D)$ is a C^2-diffeomorphism with*

$$0<b\leqslant |dg(\varphi)/d\varphi| \qquad \textit{for all } \varphi\in\partial D. \tag{2.7.53}$$

Assume furthermore that $g(\partial D)$ is strictly convex w.r.t. $u(D)$, and that we have the following estimates for the geodesic curvature of $g(\partial D)$:

$$0<a_1\leqslant \kappa_g(g(\partial D))(\varphi)\leqslant a_2 \qquad \textit{for all } \varphi\in\partial D. \tag{2.7.54}$$

Then

$$|J(u)(z)|\geqslant \delta_1^{-1} \qquad \textit{for all } z\in\partial D \tag{2.7.55}$$

where $\delta_1=\delta_1(\omega,\kappa,r,a_1,a_2,b,|g|_{C^{1,\alpha}})$ and *$-\omega^2\leqslant K\leqslant\kappa^2$ are curvature bounds as before.*

Proof: We define $h(q):=-d(q,\partial u(D))$ for $q\in u(D)$. Equations (2.7.53) and (2.7.54) imply for $q\in\partial u(D)$

$$\Delta(h\circ u)\geqslant a_1 b^2. \tag{2.7.56}$$

This will enable us to get a lower bound for the radial derivative of $h\circ u$ at boundary points with the argument of the boundary lemma of Hopf. This assertion in turn implies (2.7.55), taking (2.7.53) into account.

The constants a_2 and κ control focal points of $u(\partial D)$, and a_1 and ω then determine how long the level curves of h remain strictly convex and free of double points. By Corollary 2.5.1, we can therefore find a neighbourhood V_0 of ∂D in D with the property that h is a C^2 function with strictly convex level curves on $u(V_0)$.

Suppose $z_0\in\partial D$. Using now (2.7.56) and Theorem 2.5.2, we can choose some disc $B(z_1,\rho_1)\subset D, z_0\in\partial B(z_1,\rho_1)$, in such a way that

$$\Delta(h\circ u)(z)\geqslant \tfrac{1}{2}a_1 b^2 \qquad \text{for } z\in B(z_1,r_1). \tag{2.7.57}$$

Defining the auxiliary function $\gamma(z)$ via

$$\gamma(z) := \frac{\rho_1^2}{8} a_1 b^2 \left(1 - \frac{(z - z_1)^2}{\rho_1^2}\right)$$

we have

$$\Delta\gamma(z) = -\tfrac{1}{2} a_1 b^2$$

and consequently by (2.7.57)

$$\Delta(h \circ u + \gamma)(z) \geqslant 0 \qquad \text{on } B(z_1, \rho_1). \tag{2.7.58}$$

Moreover,

$$(h \circ u)(z_0) + \gamma(z_0) = 0$$

and

$$(h \circ u)(z) + \gamma(z) \leqslant 0 \qquad \text{on } \partial B(z_1, \rho_1)$$

since by assumption u is mapped onto the side of $\partial u(D)$, where h assumes non-positive values, and $\gamma | \partial B(z_1, \rho_1) = 0$. The maximum principle now controls the derivative of $h \circ u + \gamma$ at z_0 in the direction of the outer normal, namely

$$\frac{\partial}{\partial n}(h \circ u + \gamma)(z_0) \geqslant 0$$

and thus

$$\frac{\partial}{\partial n}(h \circ u)(z_0) \geqslant \frac{\rho_1}{4} a_1 b^2 \text{ by definition of } \gamma. \tag{2.7.59}$$

Equations (2.7.53) and (2.7.59) imply

$$|J(u)(z_0)| \geqslant \frac{\rho_1}{4} a_1 b^3 =: \delta_1^{-1}. \tag{2.7.60}$$

q.e.d.

Combining the interior and boundary estimates, we get:

Corollary 2.7.1: *Assume $u: D \to \Sigma$ is an injective harmonic map, where $u(D) \subset B(p, r)$, and $B(p, r)$ is a disk with radius $r < \pi/2\kappa$. Suppose that $g := u | \partial D \in C^{1,\alpha}$, and that* (2.7.53) *and* (2.7.54) *hold. Then for all $z \in D$*

$$|J(u)(z)| \geqslant \delta_2^{-1} \tag{2.7.61}$$

where $\delta_2 = \delta_2(\omega, \kappa, r, a_1, a_2, b, |g|_{C^{1,\alpha}})$.

Proof: Equation (2.7.61) follows from Theorem 2.7.3, Theorem 2.7.2, and Theorem 2.5.2.

q.e.d.

Remark: Another interesting estimate for the functional determinant of univalent solutions of elliptic systems in two dimensions was recently obtained by Schulz [Sz].

3 CONFORMAL REPRESENTATION

In this chapter, we shall be concerned with the conformal representation of compact surfaces (with or without boundary), equipped with Riemannian metrics, by standard models. Of course, such a representation can usually be derived by first establishing the existence of conformal parameters in the small and then applying the uniformization theorem, at least for surfaces without boundary.

We want to present a different approach, however. By a global argument, we shall directly show the existence of conformal parameters in the large and hence do not need to appeal to the uniformization theorem. Our approach will shed some preliminary light on the effect of variations of the conformal structure, a subject that will be thoroughly discussed in Chapter 6. Also it will be more applicable for minimal surfaces of higher genus, treated in Section 4.7 and in a general way in [JSt]. Our approach was solicited by the incomplete approach of [M3, Section 9.3].

In order to exhibit the conceptual ideas first, we shall start with the easiest case, namely with surfaces homeomorphic to S^2.

3.1. Conformal representation of surfaces homeomorphic to S^2

Theorem 3.1.1: *Let Σ be a surface homeomorphic to S^2 with metric tensor given in local coordinates by bounded measurable functions satisfying*

$$g_{11}g_{22} - g_{12}^2 \geqslant \lambda > 0 \qquad \text{almost everywhere.} \tag{3.1.1}$$

Then there is a homeomorphism $h: S^2 \to \Sigma$ satisfying the conformality relations

$$g_{ij}\frac{\partial h^i}{\partial x}\frac{\partial h^j}{\partial x} = g_{ij}\frac{\partial h^i}{\partial y}\frac{\partial h^j}{\partial y}$$

$$g_{ij}\frac{\partial h^i}{\partial x}\frac{\partial h^j}{\partial y}=0 \tag{3.1.2}$$

almost everywhere.

If $(g_{ij})\in C^\alpha$, *then h is a diffeomorphism of class* $C^{1,\alpha}$, *satisfying* (3.1.2) *everywhere. If S is of class* $C^{k,\alpha}$, C^∞, *or* C^ω, *then so is h.*

In the proof, we shall make crucial use of the following version of the Courant–Lebesgue lemma:

Lemma 3.1.1: *Let* $u\in H^{1,2}(D,N)$, $E(u)\leqslant K$, $x_0\in D$, *N a Riemannian manifold with distance function* $d(\cdot,\cdot)$, $\delta<1$. *Suppose* $\partial B(x_0,r)\cap D$ *is connected for all* $r\in(\delta,\sqrt{\delta})$. *Then there exists some* $r\in(\delta,\sqrt{\delta})$ *for which* $u|\partial B(x_0,r)\cap D$ *is absolutely continuous and*

$$d(u(x_1),u(x_2))\leqslant(8\pi K)^{1/2}[\log 1/\delta]^{-1/2} \tag{3.1.3}$$

for all $x_1, x_2\in\partial B(x_0,r)\cap D$, *and*

$$\int_0^{2\pi}\left|\frac{\partial u(r,\varphi)}{\partial\varphi}\right|^2\leqslant\frac{4\pi}{\log(1/\delta)}. \tag{3.1.4}$$

Proof: (The proof is the same as that of Lemma 1.1.1.) For $x_1,x_2\in\partial B(x_0,r)$ and almost all $r>0$, since u is a Sobolev function, $u|\partial B(x_0,r)$ is absolutely continuous and

$$\begin{aligned}d(u(x_1),u(x_2))&\leqslant\int_0^{2\pi}\left|\frac{\partial u(r,\varphi)}{\partial\varphi}\right|\mathrm{d}\varphi\\&\leqslant(2\pi)^{1/2}\left(\int_0^{2\pi}\left|\frac{\partial u}{\partial\varphi}\right|^2\mathrm{d}\varphi\right)^{1/2}\end{aligned}$$

(w.l.o.g. $B(x_0,r)\subset D$).

The energy of u over $B(x_0,r)$ is given by

$$E(u;B(x_0,r))=\frac{1}{2}\int_{B(x_0,r)}\left(\left|\frac{\partial u}{\partial r}\right|^2+\frac{1}{r}\left|\frac{\partial u}{\partial\varphi}\right|^2\right)r\,\mathrm{d}r\,\mathrm{d}\varphi.$$

Thus, there exists $r\in(\delta,\sqrt{\delta})$ with

$$\int_0^{2\pi}\left|\frac{\partial u(r,\varphi)}{\partial\varphi}\right|^2\mathrm{d}\varphi\leqslant\frac{2K}{\int_\delta^{\sqrt{\delta}}r^{-1}\,\mathrm{d}r}=\frac{4K}{\log(1/\delta)}.$$

All claims then follow.

q.e.d.

Proof of Theorem 3.1.1: As usual, we define the energy of a map $v\colon S^2\to\Sigma$ as

$$E(v)=\frac{1}{2}\int g_{ij}(v(z))(v_x^i v_x^j+v_y^i v_y^j)\,\mathrm{d}x\,\mathrm{d}y$$

($z = x + iy \in S^2$) where $g_{ij}(v)$ are the coefficients of the metric tensor of Σ in a chart Ψ. For the moment, we assume that Σ is differentiable and that the g_{ij} are continuous, to make $E(v)$ well defined for maps $v \in H^{1,2}$. This restriction will be removed at the end by approximation arguments.

We choose three different points z_1, z_2, z_3, in S^2 and three different points p_1, p_2, p_3 in Σ. Let $\mathscr{D}$ be the class of all diffeomorphisms $v\colon S^2 \to \Sigma$ satisfying

$$E(v) = \frac{1}{2}\int g_{ij}(v(z))(v_x^i v_x^j + v_y^i v_y^j)\,dx\,dy \tag{3.1.5}$$

and let $\bar{\mathscr{D}}$ be the closure of $\mathscr{D}$ w.r.t. weak $H^{1,2}$ and uniform convergence.

We choose an energy-minimizing sequence $(v_n)_{n\in\mathbb{N}}$ in the class $\bar{\mathscr{D}}$. W.l.o.g.

$$E(v_n) \leqslant K. \tag{3.1.6}$$

Lemma 3.1.2: *Energy-bounded subsets of $\bar{\mathscr{D}}$ are sequentially compact. In particular, if $(v_n)_{n\in\mathbb{N}} \subset \bar{\mathscr{D}}$ is a sequence satisfying* (3.1.6), *then (v_n) is equicontinuous, and a subsequence converges weakly in $H^{1,2}$ and uniformly to some $v \in \bar{\mathscr{D}}$.*

Proof: Let $(v_n) \subset \bar{\mathscr{D}}$ satisfying (3.1.6). For each $z \in S^2$ and $\epsilon > 0$, by Lemma 3.1.1 we can find $\delta > 0$ and for each $n \in \mathbb{N}$ then some $r_n \in (\delta, \sqrt{\delta})$ for which

$$\operatorname{diam}(v_n(\partial B(x, r_n)) \leqslant \epsilon.$$

Here, δ is independent of z and n, since the energy of the v_n is uniformly bounded. We can choose δ so small that $B(z, \sqrt{\delta})$ contains at most one of the points z_1, z_2, z_3. Now $v_n(\partial B(z, r_n))$ divides Σ into two parts, one of them being $v_n(B(z, r_n))$, since v_n is a uniform limit of diffeomorphisms. If ϵ is chosen small enough, then the smaller part, i.e. the one having diameter at most ϵ, contains at most one of the points p_1, p_2, p_3 and hence has to coincide with $v_n(B(z, r_n))$. In particular,

$$\operatorname{diam}(v_n(B(z, \delta)) \leqslant \epsilon$$

and the v_n are equicontinuous as claimed.

The result then follows from the Arzela–Ascoli Theorem and the fact that energy-bounded subsets of $H^{1,2}$ are sequentially weakly compact.

q.e.d.

As a consequence of Lemma 3.1.2, our energy-minimizing sequence converges (after selection of a subsequence) uniformly and weakly in $H^{1,2}$ to some $v \in \bar{\mathscr{D}}$. Since the energy is lower semicontinuous w.r.t. weak $H^{1,2}$ convergence, v minimizes the energy in $\bar{\mathscr{D}}$. Also, v is continuous and homotopic to the v_n. (We can assume that all v_n are in the same homotopy class.)

Moreover, if we have a sequence $(w_n)_{n\in\mathbb{N}}$ in the uniform and weak $H^{1,2}$-closure of the class of diffeomorphisms between S^2 and Σ, but not necessarily satisfying (3.1.5), then we still have

$$E(v) \leqslant E(w) \tag{3.1.7}$$

since the normalization (3.1.5) can always be achieved by composing w_n with a

Möbius transformation, i.e. conformal automorphism of S^2, without changing $E(w_n)$ (cf. Lemma 1.3.2).

Hence, if $\sigma_t\colon S^2 \to S^2$ is a family of diffeomorphisms, depending smoothly on t, with $\sigma_0 = \mathrm{id}$, then

$$\frac{\mathrm{d}}{\mathrm{d}t} E(v \circ \sigma_t)|_{t=0} = 0 \tag{3.1.8}$$

since $v \circ \sigma_t$ is the uniform and weak H^1_2-limit of $v_n \circ \sigma_t$.

We introduce local coordinates $z = x + iy$ on S^2 by stereographic projection and put

$$E := g_{ij}(v) v^i_x v^j_x \qquad F := g_{ij}(v) v^i_x v^j_y \qquad G := g_{ij}(v) v^i_y v^j_y$$
$$\phi := (E - G - 2iF)\,\mathrm{d}z^2.$$

Since ϕ represents a holomorphic quadratic differential, by Lemma 1.2.4, Lemma 1.4.1 implies

$$\phi \equiv 0.$$

Therefore, v satisfies the conformality relations

$$\begin{aligned} g_{ij} v^i_x v^j_x &= g_{ij} v^i_y v^j_y \\ g_{ij} v^i_y v^j_y &= 0 \end{aligned} \tag{3.1.9}$$

almost everywhere.

For notational convenience, we introduce local coordinates (v^1, v^2) on Σ. We want to exploit that v is weakly (anti)conformal and the uniform limit of diffeomorphisms in order to show that the Jacobian $v^1_x v^2_y - v^1_y v^2_x$ of v has the same sign almost everywhere in S^2 (cf. [M3, Section 9.3.7]). Here, the difficulties arise from the fact that v so far is only known to be of class $C^0 \cap H^{1,2}$.

Definition 3.1.1: Suppose G is a plane domain of class C^1, $w \in C^1(G, \mathbb{R}^2)$, $p \notin w(\partial G)$. Then $m(p, w(\partial G))$ is defined to be the winding number of the curve $w(\partial G)$ w.r.t. p. If only $w \in C^0(G, \mathbb{R}^2)$, then

$$m(p, w(\partial G)) := \lim_{n \to \infty} m(p, w_n(\partial G))$$

for any sequence $w_n \in C^1(\partial G, \mathbb{R}^2)$ which converges uniformly to w on ∂G.

That $m(p, w(\partial G))$ is well defined follows from elementary properties of winding numbers (cf. e.g. [Fe]).

Lemma 3.1.3: *Let G be a plane domain, $w \in C^0 \cap H^{1,2}(G, \mathbb{R}^2)$. Then for every $z_0 \in G$, there exists a set $C(z_0)$ with $\boldsymbol{H}^1(C(z_0)) = 0$, where $\boldsymbol{H}^1$ is one-dimensional Hausdorff measure, such that for all $R \notin C(z_0)$*

$$\int_{B(z_0,R)} J(w)\,\mathrm{d}z = \int_{w(B(z_0,R))} m(p, w(\partial B(z_0, R)))\,\mathrm{d}p \tag{3.1.10}$$

if $B(z_0, R) \subset\subset G$. (Here $J(\phi) := \phi^1_x \phi^2_y - \phi^1_y \phi^2_x$.)

Proof: We can find a sequence $w_n \in C^1(G_0)$, $G_0 \subset\subset G$, converging uniformly and strongly in $H^{1,2}$ to w, so that $w_n \to w$ strongly in $H^{1,2}(\partial B(z_0, R))$ on $\partial B(z_0, R)$, if $R \notin C(z_0)$, $H^1(C(z_0)) = 0$. Since $H^{1,2}(\partial B(z_0, R))$ functions are absolutely continuous, and the lengths of $w_n(\partial B(z_0, R))$ and $w(\partial B(z_0, R))$ are uniformly bounded, the two-dimensional measure of $w(\partial B(z_0, R))$ vanishes ($R \notin C(z_0)$). Consequently, $p \notin w(\partial B(z_0, R))$ for almost all p, and thus

$$m(p, w_n(\partial B(z_0, R))) \to m(p, w(\partial B(z_0, R))) \tag{3.1.11}$$

for these p. Now

$$\lim_{n\to\infty} \int_{w_n(B(z_0,R))} m(p, w_n(\partial B(z_0, R)))\, \mathrm{d}p = \lim_{n\to\infty} \int_{B(z_0,R)} J(w_n)\, \mathrm{d}z = \int_{B(z_0,R)} J(w)\, \mathrm{d}z.$$

Since

$$\int_I m(p, w_n(\partial B(z_0, R)))\, \mathrm{d}p \leqslant \left[\frac{\operatorname{meas} I}{\pi}\right]^{1/2} \operatorname{length}(w_n(\partial B(z_0, R)))$$

for any measurable set I, we can integrate (3.1.11), and the result follows.

q.e.d.

Lemma 3.1.4: *We suppose that $u_n\colon S^2 \to \Sigma$ converges uniformly and weakly in $H^{1,2}$ to u; and that*

$$J(u_n) \geqslant 0 \qquad \textit{almost everywhere.} \tag{3.1.12}$$

Then also

$$J(u) \geqslant 0 \qquad \textit{almost everywhere.} \tag{3.1.13}$$

Proof: We introduce coordinates on S^2 by stereographic projection. Let $B(z_0, R)$, $z_0 \in \Sigma_1$, $R \notin C(z_0)$ satisfy the assumption of Lemma 3.1.3 for u and all u_n,

$$\epsilon_n := \max_{z \in \partial B(z_0,R)} |u_n(z) - u(z)|$$

$$V_n := \{w\colon d(w, u(\partial B(z_0, R))) > \epsilon_n\}.$$

For all $w \in V_n$, $m(w, u_n(\partial B(z_0, R))) = m(w, u(\partial B(z_0, R)))$. Then, since u_n converges uniformly to u, using Lemma 3.1.3

$$\begin{aligned}
\lim_{n\to\infty} \int_{B(z_0,R)} J(u_n)\, \mathrm{d}z &= \lim_{n\to\infty} \int_{u_n(B(z_0,R))} m(w, u_n(\partial B(z_0, R)))\, \mathrm{d}w \\
&= \lim_{n\to\infty} \int_{u_n(B(z_0,R)) \cap V_n} m(w, u_n(\partial B(z_0, R)))\, \mathrm{d}w \\
&= \lim_{n\to\infty} \int_{u(B(z_0,R)) \cap V_n} m(w, u(\partial B(z_0, R)))\, \mathrm{d}w
\end{aligned}$$

$$= \int_{u(B(z_0,R))} m(w, u(\partial B(z_0, R)))\,dw$$

$$= \int_{B(z_0,R)} J(u)\,dz.$$

Since $J(u_n) \geqslant 0$ and the preceding argument is valid for almost all disks $B(z_0, R)$, we infer $J(u) \geqslant 0$ as desired.

q.e.d.

We conclude that v is weakly conformal, i.e. a weak solution of the corresponding Cauchy–Riemann equations

$$\begin{aligned} v_x^2 &= -g_{22}^{-1}(g_{12}v_x^1 + k\sqrt{g}v_y^1) \\ v_y^2 &= g_{22}^{-1}(k\sqrt{g}v_x^1 - g_{12}v_y^1) \end{aligned} \tag{3.1.14}$$

$(g = g_{11}g_{22} - g_{12}^2)$ where $k = \pm 1$ is constant by Lemma 3.1.4. Since (3.1.4) is a linear first-order elliptic system, v is regular. (We shall discuss this point more thoroughly later on.) Let us assume for the moment $g_{ij} \in C^2$ (cf. the approximation arguments below).

Lemma 3.1.5: *v is a homeomorphism.*

Proof: v is the uniform limit of diffeomorphisms u_n. We assume that v is not a homeomorphism. Then v is not injective, i.e. there must exist two points z_1, z_2, $z_1 \neq z_2$ with $v(z_1) = v(z_2)$. We choose a shortest segment γ_n joining $u_n(z_1)$ and $u_n(z_2)$. Since u_n is a homeomorphism, $\tilde{\gamma}_n := u_n^{-1}(\gamma_n)$ is a curve joining z_1 and z_2.

If $p_{n,\delta}$ is a point on $\partial B(z_1, \delta) \cap \tilde{\gamma}_n$, then for $n \to \infty$ we can find a subsequence of $(p_{n,\delta})$ converging to some point p_δ on $\partial B(z_1, \delta)$. Since the u_n converge uniformly to v, we see that $v(p_\delta) = v(z_1) = v(z_2)$. Thus, a whole continuum is mapped onto the single point $v(z_1) = v(z_2)$ by v.

At interior points, we can again choose local coordinates v^1, v^2. From (3.1.14) we conclude that v is harmonic, e.g.

$$\Delta v^1 + \Gamma_{11}^1(v_x^1 v_x^1 + v_y^1 v_y^1) + 2\Gamma_{12}^1(v_x^1 v_x^2 + v_y^1 v_y^2) + \Gamma_{22}^1(v_x^2 v_x^2 + v_y^2 v_y^2) = 0. \tag{3.1.15}$$

From (3.1.14) and (3.1.15) we obtain

$$|v_{z\bar{z}}^1| \leqslant c|v_z^1| \tag{3.1.16}$$

since $v \in C^2(B)$ (cf. Theorem 2.5.1).

If now $v_z^1(z_0) = 0$ for some $z_0 \in S^2$, Lemma 2.6.1 gives the asymptotic representation

$$v_z^1 = a(z - z_0)^n + o(|z - z_0|^n) \tag{3.1.17}$$

for some $a \in \mathbb{C}$, $a \neq 0$, and some positive integer n, unless $v_z^1 \equiv 0$ in a neighbourhood of z_0. The latter is not possible, however, since it implies that the set where $v_x^1 v_y^2 - v_y^1 v_x^2 = 0$ is non-void and open in S^2, and therefore

$v_x^1 v_y^2 - v_x^1 v_y^2 \equiv 0$ in S^2 in contradiction to the fact that v is a surjective $C^{2,\alpha}$ map onto Σ.

We can choose the local coordinates in such a way that

$$g_{ij}(v(z_0)) = \delta_{ij}. \tag{3.1.18}$$

Using (3.1.14), (3.1.18) and integrating (3.1.17), we infer that after a linear change of coordinates

$$w(z) := v^1 + iv^2 = (z - z_0)^{n+1} + o(|z - z_0|^{n+1}) + w_0$$

($w_0 = (v^1 + iv^2)(z_0)$) in a neighbourhood of z_0. This, however, is a contradiction to the consequence we have obtained from the assumption that v is not injective, namely that a whole continuum of points is mapped to a single point. This proves the lemma.

q.e.d.

Lemma 3.1.6: *v is a diffeomorphism.*

Proof: This follows from Lemma 2.7.4.

q.e.d.

So far, Lemma 3.1.6 has been proved under the assumption that $g_{ij} \in C^1$. In particular, we have proved Theorem 3.1.1 in this case. We now treat the general case by approximation.

We let (g_{ij}^n) be a sequence of $C^{2,\alpha}$ metrics converging to (g_{ij}) pointwise almost everywhere. We denote the corresponding surfaces by Σ^n and let $h_n: S^2 \to \Sigma^n$ be a conformal diffeomorphism of the form just constructed and satisfying a three-point normalization. From our preceding arguments, we see that since $E(h_n)$ is uniformly bounded, the h_n are equicontinuous (cf. Lemma 3.1.2). Hence a subsequence converges uniformly and weakly in $H^{1,2}$ towards a weak solution h of (3.1.14). (Alternatively, since the h_n satisfy a system of the type of (3.1.14), elliptic regularity theory implies uniform C^α as well as $H^{1,2}$ estimates.)

Furthermore, since the h_n are diffeomorphisms, their inverses satisfy a system of the same type, namely

$$y_{v^1}^n = \frac{g_{12}^n}{\sqrt{g^n}} x_{v^1}^n - \frac{g_{11}^n}{\sqrt{g^n}} x_{v^2}^n$$

$$y_{v^2}^n = \frac{g_{22}^n}{\sqrt{g^n}} x_{v^1}^n - \frac{g_{12}^n}{\sqrt{g^n}} x_{v^2}^n \tag{3.1.19}$$

where $g^n = g_{11}^n g_{22}^n - (g_{12}^n)^2$, and by the same argument, the h_n are equicontinuous. Therefore, the limit map h has to be invertible, i.e. a homeomorphism.

In the case $\Sigma \in C^{1,\alpha}$, the metrics (g_{ij}^n) can be chosen to converge with respect to the C^α-norm to (g_{ij}). From (3.1.19) we infer (by elliptic regularity theory) that the h_n^{-1} then satisfy uniform $C^{1,\alpha}$ estimates, and consequently the limit map h is a diffeomorphism. Thus we have found the desired conformal representation of Σ, and the proof of Theorem 3.1.1 is complete.

q.e.d.

Corollary 3.1.1: *A conformal homeomorphism $v: S \to \Sigma$ (Σ as in Theorem 3.1.1) is uniquely determined by a three-point condition (3.1.5) and the requirement that it be orientation preserving.*

Proof: Let v, w be two such maps. We want to show that $w^{-1} \circ v$ is a conformal self-map of S^2, hence the identity as it is of degree 1 with three fixed points.

From (3.1.14) we get ($k=1$)

$$\begin{aligned} v_x^1 v_y^2 - v_y^1 v_x^2 &= \frac{\sqrt{g}}{g_{22}}(v_x^1 v_x^1 + v_y^1 v_y^1) \\ &= \frac{\sqrt{g}}{g_{11}}(v_x^2 v_x^2 + v_y^2 v_y^2). \end{aligned} \tag{3.1.20}$$

We put $h := w^{-1}$ and are going to show that $h \circ v$ has weak derivatives that can be computed by the chain rule, i.e.

$$\begin{aligned} (h \circ v)_z &= \frac{\partial h}{\partial v^1} v_z^1 + \frac{\partial h}{\partial v^2} v_z^2 \\ (h \circ v)_{\bar{z}} &= \frac{\partial h}{\partial v^1} v_{\bar{z}}^1 + \frac{\partial h}{\partial v^2} v_{\bar{z}}^2. \end{aligned} \tag{3.1.21}$$

Since $h, v \in H^{1,2}$, we can approximate them in $H^{1,2}$ by sequences (h_n) and (v_n), respectively, of C^1-maps. Now

$$\int_{\mathbb{C}} \left| \frac{\partial h_m}{\partial v_1} \circ v - \frac{\partial h}{\partial v_1} \circ v \right| |v_z| \leqslant \left(\int \left| \frac{\partial h_m}{\partial v_1} \circ v - \frac{\partial h}{\partial v_1} \circ v \right|^2 \right)^{1/2} \left(\int |v_z|^2 \right)^{1/2}. \tag{3.1.22}$$

By (3.1.20) and (3.1.1)

$$\int |v_z|^2 \leqslant c \int J(v) = cA(\Sigma) \tag{3.1.23}$$

where $A(\Sigma)$ is the area of Σ. (This follows with a similar argument as Lemma 3.1.3, since v is a homeomorphism and hence $m(p, v(\partial B(z_0, R))) = 1$ for $p \in v(\mathring{B}(z_0, R))$.) Therefore, the left-hand side of (3.1.22) tends to zero as $m \to \infty$. Also

$$\int \left| \frac{\partial h_m}{\partial v_1} \circ v \cdot v_z - \frac{\partial h}{\partial v_1} \circ v_n \cdot v_{n,z} \right| \leqslant \int \left| \frac{\partial h_m}{\partial v_1} \circ v - \frac{\partial h_m}{\partial v_1} \circ v_n \right| |v_z| + \int \left| \frac{\partial h_m}{\partial v_1} \circ v_n \right| |v_{n,z} - v_z|$$

and as this tends to zero for fixed m as $n \to \infty$, v_n converges to v in $H^{1,2}$ and w.l.o.g. also uniformly. Consequently,

$$\int_{\mathbb{C}} \left| (h_m \circ v_n)_z - \frac{\partial h}{\partial v_1} \circ v \cdot v_z^1 - \frac{\partial h}{\partial v_2} \circ v \cdot v_z^2 \right|$$

becomes arbitrarily small, if m and n are sufficiently large. This implies (3.1.21). (Cf. [BJS, p. 274 ff] for the preceding argument.) On the other hand, $h = w^{-1}$ satisfies a system (3.1.19), with g_{ij} in place of g^n_{ij}, and one computes

$$(h \circ v)_{\bar{z}} = 0.$$

Thus $h \circ v$ is a conformal automorphism of S^2 with three fixed points; hence the identity.

q.e.d.

We can also derive the following version of the Riemann mapping theorem (cf. e.g. [AB]):

Corollary 3.1.2: *Let S be a compact surface with boundary, homeomorphic to the unit disk D, and a metric tensor (g_{ij}) satisfying the assumptions of Theorem* 3.1.1.

Then there is a conformal representation $h: D \to S$, satisfying the same conclusion as in Theorem 3.1.1.

Proof: Let S' be an isometric copy of S with opposite orientation; let $i: S \to S'$ be the isometry. Identifying s with $i(s)$ for $s \in \partial S$ gives a surface Σ to which we can apply Theorem 3.1.1 and find a conformal homeomorphism $h: S^2 \to \Sigma$. Then $i \circ h$ is another conformal homeomorphism, and we can find a conformal automorphism k of S^2 satisfying $h \circ k = i \circ h$. (This is clear for smooth metrics on Σ, since then $h^{-1} \circ i \circ h$ is a conformal diffeomorphism of S^2. The general case follows again by approximation.) The fixed-point set of k then is a circle and hence bounds a disk which is conformally equivalent to S.

q.e.d.

Note that our proof immediately yields the one-to-one correspondence of the boundaries, first proved by Osgood and Caratheodory.

We can again normalize the conformal map by, for example, prescribing the image of ∂S of three distinct points on ∂D.

The preceding result is due to Lichtenstein [Li] (for C^α-metrics), Lavrent'ev [Lv] (for continuous metrics), and Morrey [M1].

3.2. Conformal representation of surfaces homeomorphic to circular domains

In this section, we shall prove:

Theorem 3.2.1: *Suppose S is a surface with boundary, homeomorphic to a plane domain G bounded by k circles. Suppose the coefficients of the metric tensor of S can be defined by bounded measurable functions g_{ij} with $g_{11}g_{22} - g_{12}^2 \geq \lambda > 0$ in G. Then S admits a conformal representation $h \in H^{1,2} \cap C^\alpha(\bar{B}, \bar{G})$, where B is a plane domain bounded by k circles, and h satisfies almost everywhere the* *conformality relations*

$$g_{ij}h^i_x h^j_x = g_{ij}h^i_y h^j_y \qquad g_{ij}h^i_x h^j_y = 0 \qquad (z = x + \mathrm{i}y \in B). \tag{3.2.1}$$

h (and hence also B in case $k > 1$) can be normalized by a three-point condition, namely three points on one of the boundary curves of S can be made to correspond to three given points on the outer boundary of B which can be taken as the unit circle, or by fixing the image of an interior point.

Furthermore, concerning higher regularity, h is as regular as S, i.e. if S is of class $C^{k,\alpha}(k \in \mathbb{N}, 0 < \alpha < 1)$ or C^∞, then also $h \in C^{k,\alpha}(\bar{B})$ or $C^\infty(\bar{B})$ respectively. In particular, if S is at least $C^{1,\alpha}$, then the conformality relations are satisfied everywhere, and h is a diffeomorphism.

The case $k = 1$, of course, was already treated in Corollary 3.1.2.

Proof: By assumption, there is a diffeomorphism of some circular domain $\bar{B}$ onto $\bar{G}$. Using elementary Möbius transformations, we can assume that the outer boundary of B is the unit circle and that this diffeomorphism satisfies the three-point condition. We then define $\bar{\mathscr{D}}$ as the class of all diffeomorphisms from a circular domain bounded by k circles, the outer boundary of which is the unit circle, onto G, satisfying the three-point condition. Again, we let $\bar{\bar{\mathscr{D}}}$ be the closure of $\mathscr{D}$ w.r.t. weak $H^{1,2}$ and uniform convergence. We take an energy-minimizing sequence (v_n, B_n) in $\bar{\bar{\mathscr{D}}}$ (here, B_n is a domain bounded by k circles, and v_n: $\bar{B}_n \to \bar{G}$). The argument of Lemma 3.1.2 applies to show that the v_n are equicontinuous. Therefore, after selection of a subsequence, not only v_n converges weakly in $H^{1,2}$ and uniformly, but also the domains B_n converge to a plane domain B bounded by k circles, since by the equicontinuity no boundary circle of the B_n can collapse to a single point or become tangent to another one in the limit. We thus get an energy-minimizing v: $\bar{B} \to \bar{G}$ in $\bar{\bar{\mathscr{D}}}$. Furthermore, if B_t is another circular domain and σ_t: $B_t \to B$ a diffeomorphism, depending differentiably on t, with $B_0 = B$, $\sigma_0 = \mathrm{id}$, then again

$$\frac{\mathrm{d}}{\mathrm{d}t} E(v \circ \sigma_t^{-1})|_{t=0} = 0 \tag{3.2.2}$$

as the three-point condition can be achieved by a Möbius transformation without changing the energy (cf. (3.1.7) and (3.1.8)). We put again

$$E = g_{ij}v^i_x v^j_x \qquad F = g_{ij}v^i_x v^j_y \qquad G = g_{ij}v^i_y v^j_y \tag{3.2.3}$$

$$\varphi\, \mathrm{d}z^2 = (E - G - 2\mathrm{i}F)\,\mathrm{d}z^2 \tag{3.2.4}$$

$$\frac{\partial \sigma_t}{\partial t|_{t=0}} = v + \mathrm{i}\omega. \tag{3.2.5}$$

Lemma 1.2.5 implies that $\varphi\, \mathrm{d}z^2$ is a holomorphic quadratic differential which is real (and smooth) on ∂B. Explicitly, we get from (3.2.2) (cf. Section 1.2)

$$\int_B \{(E - G)(v_x - \omega_y) + 2F(v_y + \omega_x)\}\, \mathrm{d}x\, \mathrm{d}y = 0. \tag{3.2.6}$$

We let $\gamma_1, \ldots, \gamma_k$ be the boundary circles of B,

$$\gamma_j = \{z\colon |z - z_j| = r_j\}.$$

Integrating (3.2.6) by parts (note that φ is smooth on ∂B)

$$\int_{\gamma_j} \{(E - G)(v\,\mathrm{d}y + \omega\,\mathrm{d}x) + 2F(\omega\,\mathrm{d}y - v\,\mathrm{d}x)\} = 0 \qquad (j = 1, \ldots, k). \quad (3.2.7)$$

Choosing variations which translate the centre z_j, i.e. putting $v + \mathrm{i}\omega = a_j + \mathrm{i}b_j =$ const. near γ_j, we get from (3.2.7)

$$\int_{\gamma_j} \varphi(z)\,\mathrm{d}z = 0 \qquad (j = 1, \ldots, k). \quad (3.2.8)$$

Applying a homothetic dilation of γ_j instead, i.e. choosing $v + \mathrm{i}\omega = (e_j + \mathrm{i}f_j)\,\mathrm{e}^{\mathrm{i}\theta}$ $(\mathrm{e}_j + \mathrm{i}f_j = \text{const.})$ near γ_j, where θ parametrizes γ_j, we get from (3.2.7)

$$\int_{\gamma_j} (z - z_j)\varphi(z)\,\mathrm{d}z = 0 \qquad (j = 1, \ldots, k). \quad (3.2.9)$$

Furthermore, the fact that $\varphi\,\mathrm{d}z^2$ is real on γ_j can be expressed as

$$0 = \mathrm{Im}((z - z_j)^2\varphi(z)) \qquad \text{for } z \in \gamma_j. \quad (3.2.10)$$

Lemma 3.2.1: *Let B be a domain bounded by k circles, $\varphi\colon B \to \mathbb{C}$ be a holomorphic function satisfying* (3.2.8), (3.2.9) *and* (3.2.10). *Then $\varphi \equiv 0$.*

Proof: Assume that φ is not identically zero. Let θ parametrize γ_j. Let $f_j(\theta) = (z - z_j)^2\varphi(z)$ on γ_j. By (3.2.10), $f_j(\theta)$ is real. By (3.2.8) and (3.2.9)

$$\int_0^{2\pi} f_j(\theta)\,\mathrm{d}\theta = 0 \quad (3.2.11)$$

$$\int_0^{2\pi} f_j(\theta)\cos\theta\,\mathrm{d}\theta = \int_0^{2\pi} f_j(\theta)\sin\theta\,\mathrm{d}\theta. \quad (3.2.12)$$

Since f_j is 2π-periodic, (3.2.11) and (3.2.12) imply that f_j and hence φ has at least four zeros on γ_j. We denote the zeros of φ on ∂B by z_s, and let

$$B_\rho := B \setminus \bigcup_s (B \cap B(z_s, \rho))$$

where $\rho \geqslant 0$ is so small that each ball $B(z_s, \rho)$ contains no other zero of φ besides z_s.

The number of zeros of φ inside B_ρ is non-negative and given by

$$n = \frac{1}{2\pi\mathrm{i}} \int_{\partial B_\rho} \mathrm{d}\log\varphi. \quad (3.2.13)$$

Here, ∂B_ρ is oriented in such a way that B_ρ is to the left. As $\rho \to 0$, the contribution from $\partial B(z_s,\rho)\cap B$ to the integral in (3.2.13) is $-\frac{1}{2}$ for each z_s, and, since there are at least $4k$ boundary zeros z_s, altogether at least $-2k$. (Note that the orientation of ∂B_ρ is such that $B(z_s,\rho)$ is always to the right.) On $\gamma_j\cap B_\rho$, the contribution is given by

$$\frac{1}{2\pi i}\left(\int_{\gamma_j\cap B_\rho} d\log((z-z_j)^2\varphi(z)) - \int_{\gamma_j\cap B_\rho} d\log((z-z_j)^2\right).$$

Since $(z-z_j)^2\varphi(z)$ is real on γ_j, the first integral tends to zero as $\rho\to 0$, while the second one gives

$$\int_{\gamma_j} d\log(z-z_j)^2 = \begin{cases} 2 & \text{for } j=1 \\ -2 & \text{for } j=2,\ldots,k \end{cases}$$

(because of the orientation of ∂B).

Altogether, as $\rho\to 0$, (3.2.13) tends to

$$n = \frac{1}{2\pi i}\int_{\partial B_\rho} d\log \leqslant -2k+2(k-1)-2 \leqslant -4$$

which is impossible since n is non-negative. This contradiction shows $\varphi\equiv 0$.

q.e.d.

The remaining arguments for the proof of Theorem 3.2.1 can be taken over from the proof of Theorem 3.1.1. This proves the result.

q.e.d.

Similarly to Corollary 3.1.1, we have:

Corollary 3.2.1: *Let S be as in Theorem 3.2.1. If $k=1$, then a conformal homeomorphism h between B (which in this case is the unit disk) is uniquely determined by fixing the images of three boundary points or the image of an interior point and a tangent direction at this point (of course, this latter normalization is only possible if h is differentiable at this point).*

If $k=2$, then a conformal homeomorphism is uniquely determined by fixing the image of one boundary point, and if $k>2$, it is unique.

3.3. Conformal representation of closed surfaces of higher genus

We first recall some basic results of hyperbolic geometry: Let $H := \{z=x+iy\in\mathbb{C},\ y>0\}$ be the upper half-plane. On H, we have the hyperbolic metric

$$ds^2 = \frac{1}{y^2}\, dz\, d\bar{z}$$

with curvature $K \equiv -1$. Then the group

$$\mathrm{GL}_2^+(\mathbb{R}) := \left\{ \begin{pmatrix} a & b \\ c & d \end{pmatrix} : a, b, c, d \in \mathbb{R}, ad - bc > 0 \right\}$$

operates on H by isometries, via $z \to (az+b)/(cz+d)$, and the corresponding projective group

$$\mathrm{PL}_2^+(\mathbb{R}) := \mathrm{GL}_2^+(\mathbb{R}) \Big/ \left\{ \begin{pmatrix} a & 0 \\ 0 & a \end{pmatrix} : a \neq 0 \right\}$$

operates effectively. Since this operation is also transitive, H thus becomes a homogeneous space. The isotropy group of a point, which is isomorphic to $\mathrm{SO}(2, \mathbb{R})$, operates transitively on the unit tangent space of this point. If T is a discrete fixed point-free subgroup of $\mathrm{PL}_2^+(\mathbb{R})$, then the quotient H/T becomes a hyperbolic surface, if equipped with the quotient metric. H/T can be identified with a metric fundamental polygon. For this, one chooses any $z_0 \in H$ and puts

$$\Sigma := \{ z \in H : d(z, z_0) \leqslant d(z, \Gamma z_0) \text{ for all } \Gamma \in \mathrm{T} \} \tag{3.3.1}$$

where $d(\cdot,\cdot)$, of course, is the distance w.r.t. the hyperbolic metric. Interior points of Σ cannot be identified by elements of T, whereas $\partial\Sigma$ consists of geodesic arcs, pairwise identified by elements of T, as well as possibly some intervals on the real axis. Different pairs of sides of Σ are identified by different elements of T.

If Σ does not meet the real axis, then these identifications of the sides of Σ lead to a compact oriented surface of genus $p \geqslant 2$. Two such groups T_1, T_2 or surfaces H/T_1, H/T_2 are equivalent, if there exists $S \in \mathrm{PL}_2^+(\mathbb{R})$ with

$$\mathrm{T}_1 = S^{-1}\mathrm{T}_2 S. \tag{3.3.2}$$

S defines a conformal map between H/T_1 and H/T_2 by mapping a fundamental polygon Σ of T_1 w.r.t. z_0 into the corresponding one of T_2 w.r.t. $S(z_0)$.

We shall need the elementary 'collar lemma' of Keen [Ke], Matelski [Ma], Halpern [Hp], and Randol [Rn]; the geometric proof of Randol is also reproduced in [Ch]. The present proof is similar to the latter one.

Lemma 3.3.1: *Let $\Sigma = H/\mathrm{T}$ be a compact hyperbolic surface with a simple ($=$ non-self-intersecting) closed geodesic γ of length l. Then the focal radius of γ is at least* $\operatorname{arcsinh}(1/\sinh(l/2))$. *This means that Σ contains a 'collar', i.e. an annular region about γ whose boundary curves are at a distance from γ of at least* $\operatorname{arcsinh}(1/\sinh(l/2))$.

Proof: The focal radius of γ is

$i := \sup\{r > 0$: if $c_1, c_2 : [0, r] \to \Sigma$ are geodesic arcs parametrized by arclength with $c_1(0)$, $c_2(0) \in \gamma$, $c_1(0) \neq c_2(0)$ and if both are orthogonal to γ at their initial points, then $c_1(s) \neq c_2(s)$ for $0 \leqslant s \leqslant r\}$.

If there exist two such geodesic arcs c_1, c_2 with $c_1(r) = c_2(r)$ for some $r > 0$, then there cannot exist any homotopy

$$C: [0, r] \times [0, 1] \to \Sigma$$

with $C(0, t) \in \gamma$, $C(r, t) = c_1(r) = c_2(r)$ for all t and $C(\cdot, 0) = c_1$, $C(\cdot, 1) = c_2$. Namely, such a homotopy would lift to H, and consequently lifts of c_1 and c_2 would have the same endpoint, which is not possible as they are both perpendicular to a lift of γ. Consequently, when determining i, we can look at all piecewise geodesics c in Σ with both endpoints on γ, perpendicular to γ and not homotopic to a subarc of γ (if the two pieces were homotopic, then the whole curve would be homotopic to a subarc of γ, and vice versa). The shortest such curve then is a smooth geodesic arc c with endpoints on γ, perpendicular to γ and of length precisely $2i$. When parametrized by arclength, we thus obtain a geodesic arc

$$c: [0, 2i] \to \Sigma$$

with $c(0) =: p_1$, $c(2i) =: p_2$, $p_1, p_2 \in \gamma$, $c(i) =: q$. We parametrize γ by arclength,

$$\gamma: [0, l] \to \Sigma$$

in such a way that $p_1 = \gamma(l/2)$.

We now need two auxiliary geodesics, always parametrized by arclength. τ is the geodesic with $\tau(0) = q$ and perpendicular to c at q. σ is the geodesic with $\sigma(0) = \gamma(0)$ $(= \gamma(l))$ and perpendicular to γ at $\gamma(0)$.

We now lift γ to H to obtain a geodesic $\tilde{\gamma}$, parametrized by arclength and with $\tilde{\gamma}(l/2)$ projecting to $p_1 = \gamma(l/2)$. We let $\tilde{\sigma}_1$ and $\tilde{\sigma}_2$ be the lifts of σ through $\tilde{\gamma}(0)$ and $\tilde{\gamma}(l)$, respectively. Likewise, we let $\tilde{c}$ be the lift of c through $\tilde{\gamma}(l/2) = \tilde{c}(0)$ and $\tilde{\tau}$ the lift of τ through $\tilde{c}(i)$.

We now assume

$$i < \operatorname{arcsinh}(1/\sinh(l/2)).$$

We claim that in this case $\tilde{\tau}$ intersects $\tilde{\sigma}_1$ and $\tilde{\sigma}_2$. To verify the claim, we let g be the subarc of $\tilde{\gamma}$ between $\tilde{\gamma}(0)$ and $\tilde{\gamma}(l/2)$:

$$\text{length}(g) = l/2.$$

Likewise, s is the subarc of $\tilde{\sigma}$ between $\tilde{\sigma}(0)$ and $\tilde{\sigma}(i)$:

$$\text{length}(s) = i.$$

g and s meet at a right angle.

It is then an elementary result of hyperbolic trigonometry that if

$$\sinh(l/2) \sinh(i) < 1$$

there exists a quadrangle in H with sides s and g and right angles at the endpoints of s and g and a fourth angle φ with

$$\cos \varphi = \sinh(l/2) \sinh(i).$$

As $\tilde{\tau}$ intersects s at a right angle, and $\tilde{\sigma}_1$ interesects g at a right angle, this is equivalent to the fact that $\tilde{\tau}$ and $\tilde{\sigma}_1$ intersect with an angle φ.

In the same manner, $\tilde{\tau}$ intersects $\tilde{\sigma}_2$, and the points of intersection of $\tilde{\tau}$ with $\tilde{\sigma}_1$ and $\tilde{\sigma}_2$, $\tilde{q}_1$ and $\tilde{q}_2$, respectively, are at the same distance from g, by symmetry.

Since $\tilde{\sigma}_1$ and $\tilde{\sigma}_2$ both project to σ, $\tilde{q}_1$ and $\tilde{q}_2$ project to the same point.

We let t be the subarc of $\tilde{\tau}$ between $\tilde{q}_1$ and $\tilde{q}_2$. It follows that t projects to a closed loop τ which is homotopic to γ.

We now complete the picture by symmetry to the other side of t. Namely, through $\tilde{c}(2i)$, there is another lift $\tilde{\tau}'$ of γ, obtained by reflecting $\tilde{\gamma}$ across $\tilde{\tau}$. Likewise, we obtain $\tilde{\sigma}'_1$ and $\tilde{\sigma}'_2$ by reflecting $\tilde{\sigma}_1$ and $\tilde{\sigma}_2$ across $\tilde{\tau}$.

We observe that after projection, $\tilde{\gamma}'$ is identified with $\tilde{\gamma}$, $\tilde{\sigma}'_1$ with $\tilde{\sigma}'_2$, and $\tilde{\sigma}_1$ with $\tilde{\sigma}_2$. This means that the region bounded by these six arcs is projected to all of Σ, and Σ topologically is a torus or a Klein bottle. This is not compatible with the assumption that Σ is a quotient of the upper half-plane. This contradiction proves the lemma.

q.e.d.

We shall also need Mumford's compactness lemma [Mu]:

Lemma 3.3.2: *Let $(T_n)_{n\in\mathbb{N}}$ be a sequence of discrete fixed-point-free mutually isomorphic subgroup of* $\mathrm{PL}_2^+(\mathbb{R})$ *with compact quotients H/T_n. Suppose that the lengths of simple closed geodesics on H/T_n are bounded from below by $l_0 > 0$ (independent of n).*

Then (T_n) contains a subsequence converging to an isomorphic group T, *in the sense that the metric fundamental polygons w.r.t. $z_0 \in H$* (cf. (3.3.1)) *converge to a metric fundamental polygon of H/T, where H/T is diffeomorphic to the H/T_n.*

For the proof of Lemma 3.3.2, we shall use the following elementary result:

Lemma 3.3.3: *Let X be a compact Riemannian manifold of non-positive sectional curvature. We define* $\mathrm{diam}(X) :=$ *diameter of X*, $i(X) :=$ *injectivity radius of X*, $\mathrm{vol}(X) :=$ *volume of X, $d := \dim X$. Then*

$$\mathrm{diam}(X)\,(i(X))^{d-1} \leqslant c\,\mathrm{vol}(X) \tag{3.3.3}$$

where c depends only on d.

Proof: Let p, q be points in X of distance $\mathrm{diam}(X)$, and let γ be a shortest geodesic between them of length $\mathrm{diam}(X)$. We are going to show that no two geodesic arcs γ_1, γ_2 with initial point on γ, orthogonal to γ and of length $\leqslant i(X)/4$ can meet.

This implies that the exponential map of the normal bundle N of γ maps a tubular neighbourhood T_0 of radius $i(X)/4$ of γ injectively onto a tube T in X.

Hence

$$\mathrm{vol}(X) \geqslant \mathrm{vol}(T) \geqslant \mathrm{vol}(T_0) = \omega_{d-1}(i(X)/4)^{d-1}\,\mathrm{diam}(X)$$

(where ω_d is the volume of the d-dimensional unit ball) because X has non-positive curvature.

If γ_1 and γ_2 were to meet, their initial points p_1, p_2 on γ would have distance at most $i(X)/2$, as otherwise there would exist a connection between p and q of length $< \operatorname{diam}(X)$ contradicting their choice. Thus, if γ_1 and γ_2 meet we find a closed loop of length $\leqslant i(X)$, consisting of γ_1 and γ_2 and the portion of γ between p_1 and p_2.

This loop is not homotopic to zero as on the universal cover $\tilde{X}$ of X the exponential map of the normal bundle of (the lift of) γ is injective. Thus, there is a non-trivial closed loop of length $\leqslant i(X)$, and since this loop has corners at p_1 and p_2, there exists a closed geodesic in the same free homotopy class of length $< i(X)$, contradicting the definition of $i(X)$.

q.e.d.

Proof of Lemma 3.3.2: By assumption,

$$i(H/\mathrm{T}_n) \geqslant l_0/2. \tag{3.3.4}$$

By Gauss–Bonnet,

$$\operatorname{vol}(H/\mathrm{T}_n) = 2p - 2 \qquad (p = \text{genus } H/\mathrm{T}_n) \tag{3.3.5}$$

and hence $\operatorname{diam}(H/\mathrm{T}_n)$ is bounded from above by (3.3.3), independently of n. Therefore, also the diameter of the metric fundamental polygons

$$\Sigma_n := \{z \in H : d(z, z_0) \leqslant d(z, \Gamma z_0) \forall \Gamma \in \mathrm{T}_n\}$$

is uniformly bounded from above.

We also note that actually the length of any not necessarily simple closed geodesic on H/T_n is $\geqslant l_0$ since if there exists a closed geodesic with a self-intersection one can find a smaller one without self-intersections. The interior angles of Σ_n at the corners lie between 0 and π, and we are going to show that they are uniformly bounded away from 0 and π.

If there were to exist corner points p_n of Σ_n where the angle between the corresponding sides goes to zero, then Σ_n, being convex, is contained in the angular region formed by these sides, and since $\operatorname{diam}(\Sigma_n)$ is bounded, its area goes to zero, contradicting (3.3.5). If l_n is a side of Σ_n, then a point equivalent to z_0 under T_n is found by extending the perpendicular geodesic from z_0 onto l_n beyond l_n up to the distance $d(z_0, l_n)$. If the angle between two sides l_n^1, l_n^2 went to π, then the distance between the corresponding points equivalent to z_0 would tend to 0, contradicting (3.3.4), as the geodesic arc in H between two points equivalent under T_n corresponds to a non-trivial closed geodesic loop in H/T_n. Moreover, the number of sides (and hence also of corner points) of Σ_n is uniformly bounded. Namely, to each side there corresponds a point equivalent to z_0 under T_n, and since $\operatorname{diam}(\Sigma_n)$ is bounded, the distance of these points from z_0 is uniformly bounded, and therefore (3.3.4) again implies that their number is bounded. Therefore, after selection of a subsequence, the metric fundamental polygons Σ_n tend to a polygon Σ. Σ may have fewer sides than the Σ_n as the lengths of some of the sides may go to zero. If the length of a side goes to zero, then the number of equivalence classes (w.r.t. the operation of T_n) of corner points decreases by 1 in the limit, and the number of equivalence classes of sides also decreases by 1, for each such side. On the other hand, the transformations identifying corresponding sides of Σ_n generate T_n, and each such transformation also has a well defined limit. Altogether, Σ can be considered as a fundamental region of a surface H/T of the same Euler characteristic.

q.e.d.

Theorem 3.3.1: *Let S be a compact, oriented surface of genus $p \geqslant 2$ the metric tensor of which is represented in local coordinates by bounded measurable functions g_{ij} which satisfy almost everywhere*

$$g_{11}g_{22} - g_{12}^2 \geqslant \lambda > 0. \tag{3.3.6}$$

Then there exists a conformal map $h: H \to S$ which maps some fundamental region Σ (of a subgroup T of $PL_2^+(\mathbb{R})$ homomorphic to $\pi_1(S)$) homeomorphically onto S. h is of class $H^{1,2} \cap C^\alpha(\Sigma, S)$. Conformality of h here means that in local coordinates

$$g_{ij}\frac{\partial h^i}{\partial x}\frac{\partial h^j}{\partial x} = g_{ij}\frac{\partial h^i}{\partial y}\frac{\partial h^j}{\partial y}, \qquad g_{ij}\frac{\partial h^i}{\partial x}\frac{\partial h^j}{\partial y} = 0 \tag{3.3.7}$$

for almost all $z = x + iy \in H$.

If S is of class $C^{k,\alpha}$, C^∞, or C^ω, then so is h.

In particular, if $S \in C^{1,\alpha}$, then (3.3.7) is satisfied everywhere, and h is a diffeomorphism. h can be normalized by mapping some point $z_0 \in H$ and a tangent direction at z_0 (in the case $S \in C^{1,\alpha}$) into a prescribed point in S and a direction at that point, respectively.

Proof: We assume that the g_{ij} are continuous. This can be removed later on by approximation arguments as in Sections 3.1. We let $\mathscr{D}$ be the class of all pairs (v, T) where T is a subgroup of $PL_2^+(\mathbb{R})$ homomorphic to $\pi_1(S)$ and v is a diffeomorphism of H/T onto S. $\bar{\mathscr{D}}$ then is the class of pairs (w, T) where w is in the weak $H^{1,2}$ and uniform closure of the class of diffeomorphisms $v_n: H/T \to S$. The energy integral in this class is defined as

$$E(w) = \frac{1}{2}\int_{z = x + iy \in H/T} g_{ij}(w(z))\left(\frac{\partial w^i}{\partial x}\frac{\partial w^j}{\partial x} + \frac{\partial w^i}{\partial y}\frac{\partial w^j}{\partial y}\right) dx\, dy.$$

We note that because of (3.3.6), E is uniformly equivalent to the standard Dirichlet integral. We then choose an energy-minimizing sequence (w_n, T_n) in $\bar{\mathscr{D}}$. We can assume

$$E(w_n) \leqslant K \qquad \text{for some fixed } K. \tag{3.3.8}$$

We first want to show that the groups T_n converge after selection of a subsequence to some non-degenerate limit. We shall use the following result of Schoen and Yau [SY2].

Lemma 3.3.4: *The length of the shortest simple closed geodesic on H/T_n is bounded from below by some number $l_0 > 0$ which is independent of n.*

Proof: Let γ be a simple closed geodesic on H/T_n of length l. We introduce geodesic parallel coordinates on a collar about γ, as given by Lemma 3.3.1. This means that we introduce coordinates (r, φ) as follows (cf. Section 2.7).

For $r \equiv 0$, φ is the arclength parameter on γ, whereas the curves $\varphi \equiv$ const. are geodesics normal to γ parametrized by arclength r; consequently the curves $r \equiv$ const. are parallel curves of γ. The metric tensor satisfies

$$\gamma_{11}(r, \varphi) \equiv 1 \qquad \gamma_{12}(r, \varphi) \equiv 0 \qquad \gamma_{22}(0, \varphi) \equiv 1.$$

Also, since γ as a closed geodesic on a negatively curved surface is shorter than any of its parallel curves,

$$\gamma_{22}(r, \varphi) \geqslant 1 \qquad \text{for all } (r, \varphi).$$

By the collar Lemma 3.3.1, the above construction is possible for

$$-\operatorname{arcsinh}(1/\sinh(l/2)) < r < \operatorname{arcsinh}(1/\sinh(l/2)) \tag{3.3.9}$$

and we choose

$$0 \leqslant \varphi < l.$$

We observe that v_n as uniform limit of diffeomorphisms maps each parallel curve of γ onto a homotopically non-trivial closed curve in S. We let σ be a lower bound for the lengths of such curves. Then, for each r,

$$\int_0^l \left(g_{ij}(v_n) \frac{\partial v_n^i(r, \varphi)}{\partial \varphi} \frac{\partial v_n^j(r, \varphi)}{\partial \varphi} \right)^{1/2} \mathrm{d}\varphi \geqslant \sigma. \tag{3.3.10}$$

Hence with Hölder's inequality

$$\sigma^2 \leqslant l \int_0^l g_{ij} \frac{\partial v_n^i}{\partial \varphi} \frac{\partial v_n^j}{\partial \varphi} \mathrm{d}\varphi.$$

We integrate (3.3.10) w.r.t. r and obtain, noting the above properties of the metric tensor $(\gamma_{\alpha\beta})$,

$$2R \frac{\sigma^2}{l} \leqslant \int_{-R}^{R} \int_0^l g_{ij} \frac{\partial v_n^i}{\partial \varphi} \frac{\partial v_n^j}{\partial \varphi} \mathrm{d}\varphi \, \mathrm{d}r \leqslant 2E(v_n). \tag{3.3.11}$$

Since we may choose any $R < \operatorname{arcsinh}(1/\sinh(l/2))$, e.g. $R = 1$ if l is small, (3.3.11) yields a lower bound for l.

q.e.d.

Remark: We observe that the assertion of the collar lemma is actually stronger than we need here or in Section 4.7. below. Namely, for our purposes $R = 1$ suffices for the above proof, whereas the collar lemma implies, that the upper bound for R is $\operatorname{arcsinh}(1/\sinh(l/2))$, and this tends to infinity as l tends to 0.

Lemma 3.3.5: *After selection of a subsequence,* (T_n) *converges to a subgroup* T *of* $\mathrm{PL}_2^+(\mathbb{R})$*, isomorphic to the* T_n.

Proof: This follows from Lemmata 3.3.2 and 3.3.4.

q.e.d.

Therefore, without changing $\lim_{n\to\infty} E(w_n)$, we can assume $\mathrm{T}_n = \mathrm{T}$ for all n, and also that w_n converges weakly in $H^{1/2}$ to some v.

Lemma 3.3.6: (w_n) *is an equicontinuous sequence.*

Proof: Using (3.3.8), by Lemma 3.1.1 around each point in H/T there is a sufficiently small circle mapped under w_n onto a curve of arbitrarily small diameter. This curve divides S into two regions, one being topologically a disk and the other one being of higher genus. Since w_n is a uniform limit of diffeomorphisms, the interior of the circle has to be mapped into this disk which has arbitrarily small diameter, and equicontinuity follows.

q.e.d.

Therefore, (w_n) also converges uniformly to v, and $(v, \mathrm{T}) \in \bar{\mathscr{D}}$ and the lower semicontinuity of the energy w.r.t. weak $H^{1,2}$ convergence, together with the fact that v is the limit of an energy-minimizing sequence, then implies

$$E(v) = \inf_{w \in \bar{\mathscr{D}}} E(w). \tag{3.3.12}$$

We now want to show that v satisfies (3.3.7) almost everywhere.

Let $\sigma_t : H \to H$ be a family of diffeomorphisms depending differentiably on t, with $\sigma_0 = \mathrm{id}$. For every t, we assume for $\Gamma \in T$

$$\sigma_t \circ \Gamma = \Gamma^t \circ \sigma_t \tag{3.3.13}$$

where $\Gamma^t \in \mathrm{T}^t$, and T^t is isomorphic to T. $v \circ \sigma_t^{-1}$ then can be considered as a map from H/T^t onto S and is the uniform and weak $H^{1,2}$-limit of $(w_n \circ \sigma_t^{-1})$, and by (3.3.12) thus

$$\frac{\mathrm{d}}{\mathrm{d}t} E(v \circ \sigma_t^{-1})|_{t=0} = 0. \tag{3.3.14}$$

We put

$$E := g_{ij} \frac{\partial v^i}{\partial x} \frac{\partial v^j}{\partial x} \qquad F := g_{ij} \frac{\partial v^i}{\partial x} \frac{\partial v^j}{\partial y} \qquad G := g_{ij} \frac{\partial v^i}{\partial y} \frac{\partial v^j}{\partial y}$$

(since $v \in H^{1,2}$, E, F, G are defined almost everywhere)

$$\left.\frac{\partial \sigma_t}{\partial t}\right|_{t=0} =: \nu + i\omega.$$

With $\phi := E - G - 2iF$, as in Section 3.1,

$$\int_{H/\mathrm{T}} \phi(v + i\omega)_{\bar{z}} \, dx \, dy = 0. \tag{3.3.15}$$

We first choose v and ω in such a way that they vanish in some fundamental region Σ of T outside the neighbourhood of a point and put on Σ

$$\sigma_t(z) = x + tv(z) + i(y + t\omega(z))$$

and finally

$$\sigma_t(\Gamma z) = \Gamma\sigma_t(z) \qquad \text{for } \Gamma \in \mathrm{T}.$$

σ_t then satisfies (3.3.13) and is a diffeomorphism if t is sufficiently small. Since v and ω can be arbitrarily chosen in the neighbourhood of any point of H/T, (3.3.15) implies that ϕ is holomorphic, i.e.

$$\phi_{\bar{z}} = 0 \tag{3.3.16}$$

Since ϕ transforms via

$$\phi = (\phi \circ \Gamma)(\Gamma_z)^2 \qquad \text{for } \Gamma \in \mathrm{T} \tag{3.3.17}$$

it is a holomorphic quadratic differential on H/T. Therefore,

$$\mu(z) := \bar{\phi}(z) \cdot y^2 \tag{3.3.18}$$

transforms via

$$(\mu \circ \Gamma)\bar{\Gamma}_{\bar{z}} = \mu\Gamma_z \qquad \text{for } \Gamma \in \mathrm{T}. \tag{3.3.19}$$

Putting the metric $|dz + t\mu \, d\bar{z}|^2$ (assuming of course that $|t|$ is so small that $|t\mu(z)| < 1$ on a fundamental region of H/T) onto the upper half-plane, we get a surface denoted by H_t.

By Theorem 3.2.1, there exists a conformal map

$$\sigma_t : H \to H_t.$$

We can normalize σ_t in such a way that $\sigma_0 = \mathrm{id}$ and that it satisfies

$$\sigma_{t,\bar{z}} = t\mu\sigma_{t,z} \tag{3.3.20}$$

and since $\sigma_0 = \mathrm{id}$,

$$\frac{\partial}{\partial t}\sigma_{t,\bar{z}}\bigg|_{t=0} = \mu$$

or, using the above notations,

$$(v + i\omega)_{\bar{z}} = \mu. \tag{3.3.21}$$

Equation (3.3.19) also implies that $\sigma_t \circ \Gamma$ solves (3.3.20) for any $\Gamma \in \mathrm{T}$. Hence σ_t and $\sigma_t \circ \Gamma$ differ only by a conformal automorphism of H (cf. Corollary 3.2.1), and therefore

$$\sigma_t \circ \Gamma = \Gamma^t \circ \sigma_t$$

where $\Gamma' \in T'$ and T' is isomorphic to T. Thus (3.3.13) is satisfied. Equations (3.3.15), (3.3.21) and (3.3.18) then imply

$$\int_{H/T} |\phi|^2 y^2 \, dx \, dy = 0$$

i.e. $\phi \equiv 0$. Thus, v satisfies (3.3.7) almost everywhere. The remaining parts of Theorem 3.3 now follow as in Section 3.1.

q.e.d.

In a similar, but more elementary way, one can also show:

Theorem 3.3.2: *Suppose S is homeomorphic to a two-dimensional torus and has a metric tensor satisfying the assumptions of Theorem 3.3.1. Then there exists a conformal map* $v: \mathbb{C} \to S$, *mapping the fundamental region* Σ *of some lattice homeomorphically onto S, and* $v \in H_2^1 \cap C^\alpha(\Sigma, S)$. v *satisfies the same regularity properties as in Theorem 3.3.1. As a normalization, we can prescribe the image of a point and a tangent direction at this point and furthermore require that the area of* Σ *is* 1.

As a corollary of the preceding results, we also obtain a conformal representation of a compact oriented surface S with boundary. Let S' be an isometric copy of S with the opposite orientation. We thus have an orientation-preserving isometry $i: S \to S'$. Identifying each $p \in \partial S$ with $i(p)$ we get the Schottky double of S, a closed surface $\bar{S}$ with an orientation-reversing isometric involution i. The preceding theorems therefore imply:

Corollary 3.3.1: *Let S be a compact oriented surface with boundary, the metric of which satisfies the same assumptions as in Theorem 3.3.1. Then there is a closed surface* $\bar{\Sigma}$ ($\bar{\Sigma} = S^2$, *if S is homeomorphic to a disk,* $\bar{\Sigma}$ *is covered by* $\mathbb{C}$, *if* $\bar{\Sigma}$ *is an annular region, and by H in all other cases) with an isometric involution, the fixed-point set of which consists of closed geodesics separating it into two components both conformally equivalent to S. The corresponding conformal map is a homeomorphism up to the boundary and has the same regularity properties as in Theorem 3.3.1.*

We also observe:

Corollary 3.3.2: *Let S be a compact non-orientable surface without boundary. Then there exists a compact oriented surface S′ of constant curvature with a fixed-point-free isometric involution* τ *for which* $S'/\{\mathrm{id}, \tau\}$ *is conformally equivalent to S. The regularity results are again the same above.*

Finally, one can also represent non-orientable surfaces with boundary conformally. We omit the details.

4 EXISTENCE RESULTS

4.1. The local existence theorem for harmonic maps. An easy proof of the existence of energy-minimizing maps

We start with

Lemma 4.1.1: *Let $B_0, B_1, B_0 \subset B_1$, be closed subsets of a Riemannian manifold N. We assume that there exists a projection $\pi: B_1 \to B_0$ of class C^1, with*

$$\pi|_{B_0} = \mathrm{id}|_{B_0} \tag{4.1.1}$$

and which is distance decreasing outside B_0, i.e.

$$d(\pi(x), \pi(y)) < d(x, y) \tag{4.1.2}$$

whenever $x, y \in B_1 \backslash B_0, x \neq y$.

If $h: \Omega \to B_1$ is an energy-minimizing $H^{1,2}$ mapping with respect to fixed boundary values which are contained in B_0, i.e.

$$h(\partial\Omega) \subset B_0 \tag{4.1.3}$$

then

$$h(\Omega) \subset B_0. \tag{4.1.4}$$

Proof: Because of (4.1.2), $|d\pi(v)| < |v|$ for every non-zero $v \in T_x N, x \in B_1 \backslash B_0$. Also $\pi \circ h \in H^{1,2}(\Omega, N)$. Thus we would have

$$E(\pi \circ h) < E(h)$$

contradicting the minimality of h, unless $\mathrm{d}h = 0$ a.e. on $h^{-1}(B_1 \backslash B_0)$. Thus $\mathrm{d}h = \mathrm{d}\pi \circ h$ a.e. on Ω, and since h and $\pi \circ h$ agree on $\partial\Omega$ by (4.1.3), we conclude from the Poincaré inequality that $\pi \circ h = h$ a.e. on Ω, which easily implies the claim.

q.e.d.

Lemma 4.1.2: *Suppose that B_0 and $B_1, B_0 \subset B_1$, are compact subsets of a*

Riemannian manifold N, and that every point in $B_1 \backslash B_0$ can be joined to ∂B_0 by a unique geodesic normal to ∂B_0, and that if $\gamma_1(t), \gamma_2(t)$ are two such geodesics parametrized by arclength, $t \geqslant 0$, with $\gamma_i(0) \in \partial B_0$, $i = 1, 2$, then $d(\gamma_1(t), \gamma_2(t)) > d(\gamma_1(0), \gamma_2(0))$ for $t > 0$. *Then the same conclusion as in Lemma 4.4.1 holds.*

Proof: We project $B_1 \backslash B_0$ along normal geodesics onto ∂B_0. This projection can then be approximated by maps $\pi \in C^1$ satisfying the assumption of Lemma 4.1.1. Lemma 4.1.1 then implies the result.

q.e.d.

Another consequence of Lemma 4.1.1 is:

Lemma 4.1.3: *Suppose B_0 is a geodesic ball with radius s and centre $p, s < \frac{1}{2}\min(i(p), \pi/2\kappa)$, where κ^2 is an upper bound for the sectional curvature of N and i(p) is the injectivity radius of p. If $h: \Omega \to N$ is energy minimizing among maps which are homotopic to some map $g: \Omega \to B_0$, and if $h(\partial\Omega) \subset B_0$, then*

$$h(\Omega) \subset B_0$$

(for a suitable representative of h, again).

Proof: By assumption, we can introduce geodesic polar coordinates (r, φ) on $B(p, 2s)$ $(0 \leqslant r \leqslant 2s)$. We define a map π in the following way:

$$\begin{aligned} \pi(r, \varphi) &= (r, \varphi) && \text{if } r \leqslant s \\ \pi(r, \varphi) &= (2s - r, \varphi) && \text{if } s \leqslant r \leqslant 2s \\ \pi(q) &= p && \text{if } q \in N \backslash B(p, s). \end{aligned}$$

(Here, we have identified a point in $B(p, 2s)$ with its representation in geodesic polar coordinates.) Using the Rauch comparison theorem, it is easily seen that π can be approximated by maps satisfying the assumptions of Lemma 4.1.1, and the result follows.

q.e.d.

We now want to prove a local existence result which is due to Morrey [M2] (see also [HKW2]):

Lemma 4.1.4: *Suppose Ω is a two-dimensional bounded domain, $\partial\Omega \neq \emptyset$, $B(p, r) \subset N$ (a complete Riemannian manifold) with*

$$r < \min(i(p), \pi/2\kappa)$$

where κ^2 is an upper curvature bound on N. Let $g: \partial\Omega \to B(p, r)$ be continuous and admit an extension $\bar{g}: \Omega \to B(p, r)$ of finite energy. Then there exists a harmonic map $h: \Omega \to B(p, r)$ with boundary values g, and h minimizes the energy with respect to these boundary values. Vice versa, each such energy-minimizing map is harmonic. The modulus of continuity of h can be estimated in terms of the metric of Ω, $i(\Sigma)$, r, κ, and $E(\bar{g})$ and the modulus of continuity of g.

Proof: We are going to use arguments from the proof of Theorem 4.1 in [HW] and from [Hi2]. We note that any two points in $B(p,r)$ can be joined by a unique geodesic arc in $B(p,r)$ and this arc is free of conjugate points (cf., for example, Proposition 2.4.1 in [J6]; if one does not want to appeal to this result one can merely require r to be chosen so small that this automatically holds. This does not affect any subsequent application of Lemma 4.1.4).

We choose r^1 with

$$r < r^1 < \min(\pi/2\kappa, i(p)).$$

We take a minimizing sequence for the energy in $V := \{v \in H^{1,2}(\Omega, B(p,r^1)), v|_{\partial\Omega} = g\}$. Such a sequence has a subsequence converging weakly in $H^{1,2}$, and the limit, denoted by h, minimizes energy in its class because of the lower semicontinuity of the Dirichlet integral. Applying Lemma 4.1.2 to $B_0 = B(p,r)$, $B_1 = B(p,r^1)$, we conclude that h actually maps Ω into the smaller ball $B(p,r)$. As noted, every two points in $B(p,r)$ can be joined by a unique geodesic arc in $B(p,r)$, and this arc is free of conjugate points. Therefore, we can apply the Rauch comparison theorem in the following way. Suppose that $q \in B(p,r)$, v_1 and v_2 are unit vectors in T_qN, and c_1, c_2 are the geodesics parametrized by arclength and starting at q with tangent vectors v_1, v_2 respectively. Then

$$|v_1 - v_2|\kappa^{-1}\sin(t\kappa) \leqslant d(c_1(t), c_2(t))$$

as long as $c_1(t), c_2(t) \in B(p,r)$.

Therefore, on $B(p,r) \backslash B(q,\epsilon)$,

$$d(c_1(t), c_2(t)) \geqslant \min(d(c_1(\epsilon), c_2(\epsilon)), |v_1 - v_2|\kappa^{-1}\sin(2r\kappa)).$$

Consequently, there exists $\epsilon_0 > 0$ with the property that $B_0 := B(q,\epsilon) \cap B(p,r)$ and $B_1 := B(p,r)$ satisfy the assumptions of Lemma 4.1.2 for every $q \in B(p,r)$ and every $\epsilon \leqslant \epsilon_0$. Lemma 3.1.1 then implies that for each $x \in \Omega$ there exists a sufficiently small $\rho > 0$ with the property that

$$h(\partial B(x,\rho) \cap \Omega) \subset B(q,\epsilon)$$

for some $q \in B(p,r)$. ρ depends on ϵ, the energy of h (which is bounded by the energy of $\bar{g}$), and the modulus of continuity of g. By Lemma 4.1.2 then

$$h(B(x,\rho) \cap \Omega) \subset B(q,\epsilon)$$

and continuity of h follows.

Since h is energy minimizing in V and $h(\Omega) \subset B(p,r)$ so that h actually lies in the interior of V, we conclude that h is a weak solution of (1.2.23), i.e. weakly harmonic.

Since h is also continuous, Theorem 2.5.1 implies that h is $C^{2,\alpha}$ in the interior of Ω, and hence in particular harmonic. These arguments apply to any energy-minimizing map in the present situation.

q.e.d.

Before we proceed to free boundary conditions, we want to present a simple proof of the existence of energy-minimizing maps:

Theorem 4.1.1: *Let Σ be a compact surface without boundary, N a compact Riemannian manifold, $\varphi \in C^0 \cap H^{1,2}(\Sigma, N)$. Then there exists a harmonic map $u: \Sigma \to N$ with*

$$u_\# = \varphi_\#$$

where $\varphi_\#: \pi_1(\Sigma) \to \pi_1(N)$ is the induced map on the fundamental groups.

If $\pi_2(N) = 0$, u is homotopic to φ.

Of course, this will also be a special case of Theorem 4.2.1. The proof given here is taken from [J2]. The result is originally due to Lemaire [L1] and Sacks and Uhlenbeck [SkU1]; cf. also Schoen and Yau [SY2].

Proof of Theorem 4.1.1: Let $[\varphi] := \{v \in C^0 \cap H^{1,2}(\Sigma, N): v_\# = \varphi_\#\}$. We choose $s = \frac{1}{3}\min(i(N), \pi/2\kappa)$, where $\kappa^2 \geqslant 0$ is an upper curvature bound on N, and $i(N)$ is the injectivity radius of N. Let $\delta_0 < 1$ satisfy

$$2\pi \cdot E(\varphi)^{1/2}(\log(1/\delta_0))^{-1/2} \leqslant s/2 \tag{4.1.5}$$

where $E(\varphi)$ is the energy of φ.

Let $0 < \delta \leqslant \delta_0$. There exists a finite number of points $x_i \in \Sigma (i = 1, \ldots, m = m(\delta))$ for which the disks $B(x_i, \delta/2)$ cover Σ. We let u_n be a continuous energy-minimizing sequence in $[\varphi]$, $E(u_n) \leqslant E(\varphi)$ w.l.o.g. for all n.

Applying Lemma 3.1.1 and using (4.1.5), for every n, we can find $r_{n,1}$, $\delta < r_{n,1} < \sqrt{\delta}$, and $p_{n,1} \in N$ with the property that

$$u_n(\partial B(x_1, r_{n,1})) \subset B(p_{n,1}, s). \tag{4.1.6}$$

On the other hand, if $u_n(\partial B(x, r)) \subset B(p, s)$ for some $x \in \Sigma, r > 0, p \in N$, then by Lemma 4.1.4, there exists a solution of

$$\left.\begin{array}{l} g: B(x, r) \to B(p, s) \\ g \mid \partial B(x, r) = u_n \mid \partial B(x, r) \end{array}\right\} \quad \text{harmonic and energy minimizing} \tag{4.1.7}$$

We replace u_n on $B(x_1, r_{n,1})$ by the solution of the Dirichlet problem (4.1.7) for $x = x_1$ and $r = r_{n,1}$. We can assume $r_{n,1} \to r_1$ and, using the interior modulus of continuity estimates for the solution of (4.1.7) (cf. Lemma 4.1.4), we can assume that the replaced maps, denoted by u_n^1, converge uniformly on $B(x_1, \delta - \eta)$, for any $0 < \eta < \delta$. By Lemma 4.1.3

$$E(u_n^1) \leqslant E(u_n). \tag{4.1.8}$$

By the same argument as above, we then find radii $r_{n,2}$, $\delta < r_{n,2} < \sqrt{\delta}$, with

$$u_n^1(\partial B(x_2, r_{n,2})) \subset B(p_{n,2}, s)$$

for points $p_{n,2} \in N$.

Again, we replace u_n^1 on $B(x_2, r_{n,2})$ by the solution of the Dirichlet problem (4.1.7) for $x = x_2$ and $r = r_{n,2}$. We denote the new maps by u_n^2. Again, w.l.o.g., $r_{n,2} \to r_2$.

If we take into consideration that, by the first replacement step, u_n^1 in particular converges uniformly on $B(x_2,r_2)\cap B(x_1,\delta-\eta/2)$, if $0<\eta<\delta$, we see that the boundary values for our second replacement step converge uniformly on $\partial B(x_2,r_{n,2})\cap B(x_1,\delta-\eta/2)$. Using the estimates for the modulus of continuity for the solution of (4.1.7) at these boundary points (cf. Lemma 4.1.4) we can assume that the maps u_n^2 converge uniformly on $B(x_1,\delta-\eta)\cup B(x_2,\delta-\eta)$, if $0<\eta<\delta$. Furthermore, by Lemma 4.1.3 again and (4.1.8)

$$E(u_n^2)\leqslant E(u_n^1)\leqslant E(u_n).$$

In this way, we repeat the replacement argument, until we get a sequence $u_n^m =: v_n$, with

$$E(v_n)\leqslant E(u_n) \tag{4.1.9}$$

which converges uniformly on all balls $B(x_i,\delta/2)$, $(i=1,\ldots,m)$ and hence on all of Σ, since these balls cover Σ.

We denote the limit of the v_n by u. Since replacement on disks does not affect the induced map on the fundamental groups, and because of the uniform convergence of (v_n), we see that $u\in[\varphi]$.

Since $E(v_n)\leqslant E(\varphi)$ by (4.1.9), the v_n converge also weakly in $H^{1,2}$ to u, and, by lower semicontinuity of the energy w.r.t. weak $H^{1,2}$ convergence and since the v_n are a minimizing sequence by (4.1.9), u minimizes energy in $[\varphi]$.

In particular, u minimizes energy when restricted to small balls, and hence it is harmonic and regular by Lemma 4.1.3 and Lemma 4.1.4. Observing that if $\pi_2(N)=0$, any two maps from a disk into N are homotopic, the claim follows.

q.e.d.

Remark: With a similar argument, one can handle the case $\partial\Sigma\neq\emptyset$, and where Dirichlet or Plateau-type boundary conditions are imposed.

We now turn to the situation where a free boundary condition is prescribed. We need the following local existence result.

Lemma 4.1.5: *Let N be a complete Riemannian manifold, M a complete submanifold of class C^2. Let*

$$D^+ := \{(x,y)\in\mathbb{R}^2, y\geqslant 0, x^2+y^2\leqslant 1\}$$
$$\partial_+ D^+ := \{y\geqslant 0, x^2+y^2=1\}$$
$$\partial_0 D^+ := \{y=0, x^2<1\}.$$

For every $p\in M$, there exists $\rho>0$ with the following property. Let $g:\partial_+D^+\to B(p,\rho)\subset N$ be continuous with $g(1,0)\in M, g(-1,0)\in M$. Suppose g admits an extension $\bar g:D^+\to B(p,\rho)$ with $\bar g(\partial_0 D^+)\subset M$. Then there exists a map $u: D^+\to B(p,\rho)$ which minimizes the energy among all such maps (in particular, $u(\partial_0 D^+)\subset M$, and u satisfies the free boundary condition on $\partial_0 D^+$).

u is harmonic; in particular $u\in C^{1,\alpha}(\mathring{D}^+\cup\partial_0 D^+)$.

Vice versa, this holds for each such energy-minimizing map. If N and M satisfy uniformity conditions as in Theorem 2.3.2, then ρ is independent of p.

Proof: In order to carry over the argument of Lemma 4.1.4, we have only to establish the existence of suitable projection maps. This is done in the following lemma.

Lemma 4.1.6: *Suppose M and N are as in Lemma 4.1.5. For each $p \in M$, there exists $\rho^1 > 0$ with the following property. Each $q \in B(p, \rho^1)$ has an arbitrarily small neighbourhood V with the property that there exists a projection*

$$\pi: B(p, \rho^1) \to \bar{V}$$

with

$$\pi|_{\bar{V}} = \mathrm{id}|_{\bar{V}} \tag{4.1.10}$$

$$d(\pi(x), \pi(y)) < d(x, y) \qquad \text{for } x, y \in B(p, \rho^1)\backslash \bar{V}; x \neq y \tag{4.1.11}$$

$$\pi(M \cap B(p, \rho^1)) \subset M \cap \bar{V} \qquad \text{if } M \cap V \neq \varnothing. \tag{4.1.12}$$

Proof: Using the construction in the proof of Lemma 4.1.4 and a compactness argument it suffices to establish the existence of such a projection in the case $q \in B(p, \rho^1) \cap M$. Now if M were totally geodesic, then M would intersect all spheres $\partial B(q, \sigma)$ orthogonally, as $q \in M$. Therefore, we can find arbitrarily small neighbourhoods V of q with the following properties.

$S := \partial V$ is a Lipschitz strictly convex compact hypersurface. For each $q' \in M \cap S$, S has a tangent cone C such that if $v \in T_{q'}M$ points into the interior of V, $\|v\| = 1$; then for every $w \in C, w \notin T_{q'}M$,

$$(v, w) < \pi/2. \tag{4.1.13}$$

Because of the strict convexity, the nearest point projection $\pi: B(p, \rho^1) \to \bar{V} \cap B(p, \rho^1)$ is distance decreasing in $B(p, \rho^1)\backslash \bar{V}$, i.e. satisfies (4.1.11), for sufficiently small $\rho^1 > 0$. Moreover, because of (4.1.13), $M \cap B(p, \rho^1)$ is mapped under π onto $M \cap \bar{V}$, again for sufficiently small $\rho^1 > 0$. This proves the lemma.

q.e.d.

With a similar construction, one also extends Lemma 4.1.3 to free boundaries. With this result and Lemma 4.1.5, one can then prove an analogue of Theorem 4.1.1 for free boundaries.

Proof of Lemma 4.1.5: We choose ρ^1 as in Lemma 4.1.6 and find a neighbourhood $V_p \subset B(p, \rho^1)$ of p with the properties of Lemma 4.1.6. Let $B(p, \rho) \subset V_p$, i.e. $\rho > 0$ is chosen small enough. We then minimize the energy among all maps $v \in H^{1,2}(D^+, B(p, \rho^1))$ with

$$v|_{\partial_+ D^+} = g$$
$$v(x) \in M \text{ for almost all } x \in \partial_0 D^+.$$

We then establish the existence of a minimum u with $u(D^+) \subset V_p$ as in the proof of Lemma 4.1.4. Continuity at points in the interior of D^+ or in $\partial_+ D^+$ follows as in Lemma 4.1.4, and the proof of continuity at points $x_0 \in \partial_0 D^+$ is also similar.

Namely, by Lemma 3.1.1 again, we can find for each $\epsilon > 0$ a radius $\delta > 0$ with $u(\partial B(x_0, \delta) \cap D^+) \subset B(p_0, \epsilon)$, for some $p_0 \in N$, and w.l.o.g. we assume $u(\partial B(x_0, \delta) \cap \partial_0 D^+) \subset M$, and hence also $p_0 \in M$, again w.l.o.g. Using Lemma 4.1.6 instead of Lemma 4.1.2, we can again conclude continuity as in the proof of Lemma 4.1.4.

q.e.d.

We now turn to the situation where M is a closed Jordan curve $\gamma \subset N$. Of course, this is a special case of a free boundary, but we want to impose the additional requirement that the boundary of the domain is mapped monotonically onto γ, i.e. a Plateau boundary condition. This actually simplifies the arguments. Later on, we shall need the following local existence result of [GJ]:

Lemma 4.1.7: *Let*

$$D^+ := \{(x, y) \in \mathbb{R}^2 : y \geqslant 0, x^2 + y^2 \leqslant 1\}$$
$$\partial_0 D^+ := \{(x, 0) \in \mathbb{R}^2 : x^2 < 1\}$$
$$\partial_+ D^+ := \{(x, y) \in \mathbb{R}^2 : x^2 + y^2 = 1, y \geqslant 0\}.$$

Let $B(p, r)$ be a ball in N, disjoint from the cut locus of p, with $r < \pi/2\kappa$, κ^2 being an upper curvature bound. Let γ_0 be a C^2 Jordan arc in $B(p, r)$, with $\partial\gamma_0 \subset \partial B(p, r)$. Let $g: \partial_+ D^+ \to B(p, r)$ be continuous, with

$$g(1, 0) \in \gamma_0 \qquad g(-1, 0) \in \gamma_0.$$

Suppose there exists a map $v: D^+ \to B(p, r)$ of finite energy with $v|_{\partial_+ D^+} = g$, and that v maps $\partial_0 D^+$ monotonically onto γ_0 and is continuous on ∂D^+.

Then there also exists a harmonic map $h: D^+ \to B(p, r)$ with these properties, and h can be chosen to minimize energy in this class.

Proof: The proof of the lemma is the same as that for Lemma 4.1.4. Let us sketch the argument. Let h have the minimum of energy in the given class. Given $z \in D^+$, by the Courant–Lebesgue Lemma 3.1.1, one can always find some sufficiently small radius $\delta > 0$ so that

$$\partial^+ D(z, \delta) := \{w \in D^+ : |z - w| = \delta\}$$

is mapped into some arbitrarily small ball $B(q, \epsilon)$. One then applies a maximum principle to conclude also that

$$D(z, \delta) := \{w \in D^+ : |z - w| \leqslant \delta\}$$

is mapped under h into this ball. We have to check the validity of this procedure in the case when

$$D(z, \delta) \cap \partial_0 D^+ \neq \emptyset.$$

Since γ_0 is a Jordan arc of class C^2, we can find some $\epsilon_0 > 0$ with the property that if $0 < \epsilon < \epsilon_0, p_1, p_2 \in \gamma_0 \cap B(q, \epsilon)$ (for some q), then also the segment of γ_0

between p_1 and p_2 is contained in $B(q,\epsilon)$. If $\partial^+ D(z,\delta)$ meets $\partial_0 D^+$, then $\partial^+ D(z,\delta) \cap \partial_0 D^+$ is mapped into a small ball $B(q,\epsilon)$ (w.l.o.g. $\epsilon < \epsilon_0$) by choicc of δ, and therefore by monotonicity of the free boundary values and the above property of γ_0, also $D(z,\delta) \cap \partial_0 D^+$ is mapped into $B(q,\epsilon)$. Therefore, the projection π used in the proof of Lemma 4.1.4 does not affect the boundary condition and since

$$E(\pi \circ h) \leqslant E(h)$$

we can conclude that

$$h(D(z,\delta)) \subset B(q,\epsilon)$$

thus giving continuity, and actually an explicit estimate of the modulus of continuity in terms of the geometric data and an energy bound for h.

q.e.d.

4.2. The general existence theorem. First part of the proof

Theorem 4.2.1: *Let Σ be a compact surface, without boundary (boundary conditions will be discussed in Section 4.3), and N be a compact Riemannian manifold (as always $\partial N = \varnothing$).*

Let A be a compact parameter space (typically $A = [0,1]^n$ or S^n), and let $h_0: \Sigma \times A \to N$ be continuous. Let H be the class of all maps homotopic to h_0, and

$$\kappa := \inf_{h \in H} \sup_{t \in A} E(h(\cdot, t)) \tag{4.2.1}$$

where in the case $\partial A \neq \varnothing$, $h|_{\Sigma \times \partial A}$ is fixed in such a way that the above supremum cannot be attained on ∂A.

Then there exists a harmonic map

$$u_0: \Sigma \to N$$

and possibly also some non-trivial conformal harmonic maps

$$u_i: S^2 \to N \qquad (i = 1, \ldots, m)$$

with

$$E(u_0) + \sum_{i=1}^{m} E(u_i) = \kappa. \tag{4.2.2}$$

Here $(u_0; u_1, \ldots, u_m)$ represents a saddle point corresponding to H in the sense that there exist sequences $(h_n) \subset H$, $(t_n) \subset A$, and points $x_1, \ldots, x_k \in \Sigma, k \leqslant m$ (if $m \geqslant 1$), with

$$E(h_n(\cdot, t_n)) \to \kappa \tag{4.2.3}$$

$$h_n(\cdot, t_n) \to u_0 \quad \textit{weakly in } H^{1,2} \tag{4.2.4}$$

$$h_n(\cdot, t_n) \rightrightarrows u_0 \quad \textit{uniformly on each compact subset of } \Sigma \backslash \{x_1, \ldots, x_k\}. \tag{4.2.5}$$

Furthermore, for each $i\in\{1,\ldots,m\}$, *there exists a sequence* $(\lambda_n^i)_{n\in\mathbb{N}}\in\mathbb{R}^+$, $\lambda_n^i\to 0$ *as* $n\to\infty$ *with*

$$h_n\left(\left(\frac{\rho}{\lambda_n},\varphi\right),t_n\right)\to u_i \tag{4.2.6}$$

where (ρ,φ) *are polar coordinates centred at some* $x_{j,n}$ *with* $x_{j,n}\to x_j$ $(1\leqslant j\leqslant k)$.

Theorem 4.2.1 improves the result of Sacks and Uhlenbeck [SkU1] by obtaining equality in (4.2.2) (whereas [SkU1] has only $\leqslant$). The remaining assertions can already be deduced from [SkU1]. The result of Sacks and Uhlenbeck was reproved by Struwe [St1], but again obtaining only $\leqslant$ in (4.2.2).

Sacks and Uhlenbeck approximate the energy functional by

$$E_\alpha(v):=\frac{1}{2}\int_\Sigma (1+|dv|^2)^\alpha d\Sigma$$

with $\alpha>1$, $\alpha\to 1$, because for $\alpha>1$, E_α satisfies a Palais–Smale-type condition. Struwe uses the heat-flow method and studies the behaviour of solutions of the parabolic analogue of (1.2.10). Our method is different from either of those preceding ones. It refines the idea of [J2] and depends heavily on the local existence result Lemma 4.1.4. It is also reminiscent of the curve-shortening process used to obtain unstable closed geodesics in Riemannian manifolds, as well as of the alternating method of Schwarz and the balayage method of Poincaré for obtaining harmonic functions.

Let us now state some auxiliary results that will be needed for the proof of Theorem 4.2.1.

Lemma 4.2.1: *Let N be a Riemannian manifold of bounded geometry. Then there exists* $\mu>0$ *(depending only on the geometry of N, namely on curvature bounds, a lower bound for the injectivity radius, and the dimension) with the following property. If*

$$u: S^2\to N$$

is a conformal harmonic map, i.e. a minimal surface in N, then

$$E(u)\geqslant\mu. \tag{4.2.7}$$

Similarly, if M is a C^2*-submanifold of N with bounded second fundamental form and a uniform neighbourhood in which the nearest point projection is uniquely defined, then there exists* $\mu'>0$ *with the property that for any minimal surface* $u: D\to N$ *with free boundary M,*

$$E(u)\geqslant\mu'. \tag{4.2.8}$$

Proof: This follows directly from the monotonicity formulae (2.3.12) and (2.3.24).

q.e.d.

Furthermore, we shall need the following covering lemma.

Lemma 4.2.2: *Let X be a compact Riemannian manifold. Suppose, there are points $x_1, \ldots, x_m \in X$ with*

$$X \subset \bigcup_{i=1}^{m} B(x_i, \rho/2) \tag{4.2.9}$$

$$B(x_i, \eta\rho) \cap B(x_j, \eta\rho) = \emptyset \qquad \text{for } i \neq j \tag{4.2.10}$$

for some fixed $\rho > 0, 0 < \eta < \frac{1}{2}$.

Then $\{1, \ldots, m\}$ is the disjoint union of b sets $M_1, \ldots, M_b$ so that for all $\beta \in \{1, \ldots, b\}$ and $i, j \in M_\beta, i \neq j$,

$$B(x_i, \rho) \cap B(x_j, \rho) = \emptyset. \tag{4.2.11}$$

b is independent of m and ρ; it depends only on the geometry of X and on η.

Proof: We construct M_1 as follows. We put $x_1^1 := x_1$. Iteratively, we seek points $x_j^1 \in \{x_1, \ldots, x_m\}$ with

$$2\rho < \operatorname{dist}(x_j^1, x_i^1) \qquad \text{for all } i < j. \tag{4.2.12}$$

After a certain number of steps, we cannot find a point in $\{x_1, \ldots, x_m\}$ with (4.2.12) any more. We let M_1 be the collection of all points selected so far. If $x_k \notin M_1$, there exists $x_j^1 \in M_1$ with

$$\operatorname{dist}(x_k, x_j^1) \leqslant 2\rho. \tag{4.2.13}$$

We construct M_β iteratively as follows ($\beta \geqslant 2$). We select any point $x_k \notin \bigcup_{\alpha=1}^{\beta-1} M_\alpha$, put $x_1^\beta := x_k$, and iteratively seek points $x_j^\beta \in \{x_1, \ldots, x_m\} \backslash \bigcup_{\alpha=1}^{\beta-1} M_\alpha$ with

$$2\rho < \operatorname{dist}(x_j^\beta, x_i^\beta) \qquad \text{for all } i < j \tag{4.2.14}$$

until no such point exists anymore.

If $x_k \notin M_\beta$, for each $\alpha \leqslant \beta$, there exists $x_{j(\alpha)}^\alpha \in M_\alpha$ with

$$\operatorname{dist}(x_k, x_{j(\alpha)}^\alpha) \leqslant 2\rho. \tag{4.2.15}$$

Since all these points $x_{j(\alpha)}^\alpha$ are distinct, and their mutual distance is bounded from below by 2η because of (4.2.10), there can exist at most b_0 such points satisfying (4.2.15). The claim then follows with $b = b_0 + 1$.

q.e.d.

Proof of Theorem 4.2.1: We now begin a proof which will be completed in Section 4.3.

Lemma 4.2.2 implies that given $\rho > 0$, we can find points x_i $(i = 1, \ldots, m = m(\rho))$ in Σ with the following properties:

(1) $\Sigma \subset \bigcup_{i=1}^{m} B(x_i, \rho/2)$ $\qquad B(x_i, \rho/10) \cap B(x_j, \rho/10) = \emptyset \qquad$ for $i \neq j$.

(2) $\{1, \ldots, m\}$ is the disjoint union of b sets $M_1^1, \ldots, M_b^1$ so that for all $\beta \in \{1, \ldots, b\}$ and $i, j \in M_\beta^1, i \neq j$

$$B(x_i, \rho) \cap B(x_j, \rho) = \emptyset.$$

b can be chosen independently of m.

(3) Furthermore, we can choose the sets M^1_β in such a way that

$$\Sigma \subset \bigcup_{i \in M^1_b} B(x_i, \gamma\rho/2)$$

where $\gamma \geq 4$ again is independent of m.

(4) Having constructed sets $M^k_1, \ldots, M^k_b$, we can represent M^k_b as the disjoint union of sets $M^{k+1}_1, \ldots, M^{k+1}_b$ so that for each $\beta \in \{1, \ldots, b\}$ and $i, j \in M^{k+1}_\beta, i \neq j$,

$$B(x_i, \gamma^k\rho) \cap B(x_j, \gamma^k\rho) = \varnothing$$

and

$$\Sigma \subset \bigcup_{i \in M^{k+1}_b} B(x_i, \gamma^{k+1}\rho/2).$$

(5) We repeat this construction until

$$\gamma^{k+1}\rho \geq c_0$$

where c_0 is a fixed positive number, $c_0 \leq \frac{1}{10} i(\Sigma)$, where $i(\Sigma)$ is the injectivity radius of Σ.

For simplicity of notation, we also define

$$M^1 := \{1, \ldots, m\}$$
$$M^{k+1} := M^k_b \qquad \text{for each } k.$$

We consider

$$u: \Sigma \times A \to N.$$

We assume that u is continuous and that for some $\alpha \in (0, 1)$

$$u_t: A \to C^\alpha \cap H^{1,2}$$
$$t \to u | \Sigma \times \{t\}$$

is a continuous map.

We can find $\rho_0 > 0$ so that for each $(x, t) \in \Sigma \times A$ there exists some $p \in N$ with

$$u(B(x, \rho_0) \times \{t\}) \subset B(p, 3^{-b}s) \tag{4.2.16}$$

($s \leq \frac{1}{3}\min(i(N), \pi/2\kappa), \kappa^2 \geq 0$ is an upper curvature bound for N, and $i(N) =$ injectivity radius of N). We put $\rho = \rho_0$, choose points x_i $(i = 1, \ldots, m = m(\rho))$ and sets M^k_β as above.

For each $i \in M^1_1$, we replace (for each $t \in A$)

$$u | B(x_i, \rho) \times \{t\}$$

by the energy-minimizing harmonic map with the same boundary values on $\partial B(x_i, \rho) \times \{t\}$ (cf. Lemma 4.1.4).

Since the harmonic replacement depends continuously on the boundary values (cf. Theorem 2.2.1) and is of class $C^\alpha \cap H^{1,2}$, we do not change the regularity properties of u. We denote the new map by u^1_1. Because of (4.2.16), the convex hull property of the harmonic replacement (cf. Lemma 4.1.3) and property (2)), for each $(x,t)\in\Sigma\times A$, we can find some $p\in N$ with

$$u_1^1(B(x,\rho_0)\times\{t\})\subset B(p,3^{-b+1}s).$$

Having constructed u_1^β for $\beta\in\{1,\ldots,b-1\}$, we replace each $i\in M_{\beta+1}^1$ and $t\in A$,

$$u_1^\beta|B(x_i,\rho)\times\{t\}$$

by the energy-minimizing harmonic map with the same boundary values. We obtain a new map $u_1^{\beta+1}$ with the same regularity properties.

We proceed until we have constructed the map

$$u_1^b=:u_1.$$

For each $(x,t)\in\Sigma\times A$, we can find $p\in N$ with

$$u_1(B(x,\rho_0)\times\{t\})\subset B(p,s).$$

We put $v_{1,0}=u_1$, $\gamma_0(t)\equiv 1$, and $U_0:=\Sigma\times A$ and construct $v_{1,k}$ from $v_{1,k-1}$ as follows:

We choose some metric δ_k on A with the property that whenever $\delta_k(t,t')\leqslant 4\gamma^k\rho_0$, then

$$d(v_{k-1}(x,t),v_{k-1}(x,t'))\leqslant\tfrac{1}{4}3^{-b}s\qquad\text{for all } x\in\Sigma. \tag{4.2.17}$$

We denote by d_k the metric on $\Sigma\times A$ obtained as the product of some fixed metric on Σ and δ_k. We usually write $d_k(x,y)$ instead of $d_k((x,t),(y,t))$ if the value of t is clear. Let $U_k\subset\Sigma\times A$ be the maximal set with the following property.

Let $U_k(t)=U_k\cap(\Sigma\times\{t\})$. Assume that for every $(x,t)\in\Sigma\times A$ with $d_k(x,U_k(t))\leqslant 4\gamma^k\rho_0$ there is some $p\in N$ with $v_{1,k-1}(B(x,\gamma^k\rho_0)\times\{t\})\subset B(p,\tfrac{1}{4}3^{-b}s)$. Combining this with (4.2.17), whenever $d_k((x,t),U_k)\leqslant 4\gamma^k\rho_0$, there is some $p\in N$ with

$$v_{1,k-1}(B(x,\gamma^k\rho_0)\times\{t'\})\subset B(p,\tfrac{1}{2}3^{-b}s)\qquad\text{if } \delta_k(t,t')\leqslant 4\gamma^k\rho_0. \tag{4.2.18}$$

We choose a smooth function $\gamma_k(x,t)$ with

$$\begin{aligned}&\gamma_{k-1}(x,t)\leqslant\gamma_k(x,t)\leqslant\gamma^k &&\text{for all } (x,t)\in\Sigma\times A\\ &\gamma_k(x,t)=\gamma^k &&\text{if } d_k((x,t),U_k)\leqslant\gamma^k\rho_0\\ &\gamma_k(x,t)=\gamma_{k-1}(x,t) &&\text{if } d_k((x,t),U_k)\geqslant 2\gamma^k\rho_0\end{aligned}$$

and with the property that for every (x,t) with $\gamma_k(x,t)>\gamma_{k-1}(x,t)$ there is some $p\in N$ with

$$v_{1,k-1}(B(x,\gamma_k(x,t)\rho_0)\times\{t\})\subset B(p,3^{-b}s). \tag{4.2.19}$$

The existence of such a function γ_k follows from (4.2.18) and the continuity of $v_{1,k-1}$.

For each $i\in M_1^{k+1}$, if $d_k((x_i,t),U_k)\leqslant 2\gamma^k\rho_0$, we replace $v_{1,k-1}|B(x_i,\gamma_k(x_i,t)\rho_0)\times\{t\}$ by the energy-minimizing harmonic map with the same boundary values (cf. Lemma 4.1.4), thus obtaining a new map $v_{1,k}^1$ with the same regularity properties. Having iteratively constructed $v_{1,k}^\beta$, we obtain $v_{1,k}^{\beta+1}$ by harmonic replacement on balls

$$B(x_i, \gamma_k(x_i, t)\rho_0) \times \{t\} \qquad \text{for } i \in M^{k+1}_{\beta+1}.$$

We finally obtain a map

$$v^b_{1,k} =: v_{1,k}.$$

For each $(x,t) \in \Sigma \times A$, there exists some $p \in N$ with

$$v_{1,k}(B(x, \gamma_k(x,t)\rho_0) \times \{t\}) \subset B(p, s). \tag{4.2.20}$$

This is a consequence of the corresponding property of $v_{1,k-1}$ (4.2.19) and the covering properties of M^{k+1}; cf. property (4). If there is no such U_k, we put $v_{1,k} := v_{1,k-1}$. If property (5) is satisfied for some $k = k_0$ (depending on ρ), we obtain a map

$$v_1 := v_{1,k_0}.$$

v_1 has the property that for $(x,t) \in U_{k-1} \backslash U_k$ for some $k \in \{1, \ldots, k_0\}$, then, by (4.2.20), for some $p \in N$

$$v_1(B(x, \gamma^{k-1}\rho_0) \times \{t\}) \subset B(p, s). \tag{4.2.21}$$

On the other hand, since $(x,t) \notin U_k$, there is some $x' \in \Sigma$ and $t' \in A$ with $d_k((x,t),(x',t')) < 4\gamma^k \rho_0$ and the property that

$$v_{1,k-1}(B(x', \gamma^k \rho_0) \times \{t'\})$$

is not contained in any ball $B(p, 3^{-b}s)$. By definition of U_k, this implies

$$d_k((x',t'), U_k) > 4\gamma^k \rho_0.$$

Since $v_{1,k}$ coincides with $v_{1,k-1}$ outside a neighbourhood of U_k of width $2\gamma^k \rho_0$, in particular

$$v_{1,k} | B(x', \gamma^k \rho_0) \times \{t'\} = v_{1,k-1} | B(x', \gamma^k \rho_0) \times \{t'\}.$$

Iterating, we infer

$$v_1 | B(x', \gamma^k \rho_0) \times \{t'\} = v_{1,k-1} | B(x', \gamma^k \rho_0) \times \{t'\}.$$

Therefore, also

$$v_1(B(x, 4\gamma^k \rho_0) \times \{t\}) \tag{4.2.22}$$

is not contained in any ball $B(p, 3^{-b}s)$.

By (4.2.22) and since $\gamma \geq 4$, we infer

$$d_k((x,t), U_{k+1}) \geq 4\gamma^{k+1}\rho_0.$$

Hence, if for some $\tilde{x} \in B(x, \gamma^{k+1}\rho_0)$ and some x_i

$$\tilde{x} \in B(x_i, \gamma^{k+1}\rho_0)$$

then $v_{1,k+1} | B(x_i, \gamma^k \rho_0) \times \{t\} = v_{1,k} | B(x_i, \gamma^k \rho_0) \times \{t\}$ and iteratively

$$v_1 | B(x_i, \gamma^k \rho_0) \times \{t\} = v_{1,k} | B(x_i, \gamma^k \rho_0) \times \{t\}.$$

On the other hand, the balls $B(x_i, \gamma^{k-1}\rho_0/2)$, $i \in M^k$, cover $B(x, \gamma^{k+1}\rho_0)$ in such a way that each $\tilde{x} \in B(x, \gamma^{k+1}\rho_0)$ is contained in at most b balls $B(x, \gamma^{k-1}\rho_0)$ and hence also at most in b' balls $B(x, \gamma^k\rho_0)$, where b' is a fixed number depending only on b and the geometry of Σ. Finally, we note that v_1 was obtained in $B(x, \gamma^{k+1}\rho_0)$ from a continuous map of finite energy by harmonic replacement on balls of the form $B(x_i, \gamma_k(x_i, t)\rho_0)$ with $\gamma^{k-2} \leqslant \gamma_k(x_i, t) \leqslant \gamma^k$.

Having iteratively constructed v_{n-1}, we determine the largest $\rho_{n-1} > 0$ with the property that for each $(x, t) \in \Sigma \times A$ there exists some $p \in N$ with

$$v_{n-1}(B(x, \rho_{n-1}) \times \{t\}) \subset B(p, 3^{-b}s).$$

We repeat the above construction with v_{n-1} in place of u and ρ_{n-1} in place of ρ_0, thus obtaining the map v_n, which enjoys the same properties as described for v_1, again with ρ_{n-1} in place of ρ_0.

We now want to look at the limits of $v_n|\Sigma \times \{t\}$, for each $t \in A$, as $n \to \infty$. Since the v_n need not converge uniformly, we may have to perform some local rescalings in order to keep control of the possible splitting off of minimal 2-spheres.

If $\rho_n \geqslant \delta > 0$ for all n and some fixed δ, then it is an easy consequence of the preceding construction and standard estimates for harmonic mappings[1] that for each $t \in A$ $v_n|\Sigma \times \{t\}$ converges uniformly and strongly in $H^{1,2}$ to a harmonic map $v: \Sigma \times \{t\} \to N$, possibly after selection of a subsequence. We now investigate what happens if ρ_n tends to zero as $n \to \infty$.

The condition $\rho_n \geqslant \delta > 0$, independent of n, is implied by the following condition: for all $x \in \Sigma$, in polar coordinates (ρ, φ) on $B(x, R)$

$$\operatorname*{ess\,sup}_{\rho \leqslant R} \int_{\varphi=0}^{2\pi} \left| \frac{\partial v_n(\rho, \varphi, t)}{\partial \varphi} \right|^2 \mathrm{d}\varphi \to 0 \tag{4.2.23}$$

as $R \to 0$, independent of n and x. Namely, if, for example, $v_n(B(x_0, \rho), t)$ is not contained in any ball $B(p, 3^{-b}s)$, then there are points $x_1, x_2 \in B(x_0, \rho)$ with

$$\tfrac{1}{2} 3^{-b} s \leqslant d(v_n(x_1, t), v_n(x_2, t))$$

and denoting their middle point by x_3 and choosing polar coordinates (r, φ) centred at x_3

$$\begin{aligned} \tfrac{1}{2} 3^{-b} s &\leqslant \int_{r=\frac{1}{2}d(x_1,x_2)} \left| \frac{\partial v_n}{\partial \varphi}(r, \varphi, t) \right| \mathrm{d}\varphi \\ &\leqslant \left(2\pi \int_{r=\frac{1}{2}d(x_1,x_2)} \left| \frac{\partial v_n}{\partial \varphi} \right|^2 \mathrm{d}\varphi \right)^{1/2}. \end{aligned} \tag{4.2.24}$$

We now treat the case where (4.2.23) fails to hold. Since we are looking at a fixed t, for simplicity of notation we shall drop the dependence on t for a moment.

[1] Cf., for example, Lemma 4.2.3 below and Corollary 2.5.1.

Let $s_1 \geqslant s_2 \cdots > 0$ be a sequence of positive decreasing numbers tending to zero, $s_1 \leqslant (\frac{1}{4}\pi^{-1} 3^{-b} s)^2$.

For each n, we select the infimum r_n^1 of all numbers ρ for which there exists a point $x_n^1 \in \Sigma$ so that in polar coordinates centred at x_n^1

$$\int_{\varphi=0}^{2\pi} \left| \frac{\partial v_n}{\partial \varphi}(\rho, \varphi) \right|^2 \mathrm{d}\varphi \geqslant s_1. \tag{4.2.25}$$

We assume $r_n^1 \to 0$ as $n \to \infty$.

We then perform a homothety of Σ with centre x_n^1 and scaling factor $1/r_n^1$, i.e. we expand the ball $B(x_n^1, r_n^1)$ into the ball $B(0,1) \subset \mathbb{R}^2$. We denote the corresponding rescaled maps by w_n.

Then, for any ball $B(x,r) \subset \mathbb{R}^2$ of radius $r \leqslant 1$, polar coordinates centred at x, for all $\varphi_1, \varphi_2 \in [0, 2\pi]$

$$\begin{aligned} d(w_n(r,\varphi_1), w_n(r,\varphi_2)) &\leqslant \left(2\pi \int_{\varphi=0}^{2\pi} \left| \frac{\partial w_n(r,\varphi)}{\partial \varphi} \right|^2 \mathrm{d}\varphi \right)^{1/2} \\ &\leqslant (2\pi s_1)^{1/2} \\ &\leqslant \tfrac{1}{2} 3^{-b} s. \end{aligned}$$

Our construction of the maps v_n by harmonic replacement on balls implies that the rescaled maps w_n are obtained by replacement on balls with a uniform positive lower bound on their radius, and at each point, the map is only changed a bounded finite number of times at each step.

The regularity properties of harmonic maps (cf. Corollary 2.5.1 and Lemma 4.2.3 below) imply that, possibly after selection of a subsequence, the maps w_n converge uniformly and strongly in $H^{1,2}$ on compact sets to a harmonic map

$$w^1 \colon \mathbb{R}^2 \to N.$$

Since moreover

$$\int_{\varphi=0}^{2\pi} \left| \frac{\partial w_n}{\partial \varphi} \right|^2 \mathrm{d}\varphi$$

converges (Lemma 4.2.3 below), we infer from (4.2.25) that w^1 is not a constant map.

Since w^1 has finite energy, the holomorphic quadratic differential associated to w^1 is in L^1. Since any holomorphic function on $\mathbb{C}$ of class L^1 vanishes, we conclude that w^1 is conformal, i.e. a branched minimal immersion. By Lemma A.2 of the Appendix and Theorem 2.3.1, w^1 then extends smoothly across ∞, i.e. to a harmonic and conformal map

$$w^1 \colon S^2 \to N$$

i.e. a branched minimal immersion. By Lemma 4.2.1

$$E(w^1) \geqslant \mu > 0. \tag{4.2.26}$$

The following lemma has already been used in the preceding argument.

Lemma 4.2.3: *Let γ be any regular smooth arc in $\mathbb{R}^2$, and let t be the arclength parameter on γ. Then*

$$\int_\gamma \left|\frac{\partial w_n}{\partial t}\right|^2 dt \to \int_\gamma \left|\frac{\partial w^1}{\partial t}\right|^2 dt. \tag{4.2.27}$$

(We only need the cases where γ is a circle or a straight line.)

Proof: We need only look at the following situation: w_n is obtained by harmonic replacement on a ball B, $y_0 \in \partial B$ is a point where w_n is not Lipschitz, and $\gamma \subset B$ has an endpoint in y_0.

If $B(x_0, R) \subset B$, we have the gradient estimate

$$\begin{aligned} |dw_n(x_0)| &\leqslant c \max_{x \in B(x_0, R)} \frac{d(u(x), u(x_0))}{R} \\ &\leqslant c(\alpha) R^{\alpha - 1} \end{aligned} \tag{4.2.28}$$

by Theorem 2.5.2 and Corollary 2.5.1. Thus

$$|dw_n(x)| \leqslant c(\alpha)\,\mathrm{dist}(x, \partial B)^{\alpha - 1} \qquad \text{for every } \alpha \in (0, 1).$$

This already implies the claim if γ meets ∂B at a non-zero angle, using Lebesgue's theorem on dominated convergence. Since the only points in B where u possibly is not Lipschitz are those points in ∂B where ∂B meets the boundary of another disk where a previous replacement had been performed, we can assume that, for some $R_0 > 0$, u is Lipschitz in $B(y_0, R_0) \backslash \{y_0\}$.

By our Hölder estimate (Corollary 2.5.1), w_n maps $B(y_0, R)$ into some ball $B(p, cR^\alpha)$ for some $p \in N$ $(0 < R \leqslant R_0)$.

By Theorem 2.1.2, on $B(p, 2cR^\alpha)$, we can introduce harmonic coordinates with Christoffel symbols bounded by

$$\begin{aligned} |\Gamma^i_{jk}(q)| &\leqslant c\rho^2/\rho \qquad \text{for } q \in B(p, \rho), \rho = cR^\alpha \\ &= cR^\alpha. \end{aligned} \tag{4.2.29}$$

Let v_n be the map from which w_n is obtained by replacement on B. Let $y_1 \in \partial B$, $|y_0 - y_1| = R/2$. Then v_n is harmonic in $B(y_1, R/4)$. Furthermore, using the introduced coordinates on $B(p, 2cR^\alpha)$, the equation is

$$\Delta v_n^j = -\Gamma^j_{lk}\left(\frac{\partial v_n^l}{\partial x}\frac{\partial v_n^k}{\partial x} + \frac{\partial v_n^l}{\partial y}\frac{\partial v_n^k}{\partial y}\right). \tag{4.2.30}$$

From (4.2.29) and the gradient estimate (4.2.28), the right-hand side of (4.2.30) is bounded by $cR^{3\alpha-2}$ on $B(y_1, R/4)$.

Likewise,

$$\sup_{x \in B(y_1, R/4)} |v_n^j(x)| \leqslant cR^\alpha.$$

Therefore, for any $\beta\in(0,1)$,

$$\frac{|dv_n(y_2)-dv_n(y_3)|}{|y_2-y_3|^\beta}\leqslant\frac{c}{R^{1+\beta}}(R^\alpha+R^{3\alpha-2})$$

for $y_2, y_3\in B(y_1, R/4)$, by potential theory.

This in turn is bounded by $cR^{-\gamma}$, for any given γ with $\beta<\gamma<1$, if α is chosen accordingly. Therefore, w_n has boundary values with controlled $C^{1,\beta}$-norm on $\partial B\cap B(y_1, R/4)$.

We conclude

$$|dw_n(y)|\leqslant cR^{\beta-\gamma}$$

for $y\in B(y_1, R/8)\cap B$, again by an estimate from potential theory. Since $\gamma-\beta>0$ can be made arbitrarily small, the claim now can be deduced in any case from the theorem of dominated convergence.

q.e.d.

We shall also need ([M3, Lemma 9.4.8(b)]):

Lemma 4.2.4: *There exist $\delta>0, \lambda<\infty$, depending only on the geometry of N with the following property:*

If $v\colon\partial D\to N$ is absolutely continuous and satisfies

$$\int_{\varphi=0}^{2\pi}\left|\frac{\partial v(1,\varphi)}{\partial\varphi}\right|^2 d\varphi\leqslant\delta \tag{4.2.31}$$

(in polar coordinates (r,φ) on D), then there exists a map

$$\tilde{v}\colon D\to N$$

with

$$\tilde{v}|_{\partial D}=v$$

and

$$E(\tilde{v})\leqslant\lambda\int_0^{2\pi}\left|\frac{\partial v}{\partial\varphi}\right|^2 d\varphi. \tag{4.2.32}$$

Proof: If N is Euclidean space, the harmonic extension $\tilde{v}$ of v satisfies (4.2.32) with $\lambda=1$. This follows, for example, by Fourier expansion. Furthermore, in the general case, for $x_1, x_2\in\partial D$,

$$\begin{aligned} d(v(x_1), v(x_2)) &\leqslant\int_{\varphi=0}^{2\pi}\left|\frac{\partial v(1,\varphi)}{\partial\varphi}\right| d\varphi \\ &\leqslant\left(2\pi\int_{\varphi=0}^{2\pi}\left|\frac{\partial v(1,\varphi)}{\partial\varphi}\right|^2 d\varphi\right)^{1/2} \end{aligned} \tag{4.2.33}$$

so that for small δ, $v(\partial D)$ is contained in a small ball $B(p,s)$ in N. The result then follows from the result for Euclidean space, as the metric tensor on $B(p,s)$

in normal coordinates is uniformly equivalent to the Euclidean metric tensor (δ_{ij}). (cf. also Corollary 2.1.1.)

q.e.d.

Lemma 3.1.1 implies that there exists a sequence $r_{1,n}$ of radii with

$$r_{1,n}\to 0 \qquad r_{1,n}/r_n^1\to\infty$$

$$\int_{\varphi=0}^{2\pi}\left|\frac{\partial v}{\partial\varphi}(r_{1,n},\varphi)\right|^2 d\varphi\to 0$$

(in polar coordinates centred at x_n^1). In particular, $v_n(\partial B(x_n^1,r_{1,n}))\subset B(p_n,\eta_n)$ for some $p_n\in N$ and some sequence $\eta_n\to 0$. Let $\tilde{v}_n$ be the harmonic map (cf. Lemma 4.1.4)

$$\tilde{v}_n\colon B(x_n^1,r_{1,n})\to B(p_n,\eta_n)$$
$$\tilde{v}_n|_{\partial B(x_n^1,r_{1,n})}=v_n|_{\partial B(x_n^1,r_{1,n})}.$$

We obtain maps

$$v_n^1\colon\Sigma\to N$$
$$v_n^1|_{\Sigma\backslash B(x_n^1,r_{1,n})}=v_n|_{\Sigma\backslash B(x_n^1,r_{1,n})}$$
$$v_n^1|_{B(x_n^1,r_{1,n})}=\tilde{v}_n|_{B(x_n^1,r_{1,n})}$$

and

$$w_n^1\colon S^2\to N$$
$$w_n^1|_{B(0,r_{1,n}/r_n^1)}=w_n$$
$$w_n^1|_{S^2\backslash B(0,r_{1,n}/r_n^1)}=\tilde{v}_n|_{B(x_n^1,r_{1,n})}$$

where $S^2\backslash B(0,r_{1,n}/r_n^1)$ and $B(x_n^1,r_{1,n})$ are identified with the help of the common boundary values of w_n on $\partial B(0,r_{1,n}/r_n^1)$ and v_n and $\tilde{v}_n$ on $\partial B(x_n^1,r_{1,n})$.

Since by Lemma 4.2.4

$$E(\tilde{v}_n,B(x_n^1,r_{1,n}))\to 0 \qquad \text{as } n\to\infty$$

$$\begin{aligned}\lim_{n\to\infty}E(v_n^1)&\leqslant\lim E(v_n)-\lim E(w_n^1)\\&\leqslant\lim E(v_n)-E(w^1)\\&\leqslant\lim E(v_n)-\mu.\end{aligned}\tag{4.2.34}$$

In the next step, we select, for each n, the infimum r_n^2 of all positive numbers ρ for which there exists a point $x_n^2\in\Sigma$ so that (in polar coordinates centred at x_n^2)

$$\int_{\varphi=0}^{2\pi}\left|\frac{\partial v_n^1(\rho,\varphi)}{\partial\varphi}\right|^2 d\varphi\geqslant s_2.\tag{4.2.35}$$

If r_n^2 has a positive lower bound (independent of n), v_n^1 as before converges uniformly and strongly in $H^{1,2}$ to a harmonic map (note that by Theorem 2.4.1, the possible singularity of the limit map at $\lim_{n\to\infty} x_n^1$ is removable). In the case $r_n^2 \to 0$ as $n \to \infty$, we apply the preceding construction to v_n^1. The image of $B(x_n^1, r_{1,n})$ under the rescaling $(B(x_n^2, r_n^2) \to B(0,1))$ can either disappear at infinity as $n \to \infty$, or converge to a disk or half-space in the plane. In the first case, we can proceed precisely as before and obtain another harmonic map $w^2: S^2 \to N$ with energy at least μ.

In the second case, in the limit we obtain a non-trivial harmonic map into N, defined on the complement of a disk or a half-plane. Moreover, the boundary values are constant. By Corollary 1.4.3, this map would have to be constant. Hence this case cannot occur. Likewise, we can check whether another harmonic map from S^2 into N splits off from w_n^1 as $n \to \infty$.

We repeat this procedure.

Since every splitting off of a harmonic map from S^2 into N leads to a decrease of energy of at least μ as in (4.2.34), this can happen at most a finite number of times, say m. In other words, we obtain sequences

$$v_n^m: \Sigma \to N$$
$$w_n^1, \dots, w_n^m: S^2 \to N$$

converging uniformly to harmonic maps

$$v^m: \Sigma \to N$$
$$w^1, \dots, w^m: S^2 \to N.$$

Also, by construction

$$\begin{aligned} E(v^m) + E(w^1) + \cdots + E(w^m) &\leqslant \lim_{n\to\infty} E(v_n^m) + \lim E(w_n^1) + \cdots \lim E(w_n^m) \\ &= \lim_{n\to\infty} E(v_n). \end{aligned} \tag{4.2.36}$$

We perform this construction for each t.

Instead of taking a fixed map u, we can also take a sequence $(u_\nu)_{\nu\in\mathbb{N}}$ with

$$\lim_{\nu\to\infty} \sup_{t\in A} E(u_\nu(\cdot, t)) = \kappa.$$

We then construct maps $v_{\nu,n}$ for each ν as before, and

$$\kappa \leqslant \sup_{t\in A} E(v_{\nu,n}(\cdot, t)) \qquad \text{for all } \nu, n$$

as well as

$$E(v_{\nu,n}(\cdot, t)) \leqslant E(u_\nu(\cdot, t)) \qquad \text{for all } t, \nu, n$$

imply, for example, that

$$\lim_{n\to\infty} \sup_{t\in A} E(v_{n,n}(\cdot, t)) = \kappa.$$

This already implies the claims of Theorem 4.2.1, with the sole exception that, so far, we only get $\leqslant$ instead of equality in (4.2.2). That equality will be achieved in the next section.

We point out, however, that the arguments given so far, also imply that if $\kappa < \mu$, where μ is given in Lemma 4.2.1, then no non-trivial minimal sphere can split off. In particular, if $\Sigma = S^2$, then in this case a path $h \in H$ can be deformed into a collection of constant maps. Furthermore, the limiting path can also be made continuous by the following argument.

For simplicity, assume $A = [0,1]$, and let $h'(\cdot, t) := \lim_{n\to\infty} h(\cdot, t_n)$ and

$$\delta_0 := \operatorname{dist}(h'(\cdot,0), h'(\cdot,1))$$

where the distance is chosen w.r.t. the C^0 or $H^{1,2}$-norm.

Then for every $\delta, 0 < \delta < \delta_0$, and sufficiently large $n \in \mathbb{N}$, there exists $t_n \in [0,1]$ with

$$\operatorname{dist}(h_n(\cdot, t_n), h_n(\cdot, 0)) = \delta.$$

After selection of a subsequence, $h_n(\cdot, t_n)$ then converges uniformly and strongly in $H^{1/2}$ to some limit $\tilde{h}$ with

$$\operatorname{dist}(\tilde{h}, h'(\cdot, 0)) = \delta.$$

In general, such a limit $\tilde{h}$ may be distinct from all limits of $h_n(\cdot, t)$ for fixed t and need not even be harmonic, but in the present case it has to be constant at least. Therefore, we can obtain a path of constant maps connecting $h'(\cdot, 0)$ and $h'(\cdot, 1)$ by taking suitable limits of our original path $h_n(\cdot, t)$.

The preceding results already suffice for all corollaries stated in Section 4.4.

4.3. Completion of the proof of Theorem 4.2.1

We now want to establish equality in (4.2.2).

The only possible loss of energy in (4.2.36) can occur in the following way.

There exist $\delta > 0$, points x_n which are centres of rescaling with factor r_n in our previous construction, radii $r_{1,n}, r_{2,n}$ with

$$0 < r_n < r_{1,n} < r_{2,n} \tag{4.3.1}$$

$$r_{2,n} \to 0 \qquad \text{as } n \to \infty \tag{4.3.2}$$

$$r_{1,n}/r_n^1 \to \infty \qquad \text{as } n \to \infty \tag{4.3.3}$$

and, putting $A(r_{1,n}, r_{2,n}) := \{x \in \Sigma : r_{1,n} \leqslant d(x, x_n) \leqslant r_{2,n}\}$ and using polar coordinates centred at x_n,

$$E(v_n(\cdot, t), A(r_{1,n}, r_{2,n})) \geqslant \delta \tag{4.3.4}$$

but

$$\sup_{r_{1,n} \leqslant r_{0,n} \leqslant r_{2,n}} \int_{\varphi=0}^{2\pi} \left| \frac{\partial v_n(r_{0,n}, \varphi, t)}{\partial \varphi} \right|^2 \mathrm{d}\varphi \to 0 \qquad \text{as } n \to \infty. \tag{4.3.5}$$

Namely, in this case after rescaling with factor r_n, the annulus $A(r_{1,n}, r_{2,n})$ disappears at infinity of $\mathbb{R}^2$ as $n \to \infty$ (4.3.3), and the loss of energy (4.3.4) will not be compensated by another blow-up of Σ, because of (4.3.5).

We let

$$\kappa := \lim_{n\to\infty} \sup_t E(v_n(\cdot, t)).$$

Suppose there exist $\delta > 0$, points $x_n = x_n(t)$, sequences $r_{1,n} = r_{1,n}(t)$, $r_{2,n} = r_{2,n}(t)$, and σ_n, $0 < r_{1,n} < r_{2,n}$, $0 < \sigma_n$, $r_{2,n} \to 0$, $\sigma_n \to 0$ as $n \to \infty$, and $n_0 \in \mathbb{N}$ so that for all $n \geqslant n_0, t \in [0, 1]^d$ with

$$E(v_n(\cdot, t)) \geqslant \kappa - 2\delta \tag{4.3.6}$$

we have in the annulus centred at x_n

$$E(v_n(\cdot, t), A(r_{1,n}, r_{2,n})) > \delta \tag{4.3.7}$$

and

$$\operatorname*{ess\,sup}_{r_{1,n} \leqslant r_{0,n} \leqslant r_{2,n}} \int_{\varphi=0}^{2\pi} \left| \frac{\partial v_n(r_{0,n}, \varphi, t)}{\partial \varphi} \right|^2 \mathrm{d}\varphi < \sigma_n \tag{4.3.8}$$

(in polar coordinates centred at x_n).

We claim that δ, σ_n, n_0 can be chosen independently of t. This is justified as follows. Otherwise, for each $\delta > 0$, we can find $\epsilon > 0$ and a sequence $(t_n) \subset [0, 1]^d$ with (after suitable selections of subsequences):

$$t_n \to t_0 \qquad \text{as } n \to \infty$$
$$E(v_n(\cdot, t_n)) \geqslant \kappa - 2\delta$$

and the property that whenever, for $0 < r_1 < r_2$,

$$E(v_n(\cdot, t_n), A(r_1, r_2)) \geqslant \delta$$

then also

$$\operatorname*{ess\,sup}_{r_1 \leqslant r_0 \leqslant r_2} \int_{\varphi=0}^{2\pi} \left| \frac{\partial v_n(r_0, \varphi, t_n)}{\partial \varphi} \right|^2 \mathrm{d}\varphi \geqslant \epsilon.$$

The preceding considerations then imply the existence of harmonic maps

$$v_\delta: \Sigma \to N$$
$$w_\delta^1, \ldots, w_\delta^m: S^2 \to N \qquad \text{(possibly } m = 0)$$

so that

$$\kappa \geqslant E(v_\delta) + E(w_\delta^1) + \cdots + E(w_\delta^m)$$
$$= \lim_{n\to\infty} E(v_n(\cdot, t_n)) \geqslant \kappa - 2\delta.$$

Letting $\delta \to 0$ and using the convergence properties of harmonic maps of

Lemma 4.3.1 below, we obtain harmonic maps

$$v: \Sigma \to N$$
$$w^1, \ldots, w^m: S^2 \to N \qquad \text{(possibly } m = 0\text{)}$$

with

$$E(v) + \sum_{\mu=1}^{m} E(w^\mu) = \kappa$$

which is precisely what we want.

Therefore, we can assume, as claimed, that δ, σ_n, n_0 are independent of t.

Lemma 4.3.1: *Let $v_n: \Sigma \to N$ be a sequence of harmonic maps with*

$$E(v_n) \leqslant K.$$

Then, after selection of a subsequence, there exist harmonic maps

$$v_0: \Sigma \to N$$
$$w^1, \ldots, w^l: S^2 \to N$$

with

$$v_n \to (v_0, w^1, \ldots, w^l)$$

and in particular

$$\lim_{n \to \infty} E(v_n) = E(v_0) + \sum_{\lambda=1}^{l} E(w^\lambda).$$

Proof: We reduce the situation to the case of a sequence of minimal surfaces, using the same argument as in the proof of Theorem 2.3.4. Namely, each v_n yields a holomorphic quadratic differential

$$\varphi_n := g_{jk} \left(\frac{\partial v_n^j}{\partial r} \frac{\partial v_n^k}{\partial r} - \frac{1}{r^2} \frac{\partial v_n^j}{\partial \varphi} \frac{\partial v_n^k}{\partial \varphi} - 2\mathrm{i} \frac{1}{r} \frac{\partial v_n^j}{\partial r} \frac{\partial v_n^k}{\partial \varphi} \right)$$
$$\cdot (\mathrm{d}r^2 - r^2\, \mathrm{d}\varphi^2 + 2\mathrm{i} r\, \mathrm{d}r\, \mathrm{d}\varphi)$$

where g_{jk} is the metric of N in local coordinates v^j, and (r, φ) are conformal local polar coordinates on Σ.

Since $E(v_n)$ is uniformly bounded, the φ_n are uniformly bounded in L^1, hence also in L^∞ as they are holomorphic. We then find holomorphic functions Ψ_n on local disks in Σ with

$$\frac{\partial}{\partial z} \Psi_n = \tfrac{1}{4} \varphi_n$$

(on a local disk, we can just consider φ_n as a holomorphic function). Furthermore, we can also choose the Ψ_n to be uniformly bounded. We put

$$f_n(z) = \bar{z} - \Psi_n(z).$$

Then

$$\frac{\partial}{\partial z} f_n \frac{\partial}{\partial z} \overline{f_n} = -\tfrac{1}{4}\varphi_n$$

and

$$V_n := (v_n, f_n): B \to N \times \mathbb{R}^2$$

(B is our local disk on σ) is harmonic and conformal, i.e. a minimal surface.
Also

$$E(V_n) = E(v_n) + E(f_n) =$$
$$E(v_n) + \int (1 + \tfrac{1}{16}|\varphi_n|^2)$$

is uniformly bounded.

We are thus reduced to excluding the following. $(V_n)_{n\in\mathbb{N}}: B \to Y$ is a sequence of minimal surfaces in a Riemannian manifold Y of bounded geometry, D is the unit disk,

$$A(r_{1,n}, r_{2,n}) = \{x \in D: r_{1,n} \leqslant |x| \leqslant r_{2,n}\} \qquad 0 < r_{1,n} < r_{2,n} < \tfrac{1}{2}$$

$$E(V_n, D) \leqslant K \tag{4.3.9}$$

$$E(V_n, A(r_{1,n}, r_{2,n})) \geqslant \delta > 0 \tag{4.3.10}$$

with K, δ independent of n, and

$$\operatorname*{ess\ sup}_{r_{1,n} \leqslant r_{0,n} \leqslant r_{2,n}} \int_{\varphi=0}^{2\pi} \left| \frac{\partial V_n(r_{0,n}, \varphi)}{\partial \varphi} \right|^2 \mathrm{d}\varphi < \sigma_n \to 0 \tag{4.3.11}$$

as $n \to \infty$.

We consider two possibilities:

(1) There exist balls $B(p_n, \rho_n) \subset Y, \rho_n \to 0$, with $V_n(A(r_{1,n}, r_{2,n})) \subset B(p_n, \rho_n)$. Choosing p_n as the centre of local coordinates in Y, we have

$$\begin{aligned} E(V_n, A(r_{1,n}, r_{2,n})) &= \frac{1}{2} \int_{A(r_{1,n}, r_{2,n})} \langle \mathrm{d}V_n, \mathrm{d}V_n \rangle r \,\mathrm{d}r \,\mathrm{d}\varphi \\ &= \frac{1}{2} \int_{A(r_{1,n}, r_{2,n})} g_{ij} \frac{\partial V_n^i}{\partial x^\alpha} \frac{\partial V_n^j}{\partial x^\alpha} r \,\mathrm{d}r \,\mathrm{d}\varphi \\ &= -\frac{1}{2} \int_{A(r_{1,n}, r_{2,n})} \frac{\partial}{\partial x^\alpha} \left(g_{ij} \frac{\partial V_n^i}{\partial x^\alpha} \right) V_n^j r \,\mathrm{d}r \,\mathrm{d}\varphi \\ &\quad + \frac{1}{2} \int_{\partial A(r_{1,n}, r_{2,n})} g_{ij} \frac{\partial V_n^i}{\partial v} V_n^j r \,\mathrm{d}\varphi \end{aligned}$$

which goes to zero, as $|V_n^j| \leqslant \rho_n$, and since we have uniform gradient bounds (cf. Theorem 2.5.2), and V_n satisfies the harmonic map equation. This contradicts (4.3.10), and hence possibility (1) is ruled out.

(2) There exist $r_{3,n}, r_{4,n}, r_{1,n} \leqslant r_{3,n} < r_{4,n} \leqslant r_{2,n}$ and $\mu > 0$ (independent of n) with

$$d(V_n(r_{3,n}, \cdot), V_n(r_{4,n}, \cdot)) \geqslant \mu$$

(using polar coordinates (r, φ) in D)

$$\operatorname{diam}(V_n(r_{2,n}, \cdot)) \leqslant (2\pi\sigma_n)^{1/2} =: s_n \qquad \text{(cf. (4.3.11))}$$

$$V_n(A(r_{3,n}, r_{4,n})) \subset B(p_n, 2\mu).$$

For simplicity of notation, we put $p := p_n$ and select points

$$p_l (= p_{l,n}) \in V_n(r_{l,n}, \cdot) \qquad l = 3, 4.$$

We let $\gamma(t)$ be the geodesic connection in $B(p, 2\mu)$ between p_3 and p_4, parametrized proportionally to arclength with $\gamma(0) = p_3$, $\gamma(1) = p_4$.
W.l.o.g., μ is so small that for

$$U(\gamma, \tau) := \{q \in Y : d(q, \gamma) = \tau\}$$

if $0 < \tau \leqslant 4\mu$, the sum of any two principal curvatures (w.r.t. the inward normal) is positive.

$$V_n(A(r_{3,n}, r_{4,n})) \subset B(p, 2\mu) \subset U(\gamma, 4\mu).$$

If we then shrink τ from 4μ to s_n, the minimal surface $V_n(A(r_{3,n}, r_{4,n}))$ cannot touch any $U(\gamma, \tau)$ at an interior point by the maximum principle[2], and neither at a boundary point by choice of p_n. Hence

$$V_n(A(r_{3,n}, r_{4,n})) \in U(\gamma, s_n).$$

If μ is small enough, and σ_n (and hence s_n) small compared to μ, then the following is correct.
For every $\eta > 0$, there exists $g \in C^2([0, 1], \mathbb{R}^+)$ with $g(0) = g(1) = s_n$, $0 < g(t) < \eta$ for $t \in [\frac{1}{4}, \frac{3}{4}]$ for which $U(\gamma, g(\cdot))$, defined as the envelope of the balls $B(\gamma(t), g(t))$ has again the property that the sum of any two principal curvatures is positive.
We fix n so that s_n is small enough for this property to hold. By the same shrinking argument as before

$$V_n(A(r_{3,n}, r_{4,n})) \subset U(\gamma, g(\cdot))$$

for every such g. Letting $\eta > 0$, this means that the minimal surface $V_n(D)$ has to map an annulus onto the arc $\gamma([\frac{1}{4}, \frac{3}{4}])$ which is impossible.

Since always (1) or (2) can be achieved, if (4.3.9), (4.3.10), (4.3.11) hold, we see that this cannot happen for sequences of minimal surfaces, and hence not for sequences of harmonic maps either. This proves the lemma.

q.e.d.

[2] A minimal surface can never touch a surface with the above curvature property from the inside.

We first consider the local situation where the radii $r_{1,n}$ and $r_{2,n}$ are independent of t and the centre x_n is independent of t and n. Thus, for the moment, we have only to consider annuli centred at $x_0 \in \Sigma$.

The general case will be treated later on by a covering argument. Thus, we fix some $t_0 \in A$, with the following properties. There exists an open ball $W_0^+ \subset A$ with centre t_0, $W_0 := \partial W_0^+$, $W_0^- := A \backslash \bar{W}_0^+$ with the property that

$$E(v_n(\cdot, t)) \geqslant \kappa - 2\delta \qquad \text{for } t \in W_0^+$$

and that (4.3.7) and (4.3.8) hold for the annulus $A(r_{1,n}, r_{2,n}))$ centred at $x_0 \in \Sigma$ for all $t \in W_0^+$.

Let β be the radius of W_0^+, and let W_b^+ be the ball centred at t_0 with radius $\beta - b$, $0 \leqslant b \leqslant 2b_0 := \beta/2$, $W_b := \partial W_b^+$, $W_b^- := A \backslash \bar{W}_b^+$. For $t \in W_0$, let $v(t, b) := (1 - b/\beta)(t - t_0) \in W_b$, and conversely $\pi(v(t, b)) := t$.

We then construct a path $\tilde{v}_n$ from v_n as follows. Let σ_n be as above, w.l.o.g. $\sigma_n \leqslant s^2/100$.

For $n \leqslant n_0 : \tilde{v}_n = v_n$.

On $W_0^- : \tilde{v}_n = v_n$.

For $n \geqslant n_0$ and $t \in W_b$, $0 \leqslant b \leqslant b_0/2$:

(dropping again the dependence on t in our notation, as $t \in W_b$ is fixed for the moment).

On $\Sigma \backslash A(x_0, r_{1,n}, r_{2,n})$: $\tilde{v}_n = v_n$.

We shall use polar coordinates centred at x_0. For each r, $r_{1,n} \leqslant r \leqslant r_{2,n}$, we let $g_n(r) =$ centre of mass of $\{v_n(r, \varphi) : 0 \leqslant \varphi \leqslant 2\pi\}$. Since for φ_1, $\varphi_2 \in [0, 2\pi]$ and almost all $r \in [r_{1,n}, r_{2,n}]$

$$\begin{aligned} d(v_n(r, \varphi_1), v_n(r, \varphi_2)) &\leqslant \int_{\varphi=0}^{2\pi} \left| \frac{\partial v_n(r, \varphi)}{\partial \varphi} \right| \mathrm{d}\varphi \\ &\leqslant (2\pi)^{1/2} \left(\int_{\varphi=0}^{2\pi} \left| \frac{\partial v_n(r, \varphi)}{\partial \varphi} \right|^2 \mathrm{d}\varphi \right)^{1/2} \\ &\leqslant (2\pi)^{1/2} \sigma_n^{1/2} \leqslant s/10 \end{aligned}$$

and since v_n is continuous, $g_n(r)$ is uniquely determined for all $r \in [r_{1,n}, r_{2,n}]$. (The centre of mass of a set $\Omega \subset B(p, \rho)$ with $\rho < \min(\pi/4\kappa, i(q)/2)$ is defined as the point where $|\int_\Omega d^2(q, x)\,\mathrm{d}x|$ achieves its minimum; this minimum is unique because $d^2(\cdot, x)$ is strictly convex on $B(p, \rho)$.)

We define $w_n(r, \varphi)$ as follows. We let

$$\rho_1 = (1 + \sigma_n^{1/4}) r_{1,n}$$
$$\rho_2 = (1 - \sigma_n^{1/4}) r_{2,n}$$

w.l.o.g. $\rho_1 < \rho_2$. We note

$$\int_{\varphi=0}^{2\pi} \left| \frac{\partial v_n(\tilde{r}, \varphi)}{\partial \varphi} \right|^2 \mathrm{d}\varphi \leqslant \sigma_n \qquad \text{for } \tilde{r} = r_{1,n} \text{ and } \tilde{r} = r_{2,n}.$$

We let $\gamma_{r_{1,n},\varphi}$ be a geodesic from $v_n(r_{1,n},\varphi)$ to $g_n(r_{1,n})$ and parametrize it proportional to arclength on the interval $\{r: r_{1,n} \leqslant r \leqslant \rho_1\}$, so that $r = r_{1,n}$ corresponds to $v_n(r_{1,n},\varphi)$ and $r=\rho_1$ to $g_n(r_{1,n})$. For $r_{1,n} \leqslant r \leqslant \rho_1$ we then define $w_n(r,\varphi) = \gamma_{r_{1,n},\varphi}(r)$. We have

$$E(w_n, A(x_0, r_{1,n}, \rho_1)) \leqslant \frac{1}{2}\int_{\varphi=0}^{2\pi}\int_{r=r_{1,n}}^{\rho_1}\left(\frac{d(v_n(r_{1,n},\varphi), g_n(r_{1,n}))^2}{(\rho_1 - r_{1,n})^2} + \frac{1}{r^2}\left|\frac{\partial v_n(r_{1,n},\varphi)}{\partial\varphi}\right|^2\right) r\,\mathrm{d}r\,\mathrm{d}\varphi$$

$$\leqslant 2\pi\sigma_n\left(\frac{\rho_1^2 - r_{1,n}^2}{(\rho_1 - r_{1,n})^2} + \ln\left(\frac{\rho_1}{r_{1,n}}\right)\right) \leqslant c\sigma_n^{1/2}.$$

Similarly, we connect $g_n(r_{2,n})$ with $v_n(r_{2,n},\varphi)$ along a geodesic $\gamma_{r_{2,n},\varphi}$ parametrized on $[\rho_2, r_{2,n}]$, ρ_2 corresponding to $g_n(r_{2,n})$ and $r_{2,n}$ to $v_n(r_{2,n},\varphi)$. For $\rho_2 \leqslant r \leqslant r_{2,n}$, we put

$$w_n(r,\varphi) = \gamma_{r_{2,n},\varphi}(r)$$

and estimate as before

$$E(w_n, A(x_0, \rho_2, r_{2,n})) \leqslant c\sigma_n^{1/2}.$$

Moreover, we choose a linear map $R(r)$ from $[\rho_1, \rho_2]$ onto $[r_{1,n}, r_{2,n}]$ and define for $\rho_1 \leqslant r \leqslant \rho_2$

$$w_n(r,\varphi) = g_n(R(r),\varphi)$$

and estimate

$$E(w_n, A(x_0, \rho_1, \rho_2)) \leqslant f(\sigma_n)\left|\frac{\mathrm{d}R}{\mathrm{d}r}\right|^2 E(v_n, A(x_0, r_{1,n}, r_{2,n})).$$

We note that, as $\sigma_n \to 0$

$$f(\sigma_n) \to 1$$

and

$$\frac{\mathrm{d}R}{\mathrm{d}r} \to 1 \qquad (R \text{ depends on } \sigma_n \text{ via } \rho_1 \text{ and } \rho_2).$$

We now want to connect v_n with w_n while moving the t-parameter from W_0 to $W_{b_0/2}$. We first choose a family of continuous piecewise linear maps as follows

$$l(r,\tau) = \begin{cases} r_{1,n} & \text{for } r_{1,n} \leqslant r \leqslant r_{1,n} + \tau(\rho_1 - r_{1,n}) \\ r_{2,n} & \text{for } r_{2,n} + \tau(\rho_2 - r_{2,n}) \leqslant r \leqslant r_{2,n} \end{cases}$$

and $l(r,\tau)$ is a linear map for $r_{1,n} + \tau(\rho_1 - r_{1,n}) \leqslant r \leqslant r_{2,n} + \tau(\rho_2 - r_{2,n})$.

We then let, for $t \in W_b$, $0 \leqslant b \leqslant b_0/2$,

$$\tilde{v}_n(r,\varphi,t) = v_n(l(r, 2b/b_0), \varphi, t).$$

As in the estimate for w_n, one checks

$$E(\tilde{v}_n(\cdot,t)) \leqslant f(\sigma_n)E(v_n(\cdot,t)) + c\sigma_n^{1/4}$$

where again $f(\sigma_n)\to 1$ as $\sigma_n \to 0$.

We now connect $\tilde{v}_n$, as defined on $W_{b_0/2}$, to w_n while moving the t-parameter from $W_{b_0/2}$ to W_{b_0}.

For each $x\in\Sigma$, $t\in W_0$, we let $\gamma_{x,t}$ be the geodesic from $\tilde{v}_n(x, v(t, b_0/2))$ to $w_n(x,t)$, parametrized proportional to arclength on the interval $[b_0/2,\ b_0]$, i.e. $\gamma_{x,t}(b_0/2) = \tilde{v}_n(x, v(t, b_0/2))$, $\gamma_{x,t}(b_0) = w_n(x,t)$ and define, for $b_0/2 \leqslant b \leqslant b_0$,

$$\tilde{v}_n(x, v(t,b)) = \gamma_{x,t}(b).$$

Again, we have

$$E(\tilde{v}_n(\cdot, v(t,b)) \leqslant f(\sigma_n)E(v_n(\cdot,t)) + c\sigma_n^{1/4}$$

with $f(\sigma_n)\to 1$ as $\sigma_n \to 0$. This follows from the rescaling principle (Corollary 2.1.1) and the corresponding inequality in the Euclidean case. In the latter, we have only to check, that for $f\in H^{1,2}$, $0\leqslant\lambda\leqslant 1$, the family of functions

$$(1-\lambda)f(r,\varphi) + \lambda\frac{1}{2\pi}\int_{\delta=0}^{2\pi} f(r,\delta)\,\mathrm{d}\delta =: f_\lambda(r,\varphi)$$

satisfies $(\mathrm{d}/\mathrm{d}\lambda)E(f_\lambda) \leqslant 0$.

This in turn is implied by

$$\frac{\mathrm{d}}{\mathrm{d}\lambda}\int_{\varphi=0}^{2\pi}\int_{r_1}^{r_2}\left|(1-\lambda)\frac{\partial f}{\partial r} + \lambda\frac{1}{2\pi}\left(\int_{\delta=0}^{2\pi}\frac{\partial f}{\partial r}(r,\delta)\,\mathrm{d}\delta\right)\right|^2 r\,\mathrm{d}r\,\mathrm{d}\varphi \leqslant 0.$$

A moment's reflection shows that it suffices to establish this for $\lambda=0$, and the inequality for $\lambda=0$ in turn follows from the Cauchy–Schwarz and Hölder inequalities.

Now, for $t\in W_{b_0}$, and $\rho_1 \leqslant r \leqslant \rho_2$, we have eliminated the dependence on φ and hence can decrease the energy by stretching the r-direction.

For any $\rho<\rho_1$, we define a family $\lambda_\rho(\cdot,\tau)$ of continuous piecewise smooth maps from $[0,\rho_2]$ onto itself as follows:

$\lambda_\rho(0,\tau)=0$
$\lambda_\rho(\rho_1+\tau(\rho-\rho_1),\tau)=\rho_1$
$\lambda_\rho(\rho_2,\tau)=\rho_2$
$\lambda_\rho|[0,\rho_1+\tau(\rho-\rho_1)]\times\tau$ is linear
$(\lambda_\rho(r,\tau),\varphi)$: $A(\rho_1+\tau(\rho-\rho_1),\ \rho_2)\to A(\rho_1,\rho_2)$ is harmonic for each τ.

If $t\in W_b$, $b_0\leqslant b\leqslant 2b_0$, we put

$$\tilde{v}_n(r,\varphi,t) = w_n(\lambda_\rho(r,(b/b_0)-1),\varphi,\pi(t))$$

and for $t\in W_{2b_0}^{+}$

$$\tilde{v}_n(r,\varphi,t) = w_n(\lambda_\rho(r,1),\varphi,k^{-1}(t)).$$

We compute, for $t \in W_b$, $b_0 \leqslant b \leqslant 2b_0$

$$E(\tilde{v}_n(\cdot, t), B(x_0, \rho_1 + (b/b_0 - 1)(\rho - \rho_1))) = E(w_n(\cdot, \pi(t)), B(x_0, \rho_1))$$

by conformal invariance of the energy.

To compute the energy on the annuli, we conformally identify the annulus $\rho_1 + \tau(\rho - \rho_1) \leqslant r \leqslant \rho_2$ with a cylindrical region $Z(\tau, \rho) := [0, R_0(\tau, \rho)] \times S^1$. R_0 is increasing in τ and decreasing in ρ. Also $R_0(1, \rho) \to \infty$ as $\rho \to 0$, while $R_0(0, \rho)$ is independent of ρ. In these coordinates, λ_ρ induces a linear map from $Z(\tau, \rho)$ onto $Z(0, \rho)$, denoted by

$$R \to \mu R \qquad \mu = \mu(\tau, \rho) \qquad \mu(1, \rho) \to 0 \qquad \text{as } \rho \to 0.$$

Actually

$$\mu(1, \rho) \cdot R_0(1, \rho) = \text{const.} \tag{4.3.12}$$

We use R, δ as polar coordinates on $Z(\tau, \rho)$ and compute for $t \in W^+_{2b_0} (\tau = 1)$

$$E(\tilde{v}_n(\cdot, t), A(\rho, \rho_2)) = \int_{R=0}^{R_0(1,\rho)} \left| \frac{\partial w_n}{\partial(\mu R)} \mu \right|^2 \mathrm{d}R$$

as w_n does not depend on the angular direction,

$$= R_0(1, \rho) \mu^2(1, \rho) E(w_n(\cdot, t), A(\rho_1, \rho_2))$$
$$\to 0 \qquad \text{as } \rho \to 0 \text{ because of (4.3.12).}$$

Hence, choosing $\rho > 0$ sufficiently small, our stretching procedure makes the energy of $\tilde{v}_n$ on the annulus $A(\rho, \rho_2)$ arbitrarily small (for $t \in W^+_{2b_0}$).
Also, the increase of energy on $W_b (0 \leqslant b \leqslant 2b_0)$ caused by the interpolation between v_n and $w_n \circ \lambda_\rho$ goes to zero as $n \to \infty$. On the other hand, on $W^+_{2b_0}$, we have decreased the energy by at least δ

$$\lim_{n \to \infty} \sup_{t \in W^+_{2b_0}} E(\tilde{v}_n(\cdot, t)) \leqslant \kappa - \delta.$$

Moreover, $\tilde{v}_n$ enjoys all the convergence properties of v_n, the only difference being that one might have to choose larger radii in the rescaling at x_0.

The covering argument that is needed then is the following. For given n, we find points $t_0^1, \ldots, t_0^k \in A$ ($k = k(n)$) with the following properties. Each point t_0^i is the centre of a ball $W^+_{0,i}$ with:

- There exists $x_{0,i} \in \Sigma$, and radii $r^i_{1,n}, r^i_{2,n}$ for which (4.3.7) and (4.3.8) hold in the annulus $A(r^i_{1,n}, r^i_{2,n})$ centred at $x_{0,i}$ for all $t \in W^+_{0,i}$.
- $E(v_n(\cdot, t)) \geqslant \kappa - 2\delta$ for all $t \in W^+_{0,i}$.
- $\{t : E(v_n, (\cdot, t)) \geqslant \kappa - \delta\} \subset \bigcup_{i=1}^k W^+_{2b_0,i}$ (where $W^+_{b,i}$ is defined as above).

Using the Besicovitch covering Lemma 4.3.2 below, for each n, we can select the points $t_0^1, \ldots, t_0^k$ in such a way that each t with $E(v_n(\cdot, t)) \geqslant \kappa - \delta$ is contained in some ball $\bar{W}^+_{2b_0,i}$, and in at most Λ balls $\bar{W}^+_{0,i}$ where Λ is independent on n. In order to achieve this, for each n, we choose balls $W^+_0(t)$ with the same radius (possibly depending on n) and the above properties. Lemma 4.3.3 gives the existence of at most m disjointed subcollections of balls $W^+_{2b_0,i}$, and the argument of Lemma 4.2.2 then further divides each such subcollection into subcollections for which also the larger balls $W^+_{0,i}$ become disjoint.

We want to perform the previous energy-decreasing process on each ball $W^+_{0,i}$. We have to check what happens if the corresponding annuli intersect, e.g.

$$A(r^1_{1,n}, r^1_{2,n}) \cap A(r^2_{1,n}, r^2_{2,n}) \neq \emptyset$$

where the first one is centred at $x_{0,1}$, the second one at $x_{0,2}$. After we have performed the process on $W^+_{0,1}$, for certain t we may have increased the energy by $f(\sigma_n)$, where $f(\sigma_n) \to 0$ as $\sigma_n \to 0$. Also, let $\gamma(r^2_{0,n})$ be the image of $\{x \in \Sigma : d(x, x_{0,2}) = r^2_{0,n}\}$ under the deformation of Σ performed on $A(r^1_{1,n}, r^1_{2,n})$. The construction implies that

$$\operatorname*{ess\,sup}_{r^2_{1,n} \leqslant r^2_{0,n} \leqslant r^2_{2,n}} \int_{\gamma(r^2_{0,n})} \left| \frac{\partial \tilde{v}_n(r^2_{0,n}, \varphi, t)}{\partial \varphi} \right|^2 \mathrm{d}\varphi \leqslant g(\sigma_n)$$

where $\varphi \in [0, 2\pi]$ now parametrizes $\gamma(r^2_{0,n})$, $\tilde{v}_n$ is the map obtained from the deformation process on $W^+_{0,1}$, and $g(\sigma_n) \to 0$ as $\sigma_n \to 0$.

If n is large enough, $n \geqslant n_0$ say, we can assume

$$\gamma f(\sigma_n) + \gamma g(\sigma_n) \leqslant \delta/10.$$

From this it follows that we are able to repeat the deformation process on the deformed ring $\gamma(A(r^2_{1,n}, r^2_{2,n}))$ and decrease the energy for $t \in W^+_{2b_0,2}$, if this has not been accomplished already in the first step, i.e. the one for $t \in W^+_{0,1}$.

For large enough n, we can therefore decrease

$$E(v_n(\cdot, t)) \qquad \text{for } t \in \bar{W}_{2b_0,2}$$

by at least $\delta/2$, in the case when (4.3.7) and (4.3.8) hold. We denote the new maps by $\tilde{v}_n(\cdot, t)$.

We replace v_n by $\tilde{v}_n$ and repeat the process, i.e. look whether we can find (for $\tilde{v}_n$) a (possibly smaller) $\delta > 0$ enjoying the same properties as before (cf. (4.3.7), (4.3.8)).

Iterating, if necessary, to get sequences $v^0_n = v_n$, $v^1_n = \tilde{v}_n$, $v^2_n, \ldots,$ and taking a diagonal sequence v^n_n, we infer that for all $t_0 \in [0, 1]^d$ with

$$\lim_{n \to \infty} E(v^n_n(\cdot, t_0)) = \lim_{n \to \infty} \sup_{t \in A} E(v^n_n(\cdot, t))$$

we have, in the previous notations,

$$E(v^m(\cdot, t_0)) + E(w^1(\cdot, t_0)) + \cdots + E(w^m(\cdot, t_0)) = \lim_{n \to \infty} E(v^n_n(\cdot, t_0))$$

where

$$v^m(\cdot, t_0) : \Sigma \to N$$
$$w^1(\cdot, t_0), \ldots, w^m(\cdot, t_0) : S^2 \to N$$

are harmonic.

This completes the proof of Theorem 4.2.1.

q.e.d.

Let us now state the covering lemma of Besicovitch (cf. [F] for a proof).

Lemma 4.3.2: *Let X be a complete Riemannian manifold of bounded geometry. Let $\Omega \subset X$, $\mathscr{B} := \{B(x, r(x)) : x \in \Omega\}$ be a collection of bounded balls with centres in Ω (i.e. each $x \in \Omega$ is the centre of some $B \in \mathscr{B}$), and suppose $0 < r(x) \leqslant R$ for all $x \in \Omega$ and some fixed R. Then there are subcollections $\mathscr{B}_1, \ldots, \mathscr{B}_m$, where m depends only on the geometry of X, of $\mathscr{B}$ with the property that each $\mathscr{B}_j$ is a pairwise disjoint (or empty) collection, and*

$$\Omega \subset \bigcup_{j=1}^{m} \left(\bigcup_{B \in \mathscr{B}_j} B \right)$$

i.e. Ω is still covered by the union of the $\mathscr{B}_j$.

4.4. Corollaries and consequences of the general existence theorem. Boundary conditions

First of all, of course, Theorem 4.1.1 is a corollary of Theorem 4.2.1. We formulate it here as:

Corollary 4.4.1: *Let Σ be a compact surface without boundary, N a compact Riemannian manifold, $g : \Sigma \to N$ continuous. Then there exists a harmonic map $u : \Sigma \to N$ with*

$$u_{\#} = g_{\#} : \pi_1(\Sigma) \to \pi_1(N). \tag{4.4.1}$$

If $\pi_2(N) = 0$ then u is homotopic to g.

Proof: It is clear from Section 4.2 that the development of harmonic spheres does not change the action on the fundamental group, and if $\pi_2(N) = 0$, it does not change the homotopy class.

q.e.d.

Corollary 4.4.2 [SkU1]: *Suppose the universal cover $\tilde{N}$ of the compact Riemannian manifold N is not contractible. Then there exists a non-trivial conformal harmonic map $u : S^2 \to N$.*

Proof: Since $\tilde{N}$ is not contractible, $\pi_k(N) \neq 0$ for some $k \geqslant 2$. Hence there exists a homotopically non-trivial map

$$h_0 : S^2 \times A \to N$$

where $A := [0, 1]^{k-2}$, with $h_0(\cdot, \tau) = \text{const.}$ for every $\tau \in \partial A$ (by suspension). Let H be the class of all maps $h : S^2 \times A \to N$ homotopic to h_0. Then

$$\kappa := \inf_{h \in H} \sup_{t \in A} E(h(\cdot, t)) > 0 \tag{4.4.2}$$

as otherwise, by Theorem 4.2.1, all maps $h_0(\cdot, t)$ could be simultaneously deformed into constant maps, and h_0 would represent the zero element of $\pi_k(N)$, contrary to our assumption.

Applying Theorem 4.2.1 again, (4.4.2) then shows the existence of a non-trivial harmonic map which is conformal by Corollary 1.4.1.

q.e.d.

We shall need the following definition:

Definition 4.4.1: For a homotopy class $\alpha \in [\Sigma, N]$, we put

$$E_\alpha := \inf_{h \in \alpha} E(h). \tag{4.4.3}$$

Let $\mathbb{Z}[\pi_1(N)]$ be the integral group ring of $\pi_1(N)$. $\mathbb{Z}[\pi_1(N)]$ acts on $\pi_2(N)$ in a natural manner.

Corollary 4.4.3 ([SkU1] and also [MY2]): *Let N be a compact Riemannian manifold. Then there exist conformal harmonic maps $u_1, \ldots, u_m: S^2 \to N$ which generate $\pi_2(N)$ as a $\mathbb{Z}[\pi_1(N)]$-module, and each u_i $(i = 1, \ldots, m)$ is energy minimizing in its free homotopy class.*

Proof: Let α_i be the free homotopy classes containing energy-minimizing maps, and let M be the submodule of $\pi_2(N)$ generated by elements of the α_i. Let μ be the constant of Lemma 4.2.1. Let α_0 be a homotopy class with $\alpha_0 \notin M$ and

$$E_{\alpha_0} < \inf_{\alpha \notin M} E_\alpha + \mu/2. \tag{4.4.4}$$

We then minimize the energy in α_0. Theorem 4.2.1 shows the existence of conformal harmonic maps

$$u_1, \ldots, u_l: S^2 \to N \qquad (l \geq 2)$$

which are all non-trivial and with

$$u_1 + \cdots + u_l \in \alpha_0 \qquad \sum_{j=1}^{l} E(u_j) = E_{\alpha_0}. \tag{4.4.5}$$

By Lemma 4.2.1, and since $l \geq 2$ and because of (4.4.4) and (4.4.5) then

$$E(u_j) < \inf_{\alpha \notin M} E_\alpha. \tag{4.4.6}$$

Hence u_j is homotopic to an energy-minimizing map by definition of M, and because of (4.4.5) again, u_j itself has to be energy minimizing $(j = 1, \ldots, l)$. Thus α_0 is represented by a sum of energy-minimizing maps and therefore has to lie in M. We conclude $M = \pi_2(N)$.

q.e.d.

Corollary 4.4.4: *Let Σ and N be as in Corollary 4.4.1, $\alpha_0 \in [\Sigma, N]$ and let*

$$\alpha_{0\#}: \pi_1(\Sigma) \to \pi_1(N)$$

be the induced map on the fundamental groups. Put

$$E_0 := \inf\{E_\alpha : \alpha \in [\Sigma, N] \text{ with } \alpha_\# = \alpha_{0_\#}\}.$$

If then, for some $\beta \in [\Sigma, N]$ with $\beta_\# = \alpha_{0_\#}$,

$$E_\beta < E_0 + \mu \tag{4.4.7}$$

where μ is defined in Lemma 4.2.1, then β contains an energy-minimizing map.

Proof: This follows directly from Theorem 4.2.1 and Lemma 4.2.1.

q.e.d.

Likewise, if an appropriate analogue of (4.4.7) holds, one can also find unstable harmonic maps in a given class, as again no sphere can split off. This was studied in detail by Ding [Di].

As an application of Corollary 4.4.4, we consider the existence question for harmonic maps from a closed surface Σ into the real projective plane P, equipped with any (smooth) Riemannian metric; cf. [J8]. We let

$$\theta: \pi_1(\Sigma) \to \pi_1(P)$$

be a map of the fundamental groups. θ is called oriented if it maps orientation-preserving loops on to orientation-preserving ones, and orientation-reversing loops onto orientation-reversing ones. Otherwise, it is called non-orientable. If Σ is oriented, then any non-trivial θ is non-orientable. If Σ is non-orientable and $\pi_1(\Sigma)$ described by the relation $a_1^2 \cdots a_n^2 b_1 c_1 b_1^{-1} c_1^{-1} \cdots b_m c_m b_m^{-1} c_m^{-1} = 1$ then θ is non-orientable if it maps at least one a_i onto the trivial element of $\pi_1(P)$ or at least one b_i or c_i onto the non-trivial one.

Corollary 4.4.5 [J8]: *Let Σ and P be compact two-dimensional Riemannian manifolds, P being homeomorphic to the real projective plane. Let $\alpha \in [\Sigma, P]$ induce a non-orientable homomorphism*

$$\theta: \pi_1(\Sigma) \to \pi_1(P) = Z_2$$

which is not the zero homomorphism. Then α contains an energy-minimizing harmonic map.

In order to illuminate the significance of this result, we have to recall some details of the homotopy classification of maps from a surface into the real projective plane (due to Olum [Ol2]). While an oriented θ is always represented by infinitely many distinct homotopy classes, parametrized by their twisted degree [Ol1], in the non-orientable case it is convenient to look first at based homotopies, i.e. fixing $x_0 \in \Sigma$ and $p_0 \in P$ and requiring that all maps and homotopies map x_0 into p_0. Using the methods of obstruction theory, one sees that each non-orientable θ induces precisely two homotopy classes relative to the base points. The simplest case is where $\Sigma = P$ and θ is the zero homomorphism. One induced class contains the constant maps, and the other one is obtained from the first one by taking a small disk D in P and mapping it with degree one onto the 2-sphere S^2, ∂D going to a point, and then projecting S^2 in the standard way onto P. In other words, the second class is obtained from the first by attaching a sphere, and a geometric version of this process plays an important role in our existence proof. If we forget the base point, the two classes might become a single one, however. θ induces two distinct homotopy classes (without base point), if its degree mod 2 is 0, but only one, if the degree mod 2 is 1; cf. [Ol2].

Therefore, since, as mentioned before, θ is always induced by a minimizing map, we only obtain a new result compared to Corollary 4.4.1, if

$$\deg \theta = 0 \bmod 2$$

since then we prove that both classes, and not only one class, inducing θ can be represented (if θ is non-trivial) by an energy-minimizing map. On the other hand, it already follows from topological arguments that both these classes inducing θ can be represented by a map from Σ onto a closed geodesic in P. This was observed by Adams [Ad]. Of course, such a map can be made harmonic. It is not clear, however, whether—for an arbitrary metric on P–such a map is energy minimizing.

In contrast to the non-orientable case, as mentioned before, an oriented θ always induces infinitely many distinct homotopy classes likewise related by attaching spheres, and we can prove:

Corollary 4.4.6 [J8]: *Let Σ and P be as in Corollary 4.4.5, and let*

$$\theta: \pi_1(\Sigma) \to \pi_1(P)$$

be a non-trivial oriented homomorphism. Then there are at least two homotopy classes in $[\Sigma, P]$ inducing θ that contain an energy-minimizing harmonic map.

Note that an oriented θ is always non-trivial if Σ is non-orientable.

In contrast to Corollaries 4.4.5 and 4.4.6, there are also some non-existence results for harmonic maps into P; cf. Proposition 4.5.3 below.

There is still a number of homotopy classes where the existence question is open, most notably for oriented maps from non-orientable surfaces into P (except those covered by Corollary 4.4.6).

Finally, in the case where P is equipped with the standard metric of $\mathbb{RP}(2)$, we can improve Corollary 4.4.5 in the following way.

Corollary 4.4.7: *Let Σ be a closed surface and P be $\mathbb{RP}(2)$ equipped with its standard metric of constant positive curvature. Let $x_0 \in \Sigma$, $p_0 \in P$, and let*

$$\theta: \pi_1(\Sigma) \to \pi_1(P)$$

be a non-trivial homomorphism. Then there are at least two harmonic maps inducing θ and mapping x_0 onto p_0.

Note that these two harmonic maps may be homotopic, for example if θ is non-orientable with

$$\deg \theta = 1 \bmod 2. \tag{4.4.8}$$

They are not homotopic, however, if we fix the base points. Moreover, if θ is induced by only one class, it cannot be represented by a map onto a closed geodesic.

Proof of Corollaries 4.4.5 and 4.4.6: Let

$$E_\theta := \inf\{E_\alpha : \alpha_\# = \theta, \alpha \in [\Sigma, P]\}.$$

By Corollary 4.4.4, the minimum of energy has to be attained in any class α inducing θ that contains a map v with

$$E(v) < E_\theta + 2\,\text{Area}(P). \tag{4.4.9}$$

By definition of E_θ, we can always find at least one such class α_0 which hence contains a harmonic map $\tilde{u}$ with

$$E(\tilde{u}) = E_\theta. \tag{4.4.10}$$

We now assume that θ is non-trivial and hence $\tilde{u}$ is non-constant, and we want to add a sphere to $\tilde{u}$ and construct a map v satisfying (4.4.9). In the case when v is not homotopic to $\tilde{u}$, we therefore obtain a homotopy class different from α_0 that contains an energy-minimizing and thus harmonic map. As explained after the statement of Corollary 4.4.5, different classes inducing θ are related by attaching a sphere.

It thus remains to construct v.

Since θ is non-trivial, $\tilde{u}$ is not a constant map, and hence we can find some $x_1 \in \Sigma_1$ with

$$d\tilde{u}(x_1) \neq 0.$$

Let $k: U \to \mathbb{C}$ be a conformal chart on some neighbourhood U of $\tilde{u}(x_1)$ with $k\tilde{u}(x_1) = 0$.

By Taylor's Theorem, $k \circ \tilde{u} | \partial B(x_1, \varepsilon)$ is a linear map up to an error of order $O(\epsilon^2)$, i.e.

$$|k \circ \tilde{u}(x) - d(k \circ \tilde{u})(x_1)(x - x_1)| = O(\epsilon^2) \tag{4.4.11}$$

for $x \in \partial B(x_1, \epsilon)$.

We now look at conformal maps of the form

$$w = az + b/z \qquad a, b \in \mathbb{C}$$
$$a = a_1 + ia_2 \qquad b = b_1 + ib_2.$$

The restriction of such a map to a circle $\rho\,(\cos\theta + i \sin\theta)$ in $\mathbb{C}$ is given by

$$u = (a_1\rho + b_1/\rho)\cos\theta + (b_2/\rho - a_2\rho)\sin\theta$$
$$v = (a_2\rho + b_2/\rho)\cos\theta + (a_1\rho - b_1/\rho)\sin\theta$$

where $w = u + iv$.

Therefore, we can choose a and b in such a way that w restricted to this circle coincides with any prescribed non-trivial linear map. This map is non-singular if

$$\rho^4 \neq \frac{b_1^2 + b_2^2}{a_1^2 + a_2^2}.$$

W.l.o.g.

$$\rho^4 \leqslant \frac{b_1^2 + b_2^2}{a_1^2 + a_2^2} \tag{4.4.12}$$

(otherwise we perform an inversion at the unit circle).

Hence w can be extended as a conformal map from the interior of the circle $\rho(\cos\theta + \mathrm{i}\sin\theta)$ onto the exterior of its image. (If equality holds in (4.4.12) then this image is a straight line covered twice, and the exterior is the complement of this line in the complex plane).

We are now in a position to define v.

On $\Sigma_1 \backslash B(x_1, \epsilon)$ we put $v = \tilde{u}$.

On $B(x_1, \epsilon - \epsilon^2)$, we choose a conformal map w into the extended complex plane $\bar{\mathbb{C}}$ as above that coincides on $\partial B(x_1, \epsilon - \epsilon^2)$ with the linear map $(1-\epsilon)^{-1}\mathrm{d}(k \circ \tilde{u})(x_1)$. We identify $\bar{\mathbb{C}}$ with S^2 via stereographic projection and let

$$\pi : S^2 \to \mathbb{RP}(2)$$

be the standard projection, normalized by $\pi(0) = \tilde{u}(x_1)$. (We have, of course, identified the conformal structure on P with the standard $\mathbb{RP}(2)$ structure, so that $\pi : S^2 \to P$ is conformal). We then put $v = \pi \circ w$ on $B(x_1, \epsilon - \epsilon^2)$.

Finally, on $B(x_1, \epsilon)/B(x_1, \epsilon - \epsilon^2)$ we interpolate between $\tilde{u}$ and $\pi \circ w$. Introducing polar coordinates (r, φ), we define

$$f(\varphi) := k \circ u(\epsilon, \varphi)$$
$$g(\varphi) := \mathrm{d}(k \circ u)(x_1)(\epsilon, \varphi) = \frac{1}{1-\epsilon}\mathrm{d}(k \circ u)(x_1)(\epsilon - \epsilon^2, \varphi)$$

and

$$t(r, \varphi) := (f(\varphi) - g(\varphi))\frac{r}{\epsilon^2} + \frac{1}{\epsilon}(g(\varphi) - (1-\epsilon)f(\varphi))$$

and finally

$$v = k^{-1} \circ t \qquad \text{on } B(x_1, \epsilon) \backslash B(x_1, \epsilon - \epsilon^2).$$

$t(r, \varphi)$ coincides with $f(\varphi)$ and $g(\varphi)$, respectively for $r = \epsilon$ and $r = \epsilon - \epsilon^2$, respectively.

The energy of $t(r, \varphi)$ on the annulus $B(x_1, \epsilon) \backslash B(x_1, \epsilon - \epsilon^2)$ is given by

$$E(t) = \int_{r=\epsilon-\epsilon^2}^{\epsilon} \int_{\varphi=0}^{2\pi} \left(\frac{1}{\epsilon^4} |f(\varphi) - g(\varphi)|^2 + \frac{1}{r^2} \left| \left(\frac{r}{\epsilon^2} - \frac{1-\epsilon}{\epsilon} \right) f'(\varphi) \right.\right.$$
$$\left.\left. + \left(\frac{1}{\epsilon} - \frac{r}{\epsilon^2} \right) g'(\varphi) \right|^2 \right) r \, \mathrm{d}r \, \mathrm{d}\varphi.$$

Using (4.4.11) and $|f'(\varphi)| = O(\epsilon)$, $|g'(\varphi)| = O(\epsilon)$, we calculate

$$E(t) = O(\epsilon^3)$$

and hence also,

$$E(k^{-1} \circ t) = O(\epsilon^3). \tag{4.4.13}$$

We put $v = k^{-1} \circ t$ on the annulus $B(x_1, \epsilon) \backslash B(x_1, \epsilon - \epsilon^2)$. Therefore

$$E(v) = E(\tilde{u} | \Sigma \backslash B(x_1, \epsilon)) + E(k^{-1} \circ w | B(x_1, \epsilon - \epsilon^2)) + E(k^{-1} \circ t | B(x_1, \epsilon^2) \backslash B(x_1, \epsilon - \epsilon^2)).$$

The first term contributes (cf. (4.3.11))

$$E(\tilde{u}) - O(\epsilon^2) = E_\theta - O(\epsilon^2)$$

since $d\tilde{u}(x_1) \neq 0$, the second term at most

$$2\,\text{Area}(P)$$

since $\pi \circ w$ is conformal of degree 2, and the third term is controlled by (4.4.13).

Altogether

$$E(v) \leqslant E_\theta - O(\epsilon^2) + 2\,\text{Area}(P) + O(\epsilon^3)$$

and if we choose ϵ small enough, v satisfies (4.4.9) and the proof is complete.

q.e.d.

Proof of Corollary 4.4.7: For this proof we fix $x_0 \in \Sigma_2$ and $p_0 \in P$ and minimize the energy only among maps u with

$$u(x_0) = p_0. \tag{4.4.14}$$

Equation (4.4.14) does not constitute a restriction, since by assumption the isometry group of the image is transitive, and hence any $w: \Sigma_1 \to P$ can be transformed into a map satisfying (4.4.14) and having the same energy, simply by composition with an isometry of P that maps $w(x_0)$ onto p_0.

Therefore, we can work in the category of homotopy classes with base point, and the argument of the proof of Corollary 4.4.5 then shows that the minimum of energy is attained in at least two classes, since now attaching of a sphere leads to a new class in any case. This proves Corollary 4.4.7.

q.e.d.

We now turn to the case where $\partial\Sigma \neq \emptyset$, and a boundary condition is imposed on $\partial\Sigma$. We first treat the Dirichlet problem:

Theorem 4.4.1: *We assume that the assumptions of Theorem 4.2.1 hold, with the exception that Σ has a smooth*[3] *boundary $\partial\Sigma$. Let $g: \partial\Sigma \to N$ be continuous. Suppose*

$$h_0: \Sigma \times A \to N$$

[3] Note that smoothness can be obtained by conformal transformations, cf. chapter 3.

satisfies $h_0(y,t)=g(y)$ *for every* $y\in\partial\Sigma$, $t\in A$, *and*

$$E(h_0(\cdot,t))<\infty \text{ for every } t\in A.$$

Let H *be the class of all maps* $h:\Sigma\times A\to N$ *homotopic to* h_0 *with fixed boundary values* $h(y,t)=g(y)$ *for* $y\in\partial\Sigma$, $t\in A$.

$$\kappa:=\inf_{h\in H}\sup_{t\in A}E(h(\cdot,t)).$$

Suppose again that κ *cannot be attained on* ∂A.

Then there exist harmonic maps

$$u_0:\Sigma\to N \qquad \text{with } u_0|_{\partial\Sigma}=g$$

and possibly also some non-trivial conformal harmonic maps

$$u_i: S^2\to N \qquad (i=1,\ldots,m)$$

with

$$\sum_{i=0}^{m}E(u_i)=\kappa. \tag{4.4.15}$$

Furthermore, the same convergence properties as in Theorem 4.2.1 hold, and all blow-up points $x_1,\ldots,x_k$ *are contained in the interior of* Σ.

Proof: The proof is the same as the one of Theorem 4.2.1. We only have to note that if we want to perform a local replacement on some ball $B(x,r)$, with $B(x,r)\cap\partial\Sigma\neq\emptyset$, then we can use the Dirichlet data g on $B(x,r)\cap\partial\Sigma\neq\emptyset$ as boundary values for our replacement.

q.e.d.

Remark: Actually, for Theorem 4.4.1, we have only to assume that N is a complete Riemannian manifold of bounded geometry, but not necessarily compact. The Dirichlet data g then guarantee that any limit from a minimaxing sequence stays in a bounded subset of N. In this case, however, in general we can only expect

$$\sum_{i=0}^{m}E(u_i)\leqslant\kappa$$

as some of the minimal spheres may disappear at infinity.

Corollary 4.4.8 [L2]: *Let* Σ *be a compact surface with smooth boundary* $\partial\Sigma$, N *a complete Riemannian manifold of bounded geometry,* $g:\partial\Sigma\to N$ *be continuous with an extension* $\bar{g}\in C^0\cap H^{1,2}(\Sigma,N)$. *Then there exists a harmonic map* $u:\Sigma\to N$ *with*

$$u|_{\partial\Sigma}=g.$$

If $\pi_2(N)=0$, *then* u *is homotopic to* $\bar{g}$, *and, in general, it induces at least the same action on the fundamental group as* $\bar{g}$. *More generally, for a homotopy class* $\alpha_{\mathrm{rel}\,g}$

of maps $v: \Sigma \to N$ *with* $v|_{\partial\Sigma} = g$, *we put*

$$E_{\alpha_{\mathrm{rel}\,g}} := \inf_{v \in \alpha_{\mathrm{rel}\,g}} E(v)$$

and

$$E_g := \inf_{\alpha_{\mathrm{rel}\,g}} E_{\alpha_{\mathrm{rel}\,g}}.$$

Then any homotopy class with

$$E_{\beta_{\mathrm{rel}\,g}} < E_g + \mu \tag{4.4.16}$$

where μ *is defined in Lemma 4.2.1, contains a harmonic map* $u: \Sigma \to N$ *with* $u|_{\partial\Sigma} = g$.

Proof: Corollary 4.4.8 follows from Theorem 4.4.1 in the same way as Corollaries 4.4.1 and 4.4.4 follow from Theorem 4.2.1.

q.e.d.

As an application, we have the following result, proved in [BC] and [J3].

Corollary 4.4.9: *Let* Σ *be a compact surface with boundary* $\partial\Sigma$ *and let* S *be a Riemannian manifold homeomorphic to* S^2 *(the standard 2-sphere). Let* $g: \partial\Sigma \to S$ *be continuous, non-constant, and admitting an extension* $\bar{g} \in C^0 \cap H^{1,2}(\Sigma, S)$. *Then there are at least two homotopically different harmonic maps:*

$$u_1, u_2: \Sigma \to S \qquad \text{with } u_i|_{\partial\Sigma} = g \qquad (i = 1, 2)$$

both of them energy minimizing in their respective homotopy classes.

Proof. First of all, by Corollary 4.4.8, there is a harmonic map $u_1: \Sigma \to S$ which minimizes the energy among all maps with boundary values g. In order to conclude the existence of a second harmonic map $u_2: \Sigma \to S$ with boundary values g, we have only to check the existence of a map $v: \Sigma \to S$, $v|_{\partial\Sigma} = g$, which is not homotopic to u and satisfies

$$E(v) < E(u_1) + \mathrm{Area}(S) \qquad \text{(cf. Lemma 1.3.1)} \tag{4.4.17}$$

which is the relevant form of (4.4.16) in the present situation. As $g \neq$ const., there exists a point $x_0 \in \Sigma^0$ with

$$du_1(x_0) \neq 0. \tag{4.4.18}$$

We can then use the same construction as in the proof of Corollaries 4.4.5 and 4.4.6 to construct a map v with the required properties.

q.e.d.

We now discuss Plateau-type boundary conditions.

For this, let Σ again have a smooth boundary, and denote the components of $\partial\Sigma$ by $c_1, \ldots, c_k$. Let $\gamma_1, \ldots, \gamma_k$ be closed oriented Jordan curves of class C^2 in N. We then seek harmonic maps $u: \Sigma \to N$ which map c_i monotonically and orientation preservingly onto γ_i, and satisfy a free boundary condition there, i.e.

$$\varphi \mathrm{d}z^2 := (\langle u_x, u_x \rangle - \langle u_y, u_y \rangle - 2i\langle u_x, u_y \rangle)\,\mathrm{d}z^2$$

($z = x + \mathrm{i}y$ being a conformal parameter on Σ) is real on $\partial\Sigma$.

We want to imitate the proof of Theorem 4.2.1 to achieve this. A local harmonic replacement on a ball $B(x, r)$ this time has to be based on Lemma 4.1.7, in the case when $\mathring{B}(x, r) \cap \partial\Sigma \neq \varnothing$.

It turns out that now a blowing-up can take place not only at interior points, leading to non-trivial minimal spheres as before, but also at boundary points, e.g. at $x_0 \in c_i$, for some $i \in \{1, \ldots, k\}$. In such a case, one obtains a harmonic map of finite energy

$$v_i: H \to N \qquad (H := \{z = x + \mathrm{i}y, y > 0\})$$

which maps ∂H monotonically onto γ_i, and satisfies a free boundary condition on ∂H, i.e.

$$\left\langle \frac{\partial v_i}{\partial x}, \frac{\partial v_i}{\partial y} \right\rangle = 0 \qquad \text{on } \partial H.$$

Corollary 1.4.3 implies that v_i is conformal.

We choose a conformal transformation $k: H \to D$, with $k(\infty) = 1$. We put $u_i := v_i \circ k^{-1}$. Then u_i is harmonic and conformal in the interior of D and regular on D, *a priori* with the possible exception of the point $1 \in \partial D$. In two dimensions, weak solutions of finite energy extend through isolated points as weak solutions (see Lemma A.2 of the Appendix). Theorem 2.3.2 then implies that u_i is continuous everywhere. Moreover, by Theorems 2.5.1 and 2.5.3, $u_i \in C^{2,\alpha}(\mathring{D}, N) \cap C^{1,\alpha}(D, N)$. Therefore, u_i represents a minimal surface with Plateau boundary γ_i. After we split off the map u_i, the remaining maps map the boundary curve c_i onto a single point.

Together with these observations, the proof of Theorem 4.2.1 implies:

Theorem 4.4.2: *Let Σ be a surface with smooth boundary $\partial\Sigma$, consisting of components $c_1, \ldots, c_k$. Let N be a compact Riemannian manifold, and let $\gamma_1, \ldots, \gamma_k$ be oriented closed Jordan curves of class C^2 in N. Let A be as in Theorem 4.2.1, let*

$$h_0: \Sigma \times A \to N$$

map $c_i \times \{t\}$ monotonically and with preserved orientation onto γ_i (for every $t \in A$), and let H be the class of all such maps $h: \Sigma \times A \to N$ homotopic to h_0 (preserving the Plateau boundary condition). Let

$$\kappa := \inf_{h \in H} \sup_{t \in A} E(h(\cdot, t)).$$

Suppose again that κ cannot be attained on ∂A. Then there exists $0 \leqslant l \leqslant k$ so that, after renumbering of the curve c_i, γ_i, there exist conformal harmonic maps

$$u_1, \ldots, u_l: D \to N$$

with u_j mapping ∂D monotonically onto γ_j ($j=1,\ldots,l$)[4]*, a harmonic map*

$$u_0: \Sigma' \to N$$

mapping $c_1,\ldots,c_l$ to points on $\gamma_1,\ldots,\gamma_l$, resp., and with a Plateau condition satisfied on $c_{l+1},\ldots,c_k$ (i.e. c_p, $p=l+1,\ldots,k$, is mapped monotonically and with preserved orientation onto γ_p and with

$$\varphi\,\mathrm{d}z^2 := (\langle u_x, u_x\rangle - \langle u_y, u_y\rangle - 2\mathrm{i}\langle u_x, u_y\rangle)\,\mathrm{d}z^2$$

$z=x+\mathrm{i}y$ being a conformal parameter on Σ, real on c_p), and possibly some non-trivial conformal harmonic maps

$$v_1,\ldots,v_m: S^2 \to N$$

with

$$E(u_0) + \sum_{j=1}^{l} E(u_j) + \sum_{i=1}^{m} E(v_i) = \kappa \tag{4.4.19}$$

and similar convergence properties as in Theorem 4.2.1.

In order to draw some consequences, let us introduce the following notation.

Let $g \in C^0 \cap H^{1,2}(\Sigma, N)$ map c_i monotonically onto γ_i with preserved orientation, and put

$$\theta := g_\#: \pi_1(\Sigma) \to \pi_1(N).$$

Let

$$d(\Sigma,\theta,\gamma) := \inf\{E(u): u: \Sigma \to N,\ u_\# = \theta,\ u \text{ maps } c_i \text{ onto } \gamma_i \text{ with preserved orientation } (i=1,\ldots,k)\} \tag{4.4.20}$$

and

$$d^*(\Sigma,\theta,\gamma) := \inf\Big\{E(v) + \sum_{i=1}^{l} E(v_i): v: \Sigma \to N, \text{ where } v \text{ maps some boundary curves } c_i \text{ to points on } \gamma_i, \text{ say } c_1,\ldots,c_l, \text{ with } 1 \leqslant l \leqslant k, \text{ where } \gamma_1,\ldots,\gamma_l \text{ are homotopically trivial in } N, \text{ with } v_\# = \theta, \text{ and } v \text{ maps } c_{l+1},\ldots,c_k \text{ monotonically with preserved orientation onto } \gamma_{l+1},\ldots,\gamma_k;\ v_i: D \to N \text{ maps } \partial D \text{ monotonically onto } \gamma_i,\ i=1,\ldots,l\Big\}. \tag{4.4.21}$$

If all curves $\gamma_1,\ldots,\gamma_k$ are homotopically non-trivial in N, we put

$$d^*(\Sigma,\theta,\gamma) := \infty.$$

For a single Jordan curve $\gamma_0 \subset N$, we also put

$$d(D,\gamma_0) := \inf\{E(v): v: D \to N \text{ maps } \partial D \text{ monotonically onto } \gamma_0\} \tag{4.4.22}$$

[4] If $l=0$, no such map occurs.

and again

$$d(D, \gamma_0) := \infty \qquad \text{if } \gamma_0 \text{ is homotopically non-trivial.}$$

We then have the following result of [GJ]:

Corollary 4.4.10: *If under the assumption of Theorem 4.4.2,*

$$d(\Sigma, \theta, \gamma) < d^*(\Sigma, \theta, \gamma) \tag{4.4.23}$$

then there exists a harmonic map $u: \Sigma \to N$ with $u_\# = \theta$ which maps c_i monotonically and with preserved orientation onto γ_i $(i = 1, \ldots, l)$. In particular, if $\partial\Sigma$ is connected and γ_0 is a Jordan curve of class C^2 in N, then there exists a harmonic map $u: \Sigma \to$ with $u_\# = \theta$, mapping $\partial\Sigma$ monotonically onto γ_0 provided

$$d(\Sigma, \theta, \gamma_0) < d(D, \gamma_0). \tag{4.4.24}$$

This latter condition is in particular satisfied if γ_0 is homotopically non-trivial in N.

Proof: This follows directly from Theorem 4.4.2.

q.e.d.

We also have:

Corollary 4.4.11: *Let γ be a closed Jordan curve in N, a compact Riemannian manifold, and suppose N contains no minimal spheres (for example, if N has non-positive sectional curvature). Let $u_1, u_2: D \to N$ be homotopic minimal surfaces with boundary γ. If both of them are strict relative minima (w.r.t the C^0 or $H^{1,2}$ topology), then there exists a third minimal surface $u_3: D \to N$ with boundary γ with*

$$E(u_3) > \max(E(u_1), E(u_2)) \tag{4.4.25}$$

and u_3 is unstable.

Proof: It is not difficult to see from the arguments of Section 4.2 that if both u_1 and u_2 are strict relative minima, then the minimax value κ (as in Theorem 4.4.2, with $A = [0, 1]$, h_0 a homotopy between u_1 and u_2) satisfies

$$\kappa > \max(E(u_1), E(u_2)) \tag{4.4.26}$$

and the claim follows again from Theorem 4.4.2.

q.e.d.

Remarks: (1) We shall see a different proof of such a result in Section 4.6, where also (4.4.26) will be discussed in detail.

(2) For Corollary 4.4.10 it again suffices to assume that N is a complete Riemannian manifold of bounded geometry, not necessarily compact. For Corollary 4.4.11 we need slightly more, but it is sufficient to assume that $u_1(D)$ and $u_2(D)$ are contained in a bounded strictly convex subset of N, without

further restrictions on N. Therefore, the result applies in particular to $\mathbb{R}^d$. For the proof of this assertion, one has only to note that all replacement arguments can be carried out inside this strictly convex set. 'Convexity' here refers to the boundary of this set.

It also suffices then to assume that there are no minimal spheres inside this convex set. This is, for example, satisfied if this set is of the form $B(p,r)$, for some $p \in N$ with $r < \min(i(p), \pi/2\kappa)$, κ^2 being an upper bound for the sectional curvature of N. This latter case, as well as the case of a non-positively curved N, is due to Ströhmer [Sö1, Sö2].

Let us finally discuss the general free boundary problem, where a compact submanifold M of N of class C^2 is given. In this case, one has to use Lemma 4.1.5 for local replacements. Again, as in the case of the Plateau-type boundary conditions (which, apart from the additional requirement of monotonicity, are a special case of a free boundary condition), at the boundary minimal disks can split off. The general existence result consequently reads:

Theorem 4.4.3: *Let Σ be a surface with smooth boundary $\partial\Sigma$ with components $c_1, \ldots, c_k$, N a compact Riemannian manifold, $M \subset N$ a compact submanifold of class C^2 (not necessarily connected if $k \geqslant 2$), A as in Theorem 4.2.1. Let*

$$h_0: \Sigma \times A \to N$$

map $c_i \times \{t\}$ onto a given component of M, for all $t \in A$ ($i = 1, \ldots, k$), let H be the class of all such maps homotopic to h_0 (with preserved free boundary condition). Let

$$\kappa := \inf_{h \in H} \sup_{t \in A} E(h(\cdot t))$$

where we assume that κ cannot be attained on ∂A. Then there exists l with $0 \leqslant l \leqslant k$ so that after renumbering of the curves c_i, there exist conformal harmonic maps

$$u_1, \ldots, u_l: D \to N$$

mapping ∂D into M and with $u_j(D)$ meeting M orthogonally along $u_j(\partial D)$ ($j = 1, \ldots, l$)[5], a harmonic map

$$u_0: \Sigma \to N$$

with $u_0(\partial\Sigma) \subset M$, and $u_0(\Sigma)$ meeting M orthogonally along $u_0(\partial\Sigma)$, and with

$$\varphi\, \mathrm{d}z^2 := (\langle u_x, u_x \rangle - \langle u_y, u_y \rangle - 2\mathrm{i}\langle u_x, u_y \rangle)\mathrm{d}z^2$$

($z = x + \mathrm{i}y$ being a local conformal parameter on Σ) real on $\partial\Sigma$, as well as possibly some non-trivial conformal maps

$$v_1, \ldots, v_m: S^2 \to N$$

[5] If $l = 0$, so such map occurs.

with

$$E(u_0) + \sum_{j=1}^{l} E(u_j) + \sum_{i=1}^{m} E(v_i) = \kappa \qquad (4.4.27)$$

and similar convergence properties as in Theorem 4.2.1.

Remarks: 1) Here, we have not provided an argument to show that u_0 is smooth at those boundary points where a minimal disk splits off. If u_0 is conformal, i.e. a minimal surface, however, smoothness is a consequence of Theorem 2.3.2, and this case is the only one that will be further considered here. 2) As before, N need not to be compact; it suffices that M and the image of h_0 are contained in a bounded strictly convex subset B of N.

Let us state some consequences of Theorem 4.4.3. The first one is due to Courant ([C4]) in a special case and to Meeks and Yau [MY2]:

Corollary 4.4.12: *Let N and M be as above, and suppose there exists a map $g \in C^0 \cap H^{1,2}(D, N)$ with $g(\partial D) \subset M$ and that $g(\partial D)$ represents a non-trivial element of $\pi_1(M)$.*

Then there exists a conformal harmonic map $u: D \to N$ with $u(\partial D) \subset M$ representing a non-trivial element of $\pi_1(M)$. u minimizes the energy among all such maps, and $u(D)$ hits M orthogonally along $u(\partial D)$.

The next result is due to Tolksdorf [To2] and Ye [Ye2]:

Corollary 4.4.13: *Let M be a compact C^2-submanifold of N, and suppose M is contained in a bounded strictly convex set B (of course, we can choose $B = N$ if N is compact). Let γ be a curve in M which is non-zero in $\pi_1(M)$, but zero in $\pi_1(B)$. Then there exist minimal surfaces $u_1, \dots, u_l: D \to B$ with free boundary M with*

$$\gamma = u_1(\partial D) + \cdots u_l(\partial D) \qquad \text{as free homotopy classes in } M. \qquad (4.4.28)$$

Furthermore, if we put

$$E_\beta := \inf\{E(v): v: D \to N,\ v(\partial D) \subset M \text{ represents } \beta\}$$

for a free homotopy class of curves in M, we have

$$E_{[\gamma]} = \sum_{i=1}^{l} E_{[u_i(\partial D)]} \qquad (4.4.29)$$

where $[\gamma]$ denotes the free homotopy class of γ in M. In particular if

$$E_{[\gamma]} < E_{[\gamma_1]} + E_{[\gamma_2]} \qquad (4.4.30)$$

whenever $[\gamma_1] + [\gamma_2] = [\gamma]$ as free homotopy classes, there exists a minimal surface $u: D \to N$ with free boundary M with

$$u(\partial D) \in [\gamma].$$

Proof: The argument is similar to that of Corollary 4.4.3. First of all, splitting off of a minimal sphere does not affect the free homotopy class $v(\partial D)$ for $v: D \to N$ with $v(\partial D) \subset M$. Moreover, splitting off of a minimal disk at the boundary decomposes the free homotopy class into the sum of two others. This should be clear from the analogous behaviour when a minimal sphere splits off in the interior; cf. the argument in Section 4.2.

The remaining argument can then be taken over from the proof of Corollary 4.4.3.

q.e.d.

Remark: Obviously, Corollary 4.4.12 is a special case of Corollary 4.4.13, because under the assumption of Corollary 4.4.12 there must exist a non-trivial free homotopy class satisfying (4.4.30).

The next result generalizes a theorem of Struwe ([St2], where it is proved by the method of [SKU1]).

Corollary 4.4.14: *Let M be a compact C^2-submanifold of the three-dimensional Riemannian manifold N, contained in a bounded strictly convex subset B, and suppose M bounds a ball in N. Assume that there are no minimal spheres in B. Then there exists a non-trivial minimal surfaces $u: D \to B$ with free boundary M.*

Proof: After a diffeomorphism, $M = S^2 \subset \mathbb{R}^3$ bounds the unit ball $A = B(0, 1)$. We define

$$h_0: D \times [0, 1] \to B(0, 1)$$

in the following way.

Let $\rho(t)$ be the radius of the disk $B(0, 1) \cap \{x^3 = 2t - 1\} (x = (x^1, x^2, x^3) \in \mathbb{R}^3)$, and put

$$h_0(y^1, y^2, t) := (\rho(t)y^1, \rho(t)y^2, 2t - 1)$$

$((y^1, y^2) \in D,\ t \in [0, 1])$. In particular, $h_0(D \times \{0\})$ and $h_0(D \times \{1\})$ are points in M. Let H be the class of all such maps $h: D \times [0, 1] \to B$ (in particular, we require that $D \times \{0\}$ and $D \times \{1\}$ are always mapped to points in M) which are homotopic to h_0.

It is an easy consequence of the isoperimetric inequality that

$$\kappa := \inf_{h \in H} \sup_{t \in [0,1]} E(h(\cdot, t)) > 0.$$

Theorem 4.4.3 then yields the result.

q.e.d.

Remark: In the present monograph, the equation of the geometric regularity (immersion and embeddedness properties) of minimal surfaces is not addressed. Some references to the literature might be appropriate at this point, however.

Interior branch points for area-minimizing minimal surfaces can be ruled out by the work of Osserman [0], Gulliver [G], Alt [Alt] and Gulliver, Osserman and Royden [GOR]. Boundary branch points of area-minimizing surfaces are known to be absent only if the boundary of the surface lies on the boundary of a convex subset of N, or if the boundary is real analytic [GL]. Hardt and Simon [HS] proved the important result that a homologically trivial $C^{1,\alpha}$ Jordan curve in a three-dimensional manifold of bounded geometry always bounds an embedded oriented minimal surface, possibly of higher genus. Embedded solutions of controlled topological type were produced under suitable convexity hypotheses for Plateau and free boundary problems by Gulliver and Spruck [GS], Tomi and Tromba [TT1], Almgren and Simon [AS], Meeks and Yau [MY1, MY2, MY3], and by the author [J9 (Part I), J12]. Some of these techniques were extended to produce embedded closed minimal surfaces in three-dimensional Riemannian manifolds in [MSY] and [FHS]. All papers mentioned so far obtained their solutions by suitable minimizing techniques. Unstable embedded minimal surfaces were obtained by Pitts [P], Simon and Smith [SSm], Pitts and Rubinstein [PR], Smyth [S4] and the author [J9 (Parts II and III), J16, J17].

Also, we should point out that the physical intuition underlying free boundary problems requires that the free boundary M is impenetrable for the minimal surface. This is not reflected in the setting presented here. If one does not impose convexity conditions as in [MY2] on the free boundary, one needs to consider surfaces of varying connectivity[6], in the same way that Hardt and Simon [HS] had to consider surfaces of varying genus when considering the Plateau problem. This was carried out by Tolksdorf [To1] in the parametric context without addressing the problem of embeddedness of the solution, and by the author [J9, J12] in the context of embedded surfaces. We also mention the earlier work of Taylor [Ty] who admitted surfaces not only of varying connectivity but of varying genus as well.

4.5. Non-existence results. Existence of maps with holomorphic quadratic differentials

We already saw (cf. Corollary 1.4.2):

Proposition 4.5.1: *There is no non-constant harmonic map from the unit disk onto S^2 mapping ∂D onto a single point.*

There are also certain cases where one can only rule out the existence of an energy-minimizing map:

Proposition 4.5.2: (*a*) *Let Σ be a closed surface of positive genus, $\alpha \in [\Sigma, S^2]$ a homotopy class of degree ± 1. Then α contains no energy minimizing map.*

(*b*) *Let $g: \partial D \to S^2$ be a parametrization of a great circle by arclength. Then*

[6] of the type considered in Theorem 3.2.1

there exist precisely two energy-minimizing extensions $u_{\pm}: D \to S$ *of* g. *These maps have degree* ± 1. *In particular, there is no energy-minimizing extension of* g *of higher degree.*

Part (*a*) is due to Lemaire [L1] and part (*b*) to Brézis and Coron [BC] and Jost [J3]. The existence result contained in (*b*) follows from Corollary 4.4.9. The non-existence proofs depend on

Lemma 4.5.1: *For every* $\epsilon > 0$ *there exists a map* $k: D \to S^2$ *of degree* 1, *mapping* ∂D *onto some point* $p \in S^2$ *and satisfying*

$$E(k) \leqslant \text{Area}(S^2) + \epsilon. \tag{4.5.1}$$

Such a map k *is called* ϵ*-conformal.*

Proof of Lemma 4.5.1: We divide S^2 into $B(p, \delta)$ and $S^2 \backslash B(p, \delta)$. All the maps to follow will be understood to be equivariant w.r.t. the rotations of D and to those of S^2 leaving p fixed.

First of all, for sufficiently small δ, we can map $\{z \in \mathbb{C}: \frac{1}{2} \leqslant z \leqslant 1\}$ onto $B(p, \delta)$, $\{|z| = \frac{1}{2}\}$ going to $\partial B(p, \delta)$ and $\{|z| = 1\}$ going onto p with energy smaller than ϵ. On the other hand, $\{z \in \mathbb{C}: |z| \leqslant \frac{1}{2}\}$ can be mapped conformally onto $S^2 \backslash B(p, \delta)$, $\{|z| = \frac{1}{2}\}$ going again onto $\partial B(P, \delta)$, and the energy of this map, since conformal, equals the area of its image and is hence smaller than the area of S^2. This proves the claim.

q.e.d.

Proof of Proposition 4.5.2: (*a*) Let B be any disc in Σ and let $\epsilon > 0$. Since B is conformally equivalent to the unit disk D, Lemmata 4.5.1. and 1.3.2 imply that we can find a map $k: B \to S^2$ of degree ± 1, mapping ∂D onto some point p, and satisfying (4.5.1). If we extend k to all of Σ by mapping $\Sigma \backslash B$ onto p, then $k: \Sigma \to S^2$ still satisfies (4.5.1) and is of degree ± 1.

If there were an energy-minimizing h in α, then h would consequently have to satisfy

$$E(h) = \text{Area}(S^2)$$

by Lemma 1.3.2 and would hence have to be conformal, by Lemma 1.3.2 again. On the other hand, a conformal map of degree ± 1 has to be a diffeomorphism which is not possible since Σ is by assumption not homeomorphic to S^2.

(*b*) Let $k: D \to S^2$ be a conformal map mapping D onto the upper hemisphere with boundary values g. In particular, k is equivariant with respect to the rotations of D and S^2 (the latter ones leaving the north and south pole of S^2 fixed).

We choose the orientation of S^2 in such a way that the Jacobian of k is positive.

Let $D(0, r)$ be the plane disk with centre 0 and radius r (i.e. $D = D(0, 1)$).

Let h_r be a map from $D(0, r)$ onto S^2 which maps $\partial D(0, r)$ onto the north pole, is injective in the interior of $D(0, r)$ and has a positive Jacobian there, and is ϵ-conformal. We introduce polar coordinates (ρ, ϕ) on D and define for $0 < r < 1$

the mapping k_r by

$$k_r(\rho, \phi) = \begin{cases} k\left(\dfrac{1}{1-r}\rho + \dfrac{r}{r-1}, \phi\right) & \text{if } r \leqslant \rho \leqslant 1 \\ h_r(\rho, \phi) & \text{if } 0 \leqslant \rho \leqslant r. \end{cases}$$

Using Lemma 4.5.1 it is easy to see that the energy of k_r can be made arbitrarily close to 6π if we choose $r > 0$ sufficiently small.

On the other hand, 6π is just the area of the image of k_r, counted with multiplicity. Hence, if there is an energy-minimizing map homotopic to k_r, its energy has to be 6π, and it therefore has to be conformal. Since the boundary values are equivariant, this conformal map itself has to be equivariant (otherwise there would exist infinitely many homotopic conformal maps with the same boundary values, which is not possible). This, however, implies that it would have to collapse a circle in D to a point which is not possible for a conformal map. Hence there is no energy-minimizing map homotopic to k_r. Other homotopy classes of degree $\neq \pm 1$ are treated by a similar construction.

q.e.d.

Remark: In certain cases, however, in the situation of (*a*), there exist harmonic, but not energy-minimizing, maps in α; cf. [L1] and [Cr]. These results are only valid for special conformal structures on Σ, and the general case is unknown, as is the existence question in case (*b*).

The following non-existence results are due to Eells and Wood [EW] and Eells and Lemaire [EL3]. (The general theorem of Eells and Wood will be treated in Section 5.2)

Proposition 4.5.3: *Let T^2 be the two-dimensional torus and P^2 be the real projective plane (with any metrics), and $\pi: S^2 \to P^2$ be a covering map. Let $f: T^2 \to S^2$ be a map of degree ± 1, and let $g: P^2 \to S^2$ be a homotopically non-trivial map. Then*

(*a*) $f: T^2 \to S^2$
(*b*) $\pi \circ f: T^2 \to P^2$
(c) $g: P^2 \to S^2$
(d) $\pi \circ g: P^2 \to P^2$

(domain and image may have different metrics) are not homotopic to any harmonic map.

Proof: (*a*) Assume that $u: T^2 \to S^2$ is harmonic and homotopic to f. By Lemma 1.3.3, in Euclidean coordinates on T^2

$$|u_x|^2 - |u_y|^2 - 2i\langle u_x, u_y \rangle =: a + ib$$

is constant. $a + ib = 0$ is not possible since that would mean that u is a $\pm$ holomorphic map of degree ± 1, and hence a diffeomorphism. At a point where

$u_x = 0$ we would have $a \leqslant 0$, $b = 0$, and at a point where $u_y = 0$, $a \geqslant 0$, $b = 0$. Since $a + ib \neq 0$, either u_x or u_y cannot have a zero on T^2; w.l.o.g. $u_x \neq 0$. If J denotes multiplication by $\sqrt{-1}$ in TS^2 (in local coordinates $J(u^1 + iu^2) = -u^2 + iu^1$), u_x and Ju_x both yield non-zero sections of $u^{-1}TS^2$. Therefore, $u^{-1}TS^2$ is the trivial bundle over T^2, hence the first Chern class $c_1(u^{-1}TS^2)$ vanishes. Thus

$$\begin{aligned} 0 = c_1(u^{-1}TS^2) &= \deg(u) c_1(TS^2) \\ &= 2\deg(u) \end{aligned}$$

and $\deg(u) = 0$ which contradicts the assumption (cf. [EL4]).
(*b*) We can assume that π is a local isometry, thereby reducing (*b*) to (*a*).
(*c*) Let $p: S^2 \to P^2$ again be a local isometry. The homomorphism $(g \circ p)^*: H^2(S^2) \to H^2(S^2)$ induced by $g \circ p$ factors through $H^2(P^2) = \mathbb{Z}_2$. Hence $\deg(g \circ p) = 0$. Since any harmonic self-map of S^2 is conformal (Corollary 1.4.1), $g \circ p$ is only homotopic to constant harmonic maps. Thus g, being homotopically non-trivial, cannot be homotopic to a harmonic map.
(*d*) is again reduced to (*c*).

q.e.d.

In contrast to Proposition 4.5.3, the author showed in [J7]:

Theorem 4.5.1: *Let Σ_1 and Σ_2 be compact two-dimensional Riemannian manifolds. Then any continuous map $\phi: \Sigma_1 \to \Sigma_2$ is homotopic to a map u for which*

$$(|u_x|^2 - |u_y|^2 - 2i\langle u_x, u_y \rangle)\, dz^2$$

is a holomorphic quadratic differential.

Proof: W.l.o.g., we can assume $\Sigma_1 \neq S^2$, as otherwise only for Σ_2 diffeomorphic to S^2 or $\mathbb{RP}(2)$, non-trivial homotopy classes occur, and these are represented by conformal maps (as a consequence of Theorem 3.1.1).

If γ is a sufficiently small Jordan arc on Σ_1, it divides Σ_1 into two pieces, one of them being simply connected. We call such a piece a disk. We define, for $\epsilon > 0$

$$\begin{aligned} \Delta(\epsilon) := \{ & v \in H^{1,2} \cap C^0(\Sigma_1, \Sigma_2),\ v \text{ homotopic to } \phi, \text{ and having the} \\ & \text{following property: if } B \text{ is a disk in } \Sigma_1 \text{ with } v(\partial B) \subset B(p, \epsilon) \\ & (\text{for some } p \in \Sigma_2), \text{ then } v(B) \subset B(p, \epsilon) \}. \end{aligned}$$

One checks that for sufficiently small $\epsilon > 0$, $\Delta(\epsilon) \neq \emptyset$. We now need:

Lemma 4.5.2: *Energy-bounded subsets of $\Delta(\epsilon)$ are equicontinuous.*

Proof: This is a direct consequence of Lemma 3.1.1 and the definition of $\Delta(\epsilon)$.

q.e.d.

We can now conclude the proof of Theorem 4.5.1.

We take an energy-minimizing sequence in $\Delta(\epsilon)$. By Lemma 4.5.2, after selection of a subsequence, it converges uniformly to a map u. We can also assume that we have weak $H^{1,2}$-convergence. Thus, by the lower semicontinuity of E under weak $H^{1,2}$-convergence, u minimizes E in $\Delta(\epsilon)$.

If now $\sigma_t:\Sigma_1\to\Sigma_2$ is a family of diffeomorphisms, depending differentiably on t, with $\sigma_0=\mathrm{id}$, then also $u\circ\sigma_t\in\Delta(\epsilon)$, and hence by the energy-minimizing property of u

$$\frac{\mathrm{d}}{\mathrm{d}t}E(u\circ\sigma_t)|_{t=0}=0. \tag{4.5.2}$$

Lemma 1.2.4 then implies that

$$(|u_x|^2-|u_y|^2-2\mathrm{i}\langle u_x,u_y\rangle)\,\mathrm{d}z^2$$

is holomorphic.

q.e.d.

Remark: If, for example, $\Sigma_1=T^2$ and $\Sigma_2=S^2$ and ϕ is of degree 1, Proposition 4.5.3 implies that the map u constructed in Theorem 4.5.1 cannot be harmonic. We already saw, at the end of Section 1.3, an explicit example of a map for which the associated quadratic differential is holomorphic but which is not harmonic. Therefore, the definition of a harmonic map as one for which this quadratic differential is holomorphic, employed in [GR], [Re], [RS] is different from our definition.

4.6. Another proof of the existence of unstable minimal surfaces

In this section, we shall be concerned with:

Theorem 4.6.1: *Let $\gamma\subset\mathbb{R}^d$ be an oriented Jordan curve of class C^2, and let $u_0,u_1:D\to\mathbb{R}^d$ be minimal surfaces solving the Plateau problem for γ. We suppose that u_0 and u_1 are geometrically distinct, i.e. $u_0(D)\neq u_1(D)$.*
(a) If both u_0 and u_1 are strict local minima, in either $H^{1,2}$ or C^0, then there exists another minimal surface $u:D\to\mathbb{R}^d$ with Plateau boundary γ which is distinct from both u_0 and u_1. u is unstable.
(b) If u_0 and u_1 are both global minima, then either there exists on unstable u as above or there exists a continuum of minimal surfaces $u_t:D\to\mathbb{R}^d$, solving the Plateau problem and connecting u_0 and u_1, with $E(u_t)=\mathrm{const}$.

Proof: We let $h:D\times[0,1]$ be a continuous path in $H^{1,2}\cap C^0$ with

$$\begin{aligned} h(x,0)&=u_0(x)\\ h(x,1)&=u_1(x)\end{aligned}\qquad \text{for } x\in D \tag{4.6.1}$$

and the property that, for each t,

$$h|_{\partial D\times\{t\}}\text{ maps }\partial D\text{ monotonically onto }\gamma \tag{4.6.2}$$

(preserving the orientation of γ).

We shall perform two kinds of operations on each map $h(\cdot, t)$.

Step (1): Let $\omega \in S^{d-1} \subset \mathbb{R}^d$. For a given map $f \in C^0(D, \mathbb{R}^d)$, we put

$$f^\omega(x) := f(x) \cdot \omega \in \mathbb{R}$$

and perform the following operation.

Let $(\sigma_k)_{k \in \mathbb{N}}$ be dense in $\mathbb{R}$. For each component Ω_1 of $\{x \in D: f^\omega(x) > \sigma_1\}$ with $\Omega_1 \cap \partial D = \emptyset$, we replace f^ω by σ_1 on Ω_1 (note $f^\omega(x) = \sigma_1$ for $x \in \partial\Omega_1$), and likewise for each component Ω_1' of $\{x \in D: f^\omega(x) < \sigma_1\}$ with $\Omega_1' \cap \partial D = \emptyset$. We obtain a function g_1^ω. We repeat the construction with σ_2 instead of σ_1 and g_1^ω instead of f^ω. Iteratively, we obtain functions g_k^ω. These functions are equicontinuous. If $f^\omega \in C^\alpha$, $0 \leqslant \alpha \leqslant 1$, or $f^\omega \in H^{1,2}$, then so are all g_k^ω, and all corresponding norms are non-increasing.

As $k \to \infty$, g_k^ω converges uniformly to a function g^ω with the property that for each $B \subset D$

$$\min_B g^\omega = \min_{\partial B} g^\omega, \max_B g^\omega = \max_{\partial B} g^\omega. \tag{4.6.3}$$

Moreover,

$$g^\omega|_{\partial D} = f^\omega|_{\partial D}. \tag{4.6.4}$$

We rotate our coordinate system in $\mathbb{R}^d$ so that ω becomes the first unit vector; we then replace

$$f = (f^\omega, f^2, \ldots, f^d)$$

by

$$\tilde{g} = (g^\omega, f^2, \ldots, f^d).$$

We claim that if f_t, $t \in [0, 1]$, is a continuous path of maps in $C^0(D, \mathbb{R}^d)$, then the corresponding modified maps $\tilde{g}_t$ likewise give rise to a continuous path in C^0. This is a consequence of the following observation. Let $\epsilon > 0$ be given. If there exists, for $t_0 \in [0, 1]$, $x_0 \in D$, and a non-empty component $\Omega_{t_0,k}$ of $\{x \in D: g^\omega_{t_0,k-1} > \sigma_k\}$ with $x_0 \in \Omega_{t_0,k}$ and $\Omega_{t_0,k} \cap \partial D = \emptyset$ (here $g^\omega_{t_0,k}$ is the modification of $f^\omega_{t_0}$ at the kth step), then for every $t \in [0, 1]$ which is sufficiently close to one, there exists $\sigma_{k'}$ with $\sigma_k \leqslant \sigma_{k'} < \sigma_k + \epsilon$ and a non-empty component $\Omega_{t,k'}$ of $\{x \in D: g_{t,k'-1} > \sigma_{k'}\}$ with $x_0 \in \Omega_{t,k'}$, and with $\Omega_{t,k'} \cap \partial D = \emptyset$. (Note, however, that the corresponding functions $g^\omega_{k,t}$ for $k < \infty$ need not be continuous in t.) Similarly, if $(f_t)_{t \in [0,1]}$ is a continuous path in $C^{0,\alpha}$, $0 \leqslant \alpha \leqslant 1$, or in $C^0 \cap H^{1,2}$, then so is $(\tilde{g}_t)$.

We now choose an everywhere dense sequence $(\omega_k)_{k \in \mathbb{N}}$ in S^{d-1} and, given f as above, we first obtain a map g_1 by performing the above construction with ω_1 in place of ω. Having constructed g_k we construct g_{k+1} by performing the above construction with g_k instead of f and ω_{k+1} in place of ω_k.

The maps g_k are again equicontinuous, and hence a subsequence (g_{k_l}) converges uniformly to a map g. We claim that, for any $\omega \in S^d$, with $g^\omega(x) = g(x) \cdot \omega$, g^ω satisfies (4.6.3) and (4.6.4). The second property is clear, and for the first one we note that otherwise there would exist some $\omega_{k'} \in (\omega_k)$ with

$$g_{\omega_{k'}} \neq g$$

where $g_{\omega_{k'}}$ is the map constructed as above with g instead of f and $\omega_{k'}$ in place of ω. But then also all further such modifications of $g_{\omega_{k'}}$ are different from g, and this is not compatible with the convergence of $(g_{k_l})_{l\in\mathbb{N}}$ to g. Actually, this argument also implies that the whole sequence (g_k), and not only the subsequence (g_{k_l}), converges to g. We have proved:

Lemma 4.6.1: *The map g constructed from f as above coincides with f on ∂D and satisfies the convex hull property* (4.6.3) ($g^{\omega}(x) := g(x)\cdot\omega$ *for* $\omega \in S^{d-1}$).

Again if $(f_t)_{t\in[0,1]}$ is a continuous path in $C^{0,\alpha}$, $0 \leqslant \alpha \leqslant 1$, or in $C^0 \cap H^{1,2}$, then so is the modified path $(g_t)_{t\in[0,1]}$. We also have

$$\|g\|_{C^\alpha} \leqslant \|f\|_{C^\alpha} \tag{4.6.5}$$

and

$$E(g) \leqslant E(f) \qquad (\text{when } f\in H^{1,2}) \tag{4.6.6}$$

with strict inequality in (4.6.6) if $f \neq g$.

The construction of step (1) implies that we can assume that each $h(\cdot,t)$ of our path satisfies the convex hull property (4.6.3), noting that the minimal surfaces $u_0 = h(\cdot,0)$ and $u_1 = h(\cdot,1)$ satisfy this property automatically by the maximum principle and are hence left unchanged by our construction.

Step (2): The second operation involves harmonic replacement on small disks or half disks. We first choose $\delta > 0$, depending on the Jordon curve γ, so small that the following is correct.
Let

$$\begin{aligned} D_+ &:= \{x+iy\in\mathbb{C} : x^2+y^2 \leqslant 1, y \geqslant 0\} \\ \partial_+ D_+ &:= \{x+iy\in\mathbb{C} : x^2+y^2 = 1, y \geqslant 0\} \\ \partial_0 D_+ &:= \{x+iy\in\mathbb{C} : x^2+y^2 < 1, y = 0\}. \end{aligned}$$

For $p_0\in\mathbb{R}^d$, put $\gamma_\delta := \gamma\cap B(p_0,\delta)$. Let $f:\partial_+D_+ \to B(p_0,\delta)$ be a continuous map with $f(1,0)\subset\gamma_\delta$ and $f(-1,0)\subset\gamma_\delta$. Then there exists (at most) one harmonic map

$$u: D_+ \to \mathbb{R}^d$$

with

$$u|_{\partial_+ D_+} = f \qquad u(\partial_0 D_r) \subset \gamma_\delta$$

satisfying the free boundary condition here. The existence of such a $\delta > 0$ is a consequence of Theorem 2.2.2, as the assumption can always be satisfied on sufficiently small portions of γ by rescaling. We furthermore require that $\gamma_\delta (= \gamma\cap B(p_0,\delta))$ is connected for each $p_0\in\mathbb{R}^d$ (note $\gamma\in C^2$). We can furthermore assume

$$E(h(\cdot,t)) \leqslant K \tag{4.6.7}$$

for some fixed K, and that the same holds for all subsequent modifications of

h by the construction of (1); cf. (4.6.6). We define $\eta > 0$ via

$$2\pi K^{1/2}(\log 1/\eta)^{-1/2} = \delta \tag{4.6.8}$$

w.l.o.g. $\eta < 1$.

By Lemma 3.1.1, for each $t \in [0, 1]$, and each $x_0 \in D$, we can find

$$r_t \in (\eta, \sqrt{\eta})$$

with the property that, for all $x_1, x_2 \in \partial B(x_0, r_t) \cap D$,

$$|h(x_1, t) - h(x_2, t)| \leqslant \delta. \tag{4.6.9}$$

As we may assume that each $h(\cdot, t)$ satisfies the convex hull property, (4.6.9) then also holds for each $r \leqslant r_t$, in particular for $r = \eta$. We cover D by a finite number of disks of radius $\eta/2$, with centres $x_1, \ldots, x_m$, say. We then replace each $h(\cdot, t)$ on the disk $B(x_1, \eta) \cap D$ by the harmonic map (cf. Lemma 4.1.5) $u: B(x_1, \eta) \cap D \to B(p_1, \delta)$ (because of (4.6.9), and the convex hull property, there exists $p_1 \in \mathbb{R}^d$ with $h(x, t) \subset B(p_1, \delta)$ for $x \in B(x_1, \eta) \cap D$) with

$$u|_{\partial B(x_1, \eta) \cap D} = h(\cdot, t)|_{\partial B(x_1, \eta) \cap D}$$

and

$$u|_{\partial D \cap B(x_1, \eta)} \subset \gamma_\delta$$

satisfying the free boundary condition here (in the case when $\partial D \cap B(x_1, \eta) \neq \emptyset$).

By choice of δ, this harmonic replacement is unique. We denote the new map by $\tilde{h}_1(\cdot, t): D \to \mathbb{R}^d$. By the stability result of Theorem 2.2.2, $\tilde{h}_1(\cdot, t)$ still depends continuously on t,

Performing again the construction of step (1), we obtain $h_1(\cdot, t)$ satisfying the convex hull property, without destroying regularity properties.

We repeat the construction, with h_1 instead of h, and x_2 instead of x_1, and so on. After the replacement on $B(x_l, \eta)$ and subsequent application of step(1), we obtain

$$h^1(x, t) := h_l(x, t).$$

We note

$$h^1(\cdot, t) \in C^\alpha \qquad \text{for every } \alpha < 1. \tag{4.6.10}$$

This follows from the facts that a harmonic map, defined on $B(x, \eta) \cap D$, is of class $C^{1,\alpha}$ on $B(x, \eta/2) \cap D$ (cf. Theorems 2.5.1 and 2.5.3) and that a harmonic map with C^α boundary values is itself C^α (cf. Corollary 2.5.3), and because the disks $B(x_1, \eta/2), \ldots, B(x_l, \eta/2)$ already cover D, and because the C^α-property is not destroyed by application of step (1).

By uniqueness of the harmonic replacement and because of the estimates of Corollary 2.5.3 h^1 represents a continuous path in $C^\alpha \cap H^{1,2}$ for every $\alpha < 1$. By construction h^1 already satisfies the conclusion of step (1).

We then repeat step (2) with h^1 instead of h (note (4.6.6) which implies that the same η as before suffices), obtaining h^2.

Iteratively, we obtain

$$h^n: D \times [0,1] \to \mathbb{R}^d \tag{4.6.11}$$

representing a path in $C^\alpha \cap H^{1,2}$ $(0<\alpha<1)$ with

$$h^n(\partial D \times \{t\}) \subset \gamma \qquad \text{for each } t \tag{4.6.12}$$

and

$$E(h^n(\cdot,t)) \leqslant K \qquad \text{for each } t \tag{4.6.13}$$

and

$$h^n(\cdot,0) = u_0, h^n(\cdot,1) = u_1. \tag{4.6.14}$$

In order to get rid of the action of the conformal group on D, we can also achieve a three-point normalization. We select three distinct points $z_1, z_2, z_3 \in \partial D$ and three distinct points $a_1, a_2, a_3 \in \gamma$ so that the numberings are compatible with the orientations of ∂D and γ. We then require first

$$h(z_i, t) = q_i \qquad \text{for } i = 1,2,3;\ t \in [0,1]$$

for our original path h.

This property may be destroyed by the application of step (2), but w.l.o.g. we can assume that η is so small that $\operatorname{dist}(z_i, z_j) > 2\eta$ for $i \neq j$. If then $h_1(z_i, t) \neq h_1(z_j, t)$ for $i \neq j$, there exist conformal transformations $k_t: D \to D$, depending continuously on t, with

$$h_1(k_t(z_i), t) = p_i \qquad \text{for } i = 1,2,3;\ t \in [0,1].$$

Consequently, we can assume w.l.o.g. that h^1, and iteratively also h^n satisfies the normalization

$$h^n(z_i, t) = q_i \qquad \text{for } i = 1,2,3;\ t \in [0,1]. \tag{4.6.15}$$

We also note that all constructions in step (1) and step (2) are energy non-increasing (for step (2), this follows from the fact that by uniqueness we always replace by the energy-minimizing map). Thus

$$E(h^{n+1}(\cdot,t)) \leqslant E(h^n(\cdot,t)) \qquad \text{for all } t \in [0,1]. \tag{4.6.16}$$

Normalization (4.6.15), together with Lemma 3.1.1 (note (4.6.13)) and the convex hull property implies the equicontinuity of $(h^n(\cdot,t))_{n\in\mathbb{N}}$ for each fixed t.

After selection of a subsequence, $(h^n(\cdot,t))$ therefore converges uniformly to a continuous map $u(\cdot,t)$. Moreover, this convergence is also strong in $H^{1,2}$. There are several ways to see this. One possibility is to use the argument of Section 4.1, namely convergence of integrals of the form

$$\int_{\varphi=0}^{2\pi} \left| \frac{\partial h^n(r,\varphi,t)}{\partial \varphi} \right|^2 \mathrm{d}\varphi$$

in polar coordinates (r,φ), for every centre and every $r>0$. Another possibility is to note that the $H^{1,2,2}$-norms of $h^n(\cdot,t)|_{\partial B(x_\lambda,\eta)\cap D}$ $(\lambda = 1,\ldots,l)$ are convergent, because of the C^α-estimates for all $\alpha<1$, and from this one can infer

$H^{1,2}$-convergence of the corresponding harmonic replacements on $B(x_\lambda, \eta) \cap D$ (this is also true, if $B(x_\lambda, \eta) \cap \partial D \neq \emptyset$, as we then require a free boundary condition, and one can check $H^{1,2}$ convergence by using similar perturbation arguments as in the proof of Theorem 2.2.2 and at the end of Section 2.5).

By construction $u(\cdot, t)$ then has to be a stationary point for the application of step (2). Therefore $u(\cdot, t)$ is a harmonic map: $D \to \mathbb{R}^d$ with a Plateau condition on ∂D. By Corollary 1.4.1, $u(\cdot, t)$ represents a minimal surface with Plateau boundary γ.

We have to show that for some $t \in (0, 1)$,

$$u_t := u(\cdot, t)$$

is different from both u_0 and u_1. We distinguish two cases:

Case (a) $$m := \inf_{n \in \mathbb{N}} \sup_{t \in [0,1]} E(h_n(\cdot, t)) > \max(E(u_0), E(u_1)).$$

Let

$$M_n := \{t \in [0, 1]: E(h_n(\cdot, t)) \geqslant m\}.$$

M_n is a compact set, and since (cf. (4.6.16))

$$E(h^{n+1}(\cdot, t)) \leqslant E(h^n(\cdot, t)) \qquad \text{for every } t$$

we have $M_{n+1} \subset M_n$.

Hence, there exists $t_0 \in \bigcap_n M_n$, and by strong $H^{1,2}$-convergence

$$E(u_{t_0}) = \lim_{n \to \infty} E(h_n(\cdot, t_0)) = m.$$

Since $m > \max(E(u_0), E(u_1))$, u_{t_0} is different from both u_0 and u_1.

Case (b) $$m := \inf_{n \in \mathbb{N}} \sup_{t \in [0,1]} E(h_n(\cdot, t)) = E(u_0) \qquad \text{(say)}.$$

In case (a), u_0 is a strict local minimum, i.e. there exists $\rho > 0$ with the property that for every

$$u \in N(u_0, \rho) := \{v: D \to \mathbb{R}^d,\ 0 < \| u_0 - v \| \leqslant \rho,\ v \text{ maps } \partial D \text{ monotonically onto } \gamma\}$$

($\| \cdot \|$ is either the C^0- or the $H^{1,2}$-norm) we have

$$E(u) > E(u_0).$$

Since h_n represents a continuous path in $C^0 \cap H^{1,2}$, for every $n \in \mathbb{N}$ we can find $t_n \in (0, 1)$ with

$$\| h_n(\cdot, t_n) - u_0 \| = \rho/2.$$

By the above compactness properties, $(h_n(\cdot, t_n))$ contains a subsequence $(\tilde{h}_n(\cdot))$, converging uniformly and strongly in $H^{1,2}$ towards some $u \in N(u_0, \rho)$ with

$$E(u) = \lim_{n \to \infty} E(\tilde{h}_n) = m.$$

This contradicts the assumption that u_0 is a strict local minimum.

In case (*b*), for every ρ with

$$0 < \rho < \| u_0 - u_1 \|$$

we find a sequence $(h_n(\cdot, t_n))$ with

$$\| h_n(\cdot, t_n) - u_0 \| = \rho$$

and a subsequence $(\tilde{h}_n)_{n \in \mathbb{N}}$ converging uniformly and strongly in $H^{1,2}$ to some u_ρ with

$$\| u_\rho - u_0 \| = \rho$$

and

$$E(u_\rho) = \lim_{n \to \infty} E(\tilde{h}_n) = m.$$

Since u_0 and u_1 are global minima, necessarily

$$m = E(u_0) = E(u_1)$$

and therefore u_ρ is also a global minimum, and therefore in particular a minimal surface with Plateau boundary γ. A modification of this argument also shows that u_0 and u_1 can be connected by a continuum of minimal surfaces with boundary γ, all having the same energy.

This completes the proof of Theorem 4.6.1.

q.e.d.

Remarks: (1) For the proof of Theorem 4.6.1, it would actually have been sufficient instead of step (1) to carry out the projections only in d orthogonal directions as in the proof of Theorem 1.1.1. This is technically somewhat simpler as orthogonal projections do not affect each other so that if (4.6.3) is achieved in the direction e_1, say, then projections in the remaining orthogonal directions $e_2, \ldots, e_d$ leave this property invariant. The present proof, however, has the advantage that it immediately carries over to the situation where instead of $\mathbb{R}^d$ we have as ambient space a non-positively curved simply-connected complete Riemannian manifold N, as in N we have similar energy and distance decreasing projections as in $\mathbb{R}^d$, using distance spheres instead of planes.

More generally, the same procedure still works if γ and $u_0(D)$, $u_1(D)$ are contained in a ball $B(p, r)$, $r < \min(\pi/2\kappa, i(p))$, where κ^2 is an upper curvature bound, in some Riemannian manifold N, without further restrictions on N; cf. Lemma 4.1.2 and the proof of Lemma 4.1.4.

(2) Theorem 4.6.1 is due to Morse and Tompkins [MT], Shiffman [Sf1] [Sf2], Courant [C3] and Struwe [St3].

(3) It is not difficult to adapt the arguments of Struwe [St3, St4] to the present situation in order to carry out a Morse theory for disk-type minimal surfaces with boundary γ. By the first remark, this is valid not only in $\mathbb{R}^d$ but also in those Riemannian situations mentioned there.

4.7. The Plateau–Douglas problem in Riemannian manifolds

This section is based on [J5].

In Euclidean space, the problem of the existence of minimal surfaces of higher genus spanning a given contour had been raised by Douglas and was treated in papers of Douglas [D2], Courant [C2], and Shiffman [Sf3]. For an evaluation and criticism of these papers, see Tromba [Tr1], where the problem is again taken up and a partial solution obtained. The work of Courant and Shiffman was generalized by Luckhaus [Lu] to surfaces of constant mean curvature. The problem in Riemannian manifolds was formulated and solved for the genus 0 case by Morrey [M2]. The general problem in Euclidean space was also treated in [TT3] by combining the approach of [Tr1] with an argument first obtained in [J5].

We now start describing the setting of the problem: N is a complete Riemannian manifold of bounded geometry, i.e. with a positive lower bound for the injectivity radius and an upper bound for the absolute value of the sectional curvature.

(Actually, lesser regularity assumptions suffice: N need only be of class C^1 and homogeneously regular in the sense that each $p \in N$ corresponds to the centre of a coordinate path with domain $B(0,1)$ for which

$$\lambda|\xi|^2 \leqslant g_{ij}(x)\xi^i\xi^j \leqslant \mu|\xi|^2 \qquad \text{for all } x \in B(0,1) \text{ and } \xi \in \mathbb{R}^m$$

where λ and μ are positive constants independent of p. Our reasoning can be adapted to these assumptions, but for the sake of unity of presentation, we shall assume that N is of bounded geometry.)

Let $\gamma = (\gamma_1, \ldots, \gamma_k)$ be a system of disjoint oriented Jordan curves in N. In the following, it will be technically convenient to have a definition of area for a rather general class of comparison surfaces. To this end, let S be a compact oriented surface of class C^1 with boundary curves $c_1, \ldots, c_k$. If $h: S \to N$ is a map of class C^1, then of course the area of $h(S)$ is

$$A(h, S) := \int_S |\det(\nabla h)| \, dS.$$

We now define the area of $h(S)$ in the sense of Lebesgue, if $h: S \to N$ is continuous and maps each c_i monotonically and with preserved orientation onto γ_i, as

$$A(h, S) := \inf\left(\liminf_{n \to \infty} A(h_n, S) \right)$$

for all sequences $(h_n)_{n \in \mathbb{N}} \subset C^1(S, N)$ converging uniformly to h. Later on, sometimes it will be convenient to perturb the boundary curves $\gamma_1, \ldots, \gamma_k$ slightly. We say that for a sequence of maps $\eta_n: S^1 \to N$, the image curves $\gamma_n := \eta_n(S^1)$

converge in the sense of Fréchet to an oriented Jordan curve $\gamma \subset N$, if

$$\liminf_{n\to\infty} \inf_{\tau} \left(\max_{x\in S^1} d(\eta_n(x), \tau(x)) \right) = 0 \tag{4.7.1}$$

where the infimum is taken over all orientation-preserving monotonic maps $\tau: S^1 \to \gamma$.

If now S is a surface of genus p, and if the continuous map $h: S \to N$ induces the map

$$\alpha: \pi_1(S) \to \pi_1(N)$$

on fundamental groups, we put

$$a(\gamma, p, \alpha) := \inf\left(\liminf_{n\to\infty} A(h_n, S_n) \right) \tag{4.7.2}$$

for all sequences (h_n, S_n) where S_n is of genus p and class C^1 and $h_n: S_n \to N$ is continuous, induces the map α on fundamental groups and maps the k boundary curves $c_{1,n}, \ldots, c_{k,n}$ of S_n onto curves $\gamma_{1,n}, \ldots, \gamma_{k,n}$ in N which converge towards $\gamma_1, \ldots, \gamma_k$ in the sense of Fréchet.

In the following, it will be more convenient to consider the energy functional E instead of the area functional A. We therefore define

$$d(\gamma, p, \alpha) := \inf\left(\liminf_{n\to\infty} E(h_n, S_n) \right) \tag{4.7.3}$$

where we require in addition $h_n \in H^{1,2}(S_n, N)$.

In order to formulate the so-called Douglas condition, we need the notion of a primary reduction of a surface S. A primary reduction of S leads to a surface S' of lower topological type than S. There are two possibilities for a primary reduction of S. (As before $h: S \to N$ is continuous and induces the map

$$\alpha: \pi_1(S) \to \pi_1(N)$$

on fundamental groups, and maps the ith boundary curve c_i onto γ_i.)

(1) if β is a non-trivial element of $\pi_1(S)$ which can be represented by a simple (i.e. non-self-intersecting) closed geodesic on S, and if

$$\alpha(\beta) = 0 \in \pi_1(N)$$

we cut S along a simple closed loop in the interior of S representing β. The two curves resulting from this cut are then contracted to points—for example, by glueing in disks.

(2) Let $\eta \in S$ be an arc joining two points on one of the boundary curves of S, say c_i, and assume that $\alpha(\eta)$ is homotopic to an arc contained in γ_i. We cut S along η. This possibly disconnects S. The resulting surface has two new boundary curves each formed by a portion of c_i and a copy of η after the cut. The image of one of these curves is homotopically trivial by assumption. This curve is again contracted to a point. If the cut leads to a disconnected surface, and if

one of the components is of disk type, we assume that the new boundary curve of the other component has a homotopically trivial image, and that boundary curve then is shrunk to a point.

We observe that a primary reduction either reduces the genus of S by 1 or separates S into two components, the sum of their genera not exceeding the genus of S. If one of these components has the same genus as S (and the other one is consequently a disk), then the first one has one boundary curve less than S.

At the moment the notion of primary reduction is a purely topological one. In the considerations below, all primary reductions will also carry a metric, to which Corollary 3.3.1 applies, and hence in particular a conformal structure.

We finally define

$$a^*(\gamma, 0, \alpha) := d^*(\gamma, 0, \alpha) := \infty \qquad \text{if } k = 1$$

and define $\alpha^*(\gamma, p, \alpha)$ and $d^*(\gamma, p, \alpha)$ otherwise by taking the inf in (4.7.2) or (4.7.3) respectively over surfaces S_n which are homeomorphic to primary reductions of S. If there are no primary reductions of S, we again put

$$a^*(\gamma, p, \alpha) := d^*(\gamma, p, \alpha) := \infty.$$

Lemma 4.7.1: (*a*) *In the definitions of a, d, a^*, and d^* we can assume that each S_n is of class $C^{1,\alpha}$ and h_n is an immersion of class $C^{1,\alpha}$*, $0 < \alpha < 1$.
(*b*) $d(\gamma, p, \alpha) = a(\gamma, p, \alpha) \qquad d^*(\gamma, p, \alpha) = a^*(\gamma, p, \alpha)$
(*c*) $d(\gamma, p, \alpha) \leqslant d^*(\gamma, p, \alpha)$.

Proof (*cf.* [M3], *Lemma 9.4.1*]): (*a*) It follows from standard approximation arguments that h_n and S_n can be chosen of class $C^{1,\alpha}$. In order to obtain immersions, for each $n \in \mathbb{N}$, we choose a differentiable embedding

$$i_n = (i_n^1, \ldots, i_n^l): S_n \to \mathbb{R}^l$$

for some fixed l. This is possible by Whitney's embedding theorem. We also choose a sequence $(\epsilon_n)_{n \in \mathbb{N}}$, $\epsilon_n > 0$, with

$$\lim_{n \to \infty} \epsilon_n E(i_n) = 0.$$

We then define $l_n: S_n \to N \times \mathbb{R}^l$ as

$$l_n(x) := (h_n(x), \epsilon_n i_n(x)).$$

l_n then is an embedding.
(*b*) This proof then follows from the conformal representation theorems of Chapter 3 (Corollary 3.3.1). Namely, we can choose all maps $h_n: S_n \to N$ to be conformal maps onto their images, since those can be assumed to be immersed surfaces of class $C^{1,\alpha}$ by (*a*). Then

$$A(h_n, S_n) = E(h_n, S_n)$$

for all n.

(*c*) By (*b*), we have only to show $a(\gamma, p, \alpha) \leqslant a^*(\gamma, p, \alpha)$ but this is obvious since, for example, additional handles can always be mapped onto curves without using area.

Theorem 4.7.1: *Let N be a Riemannian manifold of bounded geometry, $\gamma = (\gamma_1, \dots, \gamma_k)$ a system of k disjoint oriented Jordan curves in N, Σ a compact oriented surface with k boundary curves $\bar{c}_1, \dots, \bar{c}_k$ and of genus p, $\bar{h}: \Sigma \to N$ be a continuous mapping which maps $\bar{c}_i$ monotonically and with preserved orientation onto γ_i, and $\alpha: \pi_1(\Sigma) \to \pi_1(N)$ be the induced map on fundamental groups. If*

$$d(\gamma, p, \alpha) < d^*(\gamma, p, \alpha) \tag{4.7.4}$$

(or equivalently $a(\gamma, p, \alpha) < a^*(\gamma, p, \alpha)$) then there exists a compact oriented surface S of genus p with k boundary curves $c_1, \dots, c_k$ and a continuous map $h: S \to N$ which induces the map α on fundamental groups and maps each c_i monotonically and with preserved orientation onto γ_i. h is conformal and maps S onto a surface of least area (in its topological class) bounded by γ, i.e.

$$A(h, S) = a(\gamma, p, \alpha).$$

If $\pi_2(N) = 0$, then h is homotopic to $\bar{h}$.

In the case $N = \mathbb{R}^d$, Theorem 4.7.1 implies the solution of the Plateau–Douglas problem.

Corollary 4.7.1: *Let γ be a system $(\gamma_1, \dots, \gamma_k)$ of oriented Jordan curves in $\mathbb{R}^d$, and $p \in \mathbb{N}$. If the infimum of area of surfaces of genus p spanning γ is strictly less than the infimum of surfaces either of genus less than p or consisting of more than one component, the sum of the genera of which does not exceed p, then there exists a minimal surface of genus p spanning γ, and this surface is of least area among all such surfaces.*

Proof of Theorem 4.7.1: Let $h_n: S_n \to N$ be a minimizing sequence for $a(\gamma, p, \alpha)$. By Lemma 4.7.1, we can assume that each S_n is of class $C^{1,\alpha}$ and each h_n is an immersion of class $C^{1,\alpha}$. We may also assume that h_n is conformal, i.e. that S_n and $h_n(S_n)$ are conformally equivalent. We let Σ_n be the Schottky double of S_n. By the conformal representation theorems of Chapter 3, Σ_n is conformally equivalent to S^2 or to a quotient of $\mathbb{C}$ or H. We shall present the details only in the most difficult case where the Schottky double of S_n is covered by H, since the remaining cases where S_n is either a disk or an annular region can be handled by similar but more elementary considerations. (In the case of a disk, the Douglas condition is trivial, and if $N = \mathbb{R}^d$ this case was already treated in Theorem 1.1.1.)

Since h_n is conformal, $E(h_n, S_n) = A(h_n, S_n)$, and hence

$$E(h_n, S_n) \to d(\gamma, p, \alpha).$$

Hence also

$$E(h_n, S_n) \leqslant K \qquad \text{for some constant } K. \tag{4.7.5}$$

We want to show that (4.7.4) implies convergence of a subsequence of (S_n) and of the boundary maps $h_n|\partial S_n$.

By Lemma 3.3.2, for the convergence of a subsequence of (S_n), it suffices to show that the lengths of simple closed geodesics on the Schottky doubles Σ_n have a positive lower bound. We note that the boundary curves of S_n correspond to simple closed geodesics on Σ_n because they are invariant under an isometric involution.

We shall have to consider four types of geodesics, namely interior geodesics of S_n, boundary curves of S_n, geodesics of Σ_n that intersect one boundary curve of S_n twice, and geodesics that intersect two boundary curves of S_n. For each type, one has to consider two cases: (i) where the length of the image of the corresponding geodesic and its parallel curves has a positive lower bound, for example if the image curve is homotopically non-trivial, and (ii) where the length of the image curves tends to 0. In the first cases, one uses the collar lemma as in the proof of Lemma 3.3.4 to derive a contradiction, whereas in the second case, one constructs suitable primary reductions so that (4.7.4) is violated.

Let us consider the first case:

Let $i\colon \Sigma_n \to \Sigma_n$ be the orientation-reversing isometric involution fixing ∂S_n. We extend h_n to Σ_n by putting

$$h_n(x) = h_n(i(x)) \qquad \text{for all } x \in \Sigma_n. \tag{4.7.6}$$

Then

$$E(h_n, \Sigma_n) = 2E(h_n, S_n). \tag{4.7.7}$$

Let c_n be a simple closed geodesic on Σ_n of length l_n, and assume

$$l(h_n(c_n)) \geqslant \sigma \quad (l \text{ denotes length}).$$

Σ_n contains a collar about c_n of width $2\operatorname{arcsinh}(1/\sinh(l/2))$ by Lemma 3.3.1. As in the proof of Lemma 3.3.4, we let r denote the distance from c_n, with a negative sign on one side, and we parametrize all parallel curves by $\varphi \in [0, l_n)$.

Then, for each r with

$$-\operatorname{arcsinh}(1/\sinh(l_n/2)) < r < \operatorname{arcsinh}(1/\sinh(l_n/2))$$

$$\sigma \leqslant \int_0^{l_n} \left(g_{ij} \frac{\partial h_n^i(r,\varphi)}{\partial \varphi} \frac{\partial h_n^j(r,\varphi)}{\partial \varphi} \right)^{1/2} \mathrm{d}\varphi$$

where (g_{ij}) is the metric of N. Thus

$$\sigma^2 \leqslant l_n \int_0^{l_n} \left(g_{ij} \frac{\partial h_n^i}{\partial \varphi} \frac{\partial h_n^j}{\partial \varphi} \right) \mathrm{d}\varphi$$

and integrating w.r.t $r \in [-R, R]$

$$\frac{2R\sigma^2}{l_n} \leqslant \int_{-R}^{R} \int_0^{l_n} \left(g_{ij} \frac{\partial h_n^i}{\partial \varphi} \frac{\partial h_n^j}{\partial \varphi} \right) \mathrm{d}\varphi\, \mathrm{d}r \leqslant 2E(h_n, \Sigma_n) = 4E(h_n, S_n) \tag{4.7.8}$$

by (4.7.7).

Again, if l_n is small, we may choose $R=1$, and (4.7.8) then yields a lower bound for l_n, noting (4.7.5). This handles the first case. In particular, we have thus handled those geodesics that intersect two boundary curves of S_n. Namely their image curves have length bounded from below by the minimal distance between the various curves $\gamma_1, \ldots, \gamma_k$, as those image curves have endpoints on two different such curves.

We now treat the second case.

We first exclude that a boundary curve $c_{j,n}$ of S_n can shrink to a point in the limit. Let l_n be the length of $c_{j,n}$. S_n then contains a half-collar about $c_{j,n}$, parametrized in the same manner as before by

$$0 \leqslant \varphi \leqslant l_n$$
$$0 \leqslant r \leqslant \operatorname{arcsinh}(1/\sinh(l_n/2)).$$

We also parametrize the unit disk D by polar coordinates (ρ, θ). We map the boundary $(1, \theta)$ via

$$(1, \theta) \rightarrow \left(r_n, \frac{l_n}{2\pi}\theta\right) \tag{4.7.9}$$

onto a parallel curve of $c_{j,n}$; the value of r_n will be chosen in a moment.

We also define

$$u_n(1, \theta) := h_n\left(r_n, \frac{l_n}{2\pi}\theta\right).$$

Then

$$\int_{\theta=0}^{2\pi} \left|\frac{\partial}{\partial\theta} u_n(1, \theta)\right|^2 \mathrm{d}\theta = \int_{\varphi=0}^{l_n} \left|\frac{\partial}{\partial\varphi} h_n(r_n, \varphi)\right|^2 \frac{l_n}{2\pi} \mathrm{d}\varphi. \tag{4.7.10}$$

Again, we may assume that l_n is so small that

$$1 \leqslant \operatorname{arcsinh}(1/\sinh(l_n/2) \tag{4.7.11}$$

as otherwise there is nothing to prove.

Since

$$\int_0^1 \int_0^{l_n} \left|\frac{\partial}{\partial\varphi} h_n(r, \varphi)\right|^2 \mathrm{d}\varphi\, \mathrm{d}r \leqslant 2E(h_n, S_n)$$

(4.7.10) implies that there exists some $r = r_n$ with, when choosing this r_n in (4.7.9),

$$\int_{\theta=0}^{2\pi} \left|\frac{\partial}{\partial\theta} u_n(1, \theta)\right|^2 \mathrm{d}\theta \leqslant \frac{K}{\pi} l_n. \tag{4.7.12}$$

Using Lemma 4.2.4, it follows that there exists a continuous extension of its boundary values,

$$u_n : D \rightarrow N$$

satisfying

$$E(u_n) \leqslant c_1 l_n \tag{4.7.13}$$

where c_1 is independent of n, provided l_n is small enough. We now cut S_n along the curve $\{(r_n, \varphi),\ 0 \leqslant \varphi \leqslant l_n\}$, the curve parallel to $c_{j,n}$ at distance r_n. One of the two resulting components is an annulus, and the other one is diffeomorphic to S_n. We then identify each copy of the cut curve via (4.7.9) with the boundary of D and then attach a copy of D to each curve. In this way, we obtain two new surfaces, one being homeomorphic to D and the other one having the same genus but one boundary curve less than S_n. By Corollary 3.1.1, we can assume that they are regular. We call those surfaces S_n^1 and S_n^2. On each S_n^i we take the map h_n^i which corresponds on the part of S_n^i coming from S_n with h_n and on the inserted disk with the function u_n constructed above. Equation (4.7.13) implies

$$E(h_n^1, S_n^1) + E(h_n^2, S_n^2) \leqslant E(h_n, S_n) + c_2 l_n. \tag{4.7.14}$$

Since $l_n \to 0$, $E(h_n, S_n) \to d(\gamma, p, \alpha)$, and (S_n^1, S_n^2) is a primary reduction of S_n, (4.7.14) implies

$$d^*(\gamma, p, \alpha \leqslant d(\gamma, p, \alpha)$$

contradicting (4.7.4).

Hence the lengths of the boundary curves of S_n have to remain bounded below by some positive number which is independent of n.

With a similar argument we show that the lengths l_n of interior closed geodesics c_n of S_n cannot approach 0. Namely, otherwise we again find a curve parallel to c_n along which we can cut S_n and insert two disks in such a way that (4.7.12) and (4.7.13) again are satisfied.

It remains to show that the lengths of closed geodesics c_n on Σ_n intersecting one boundary curve of S_n twice, say $c_{1,n}$, cannot approach 0. If the image of the intersection of this geodesic with S_n is not homotopic to a subarc of γ_1, we are in the first case already treated above. Thus, we assume that this image is homotopic to a subarc of γ_1. Again, we want to derive a contradiction to (4.7.4), if $l_n \to 0$.

We again use the existence of a collar of width $2 \operatorname{arc} \sinh(1/\sinh(l_n/2))$ about c_n, with coordinates (r, φ) as before. We also extend h_n to all of Σ_n again as in (4.7.6). Then for each r with

$$-\operatorname{arcsinh}(1/\sinh(l_n/2)) < r < \operatorname{arcsinh}(1/\sinh(l_n/2)),$$

we estimate the length of $h_n(r, \cdot)$ as before

$$l(h_n(r, \cdot)) \leqslant \int_0^{l_n} \left(g_{ij} \frac{\partial h_n^i(r, \varphi)}{\partial \varphi} \frac{\partial h_n^j(r, \varphi)}{\partial \varphi} \right)^{1/2} \mathrm{d}\varphi$$

and using Hölder's inequality and integrating w.r.t. $r \in [-R, R]$ as in (4.7.8)

$$\int_{-R}^{R} l(h_n(r, \cdot))^2 \, \mathrm{d}r \leqslant 4 l_n E(h_n, S_n) \leqslant 4 l_n K \qquad \text{by (4.7.5).} \tag{4.7.15}$$

Again w.l.o.g. $R = 1$, and thus there exists some $r_n \in [-1, 1]$ with

$$l(h_n(r_n, \cdot)) \leqslant (2 l_n K)^{1/2}. \tag{4.7.16}$$

Denoting the two points of intersection of the parallel curve $r \equiv r_n$ of c_n with the boundary curve $c_{1,n}$ by x_n and y_n, we conclude in particular

$$d(h_n(x_n), h_n(y_n)) \leqslant (2l_nK)^{1/2}. \tag{4.7.17}$$

$h_n(x_n)$ and $h_n(y_n)$ both lie on $\gamma_{1,n} = h_n(c_{1,n})$. Also, $h_n|c_{1,n}$ becomes arbitrarily close to a monotonic map from $c_{1,n}$ onto the Jordan curve γ_1, if n is sufficiently large, by (4.7.1). Therefore, there exists a point $p \in N$ with

$$d(h_n(x), p) \leqslant \delta_n/2 \tag{4.7.18}$$

and $\delta_n \to 0$ as $\epsilon_n \to 0$, whenever $x \in g_n$, where g_n is the closed curve formed by the intersection of the parallel curve $r \equiv r_n$ with S_n and by one of the two subarcs of $c_{1,n}$ with endpoints x_n and y_n. We point out that here we make essential use of the assumption that γ_1 is a Jordan curve.

We now need the following auxiliary result (cf. [C5; p. 154ff.]); the proof is an easy computation and can be omitted. (See also the proof of Lemma A.2 of the Appendix.)

Lemma 4.7.2: *Let $\delta < 1$,*

$$\lambda(\rho) := \begin{cases} 1 & \text{if } \rho \geqslant \sqrt{\delta} \\ 1 + \dfrac{\log(\sqrt{\delta}/\rho)}{\log\sqrt{\delta}} & \text{if } \delta \leqslant \rho \leqslant \sqrt{\delta} \\ 0 & \text{if } \rho \leqslant \delta \end{cases}$$

and

$$u(z) := \begin{cases} h(z) & \text{if } d(h(z), p) \geqslant \sqrt{\delta} \\ p + \lambda(d(h(z), p))(h(z) - p) & \text{otherwise} \end{cases}$$

where $h(z) - p$ is defined in local coordinates centred at p. Then $E(u) \to E(h)$ as $\delta \to 0$.

We apply this transformation to $h = h_n$ and $\delta = \delta_n$, and thus get maps u_n with

$$\lim_{n\to\infty} E(u_n) = \lim_{n\to\infty} E(h_n). \tag{4.7.19}$$

Because of (4.7.18), u_n maps a neighbourhood of the curve g_n into p. (Note that in general, u_n does not map $c_{1,n}$ onto $\gamma_{1,n}$ but only onto a curve $\gamma_{1,n}^*$ for which (4.7.1) still holds.)

We now cut S_n along our curve $r \equiv r_n$. $c_{1,n}$ and the cut line give rise to new boundary curves, one of them being g_n and the other one denoted by $\tilde{g}_n \cdot g_n$ again is identified with ∂D, and D is then inserted. Furthermore, that part of $\tilde{g}_n$ which corresponds to the cut line $r \equiv r_n$ is identified with $\{\operatorname{Im} z \geqslant 0, |z| = 1\}$ and the half-disk $\{z \in \mathbb{C}: \operatorname{Im} z \geqslant 0, |z| \leqslant 1\}$ is inserted. Altogether we get a surface S_n' of genus $p-1$ which is a primary reduction of S_n. Corollary 3.1.1 again applies. We then take the map $u_n': S_n' \to N$ coinciding with u_n on S_n and mapping the two inserted pieces onto p. In this way, the Douglas condition is violated again.

We have now proved that the lengths of simple closed geodesics on Σ_n are bounded away from 0. Thus, the assumption of Lemma 3.3.2 is satisfied, and, after selection of a subsequence (Σ_n) converges to some surface Σ of the same genus, and Σ is the Schottky double of some surface $S = \lim S_n$. We may thus suppose $S_n = S$ for all n, without changing $\lim E(h_n, S_n)$.

We denote the boundary curves of S by $c_1, \ldots, c_k$. We shall also parametrize the unit circle S^1 by $0 \leqslant \varphi < 2\pi$, and each curve c_i by $0 \leqslant \theta < 2\pi$.

The next step of the proof consists of showing that after selection of a subsequence, the boundary maps $h_n|c_i$, $i = 1, \ldots, k$, converge uniformly to monotonic maps from c_i onto γ_i.

First of all, since $(\gamma_{i,n})$ converges to γ_i in the sense of Fréchet (cf. (4.7.1)), there exist continuous nowhere-constant maps $\tau_{i,n}\colon S^1 \to \gamma_{i,n}$ that converge uniformly to some homeomorphism $\tau_i\colon S^1 \to \gamma_i$. Then there also exist strictly monotonic, continuous functions $\phi_{i,n}\colon c_i \to S^1$ with

$$\tau_{i,n}(\phi_{i,n}(\theta)) = h_n(\theta).$$

After selection of a subsequence, $(\phi_{i,n})$ converges to a monotonic function $\phi_i\colon c_i \to S^1$.

Lemma 4.7.3: *Let $\theta_0 \in c_i$. Then either ϕ_i is continuous at θ_0 or*

$$\phi_i(\theta_{0^+}) - \phi_i(\theta_{0^-}) = 2\pi.$$

Proof: If $0 < \phi_i(\theta_{0^+}) - \phi_i(\theta_{0^-}) < 2\pi$, then

$$\tau_i(\phi_i(\theta_{0^+})) \neq \tau_i(\phi_i(\theta_{0^-})$$

i.e.

$$d := d(\tau_i(\phi_i(\theta_{0^+})), \tau_i(\phi_i(\theta_{0^-})) > 0.$$

As $(\tau_{i,n}(\phi))$ converges uniformly to $\tau_i(\phi)$ and $(\phi_{i,n}(\theta))$ pointwise to $\phi_i(\theta)$ and the $\phi_{i,n}$ are monotonic, we can find $\delta > 0$ and $N \in \mathbb{N}$ with the following property. Putting $c(r) := \partial B(\theta_0, r) \cap S$, we have

$$l(h_n(c(r))) \geqslant d/2 \qquad \text{for } 0 < r < \delta \text{ and } n \geqslant N.$$

Thus, we conclude that for each $\epsilon > 0$, $\epsilon < \delta$

$$\begin{aligned} 2K &\geqslant \int_\epsilon^\delta \frac{1}{r} \int_{\psi \in c(r)} \left| \frac{\partial}{\partial \psi} h_n(r, \psi) \right|^2 \mathrm{d}\psi \, \mathrm{d}r \\ &\geqslant \int_\epsilon^\delta \frac{1}{2\pi r} \left(\int_{c(r)} \left| \frac{\partial}{\partial \psi} h_n(r, \psi) \right| \mathrm{d}\psi \right)^2 \mathrm{d}r \\ &\geqslant \frac{d^2}{8\pi} \log\left(\frac{\delta}{\epsilon}\right) \end{aligned}$$

which is impossible.

q.e.d.

By Lemma 4.7.3, in order to prove the continuity of the boundary maps, it only remains to exclude

$$\phi_i(\theta_0^+) - \phi_i(\theta_0^-) = 2\pi \qquad \text{for some } \theta_0 \in c_i. \tag{4.7.20}$$

In the case when S is the unit disk, a standard three-point normalization of the boundary values excludes (4.7.20) as in Section 1.1. In the other cases, such a normalization is no longer possible, and we have to use the assumption (4.7.4). Again, we only treat the case where the Schottky double of S is covered by H, as the case of an annulus is similar.

By Lemma 3.1.1, for every sufficiently small $\epsilon > 0$ there is some $r_n \in (\epsilon, \sqrt{\epsilon})$ with

$$l(h_n(\partial B(\theta_0, r_n) \cap S)) \leqslant \left(\frac{8\pi K}{\log(1/\epsilon)}\right)^{1/2}. \tag{4.7.21}$$

$\partial B(\theta_0, r_n)$ divides c_i into two parts, and we let $\tilde{c}_n$ be the curve consisting of $\partial B(\theta_0, r_n) \cap S$ and that part of c_i which does not contain θ_0. If (4.7.20) holds, (4.7.21) implies, letting $\epsilon = \epsilon_n \to 0$,

$$l(h_n(\tilde{c}_n)) \leqslant \delta_n$$

$\delta_n \to 0$. We thus again find some $p \in N$ with

$$d(p, h_n(x)) \leqslant \delta_n/2 \qquad \text{for all } x \in \tilde{c}_n.$$

As in the argument excluding the degeneration of a geodesic intersecting a boundary curve twice we replace h_n by a map u_n which maps a neighbourhood of $\tilde{c}_n$ into p. Again

$$\lim_{n \to \infty} E(u_n) = \lim_{n \to \infty} E(h_n). \tag{4.7.22}$$

We cut S along $\partial B(\theta_0, r_n) \cap S$ into two pieces. One of them has $\tilde{c}_n$ as a boundary curve, and inserting a disk into $\tilde{c}_n$ we get a surface S' with $(k-1)$ boundary curves to which Corollary 3.3.1 again applies, and we define u'_n again as above. The second piece S'' is homeomorphic to a disk, and here we put $u'_n = u_n$ again. Thus, using (4.7.22)

$$\begin{aligned} d^*(\gamma, p, \alpha) &\leqslant \lim_{n \to \infty} (E(u'_n, S') + E(u'_n, S'')) \\ &= \lim_{n \to \infty} E(u_n, S) = \lim_{n \to \infty} E(h_n, S) = d(\gamma, p, \alpha) \end{aligned}$$

again contradicting (4.7.4).

This completes the proof of the assertion that after selection of a subsequence, $h_n | \partial S$ converges uniformly to a monotonic map $h: \partial S \to \gamma$.

The final step consists of harmonic replacement of h_n. By Theorem 4.1.1 there exists a map $u_n: s \to N$ which minimizes energy among all maps inducing $\alpha: \pi_1(S) \to \pi_1(N)$ and satisfying $u_n | \partial S = h_n | \partial S$. We replace h_n by u_n. Thus

$$E(u_n, S) \leqslant E(h_n, S). \tag{4.7.23}$$

Since $(u_n | \partial S)$ converges uniformly, (u_n) also converges uniformly (after selection of a subsequence) by the estimates of Section 2.5. By (4.7.23) (u_n) also has to converge weakly in $H^{1,2}$ and since E is lower semicontinuous with respect to weak $H^{1,2}$-convergence, with $u = \lim_{n\to\infty} u_n$,

$$d(\gamma, p, \alpha) \leqslant E(u) \leqslant \liminf_{n\to\infty} E(u_n) \leqslant \lim_{n\to\infty} E(h_n) = d(\gamma, p, \alpha).$$

Lemma 4.7.1 and Lemma 1.3.1 then imply that u is conformal and that $u(S)$ has least area in its class. Theorem 4.1.1 yields the statement for $\pi_2(N) = 0$.

This completes the proof of Theorem 4.7.1.

q.e.d.

We can also obtain a similar result about closed minimal surfaces, if again a Douglas type condition is satisfied. Thus, let now S be a closed surface of class C^1 and genus p, and $h: S \to N$ a continuous map. The induced map on fudamental groups is again denoted by $\alpha: \pi_1(S) \to \pi_1(N)$. We can then define

$$a(p, \alpha) := \inf A(h, S)$$

for all such pairs (h, S), and

$$d(p, \alpha) := \inf E(h, S)$$

where in addition $h \in H^{1,2}(S, N)$, and $a^*(p, \alpha)$ and $d^*(p, \alpha)$ are the corresponding quantities for primary reductions of S as in the beginning of this section.

The preceding considerations yield the following generalization of a result of Sacks and Uhlenbeck [SkU2] and Schoen and Yau [SY2].

Corollary 4.7.1: *Let N be a compact Riemannian manifold, Σ a closed surface of genus p, $\bar{h}: \Sigma \to N$ a continuous map, and $\alpha: \pi_1(\Sigma) \to \pi_1(N)$ the induced map on fundamental groups. If*

$$d(p, \alpha) < d^*(p, \alpha)$$

or equivalently (4.7.24)

$$a(p, \alpha) < a^*(p, \alpha)$$

then there exists a closed surface S of genus p and a continuous map $h: S \to N$ which induces the map α on fundamental groups, is conformal, and maps S onto a surface of least area (in its topological class), i.e.

$$A(h, S) = a(p, \alpha).$$

If $\pi_2(N) = 0$, then h is homotopic to $\bar{h}$.

The compactness of N is needed, because we no longer have any boundary in order to prevent a minimizing sequence from disappearing at infinity. More generally, one can require a suitable growth condition for the metric of N at infinity or a condition on $\alpha(\pi_1(\Sigma))$.

If α is injective, then there are no primary reductions of Σ, hence

$d^*(p, \alpha) = \alpha^*(p, \alpha) = \infty$, and (4.7.24) is trivially satisfied. This case, in which the argument simplifies and reduces to that of [SY2], is the theorem of Sacks and Uhlenbeck and Schoen and Yau.

Remark: An existence result for minimal surfaces of higher genus in $\mathbb{R}^3$ under different conditions was obtained in [TT2, TT3].

5 HARMONIC MAPS BETWEEN SURFACES

5.1. The existence of harmonic diffeomorphisms

We start with the result of [J1], which gives a local existence result for harmonic diffeomorphisms. The global result (Theorem 5.1.2 below) will then be obtained with the help of local replacements based on Theorem 5.1.1.

Theorem 5.1.1: *Suppose Ω is a bounded domain with C^2 boundary $\partial\Omega$ on some surface, and that Σ is another surface. We assume that $\Psi:\bar{\Omega}\to\Sigma$ maps $\bar{\Omega}$ homeomorphically onto its image, that $\Psi(\partial\Omega)$ is contained in some disk $B(p,r)$ with radius $r<\pi/2\kappa$ (where $\kappa^2\geqslant 0$ is an upper curvature bound on $B(p,r)$) and that the curves $\Psi(\partial\Omega)$ are of class C^2 and convex w.r.t. $\partial(\Omega)$.*

Then there exists a harmonic mapping $u:\bar{\Omega}\to B(p,r)$ with the boundary values prescribed by Ψ which is a homeomorphism between $\bar{\Omega}$ and its image, and a diffeomorphism in the interior.

Moreover, if $\Psi|\partial\Omega$ is even a C^2-diffeomorphism then u is a diffeomorphism up to the boundary.

Theorems 5.1.1. and 2.1.1. imply:

Corollary 5.1.1: *Under the assumption of Theorem 5.1.1, each harmonic map which solves the Dirichlet problem defined by Ψ and which maps Ω into a geodesic disk $B(p,r)$ with radius $r<\pi/2\kappa$, is a diffeomorphism in Ω.*

Proof of Theorem 5.1.1: First of all, $\partial\Omega$ is connected. Otherwise, $\Psi(\partial\Omega)$ would consist of at least two curves, both of them convex w.r.t. $\Psi(\Omega)$. Therefore, we could find a non-trivial closed geodesic γ in $\Psi(\Omega)\subset B(p,M)$ with an easy Arzela–Ascoli argument. This is impossible, because by virtue of $r<\pi/2\kappa$, $d^2(\cdot,p)$ is strictly convex on $B(p,r)$.

Therefore, we have to prove the theorem only for the case when Ω is the plane unit disk D, taking the existence (cf. Corollary 3.1.2) of a conformal map $k: D \to \Omega$ and the composition property Lemma 1.3.2 into account.

For the moment, we assume that $\Psi: \partial D \to \Psi(\partial D)$ is a C^2-diffeomorphism between curves of class $C^{2,\alpha}$, that $\Psi(\partial D)$ is not only convex, but strictly convex, and that we have the following quantitative bounds:

$$\left|\frac{d^2}{d\phi^2}\Psi(\phi)\right| \leqslant b_1 \tag{5.1.1}$$

and

for $\phi \in \partial D$

$$\left|\frac{d}{d\phi}\Psi(\phi)\right| \geqslant b_2^{-1} \tag{5.1.2}$$

and

$$0 < a_1 \leqslant \kappa_g(\Psi(\partial D)) \leqslant a_2 \qquad (\kappa_g \text{ is the geodesic curvature}). \tag{5.1.3}$$

These assumptions can be removed later on by approximation arguments which we shall indicate below.

By virtue of Corollary 3.1.2 again, there is a conformal map $k: D \to \Psi(D)$. By a variation of boundary values, we now want to deform this conformal map into a harmonic diffeomorphism u.

Without loss of generality, we may assume that the boundary value map preserves the orientation of ∂D. Now let γ be the parametrization of the boundary curve of $\Psi(D)$ by arclength. We set

$$\omega(\phi, \lambda) := \gamma(\lambda\gamma^{-1}(k(\phi)) + (1-\lambda)\gamma^{-1}\Psi(\phi)) \qquad \phi \in \partial D, \lambda \in [0, 1] \tag{5.1.4}$$

where ω deforms the boundary values of k into the boundary values prescribed by Ψ.

Since we assumed that (5.1.1) and (5.1.2) hold and that $\Psi(\partial D) \in C^{2,\alpha}$, Theorem 2.6.1 implies that

$$\omega(\phi, \lambda) \qquad \frac{\partial}{\partial\phi}\omega(\phi, \lambda) \qquad \frac{\partial^2}{\partial\phi^2}\omega(\phi, \lambda) \tag{5.1.5}$$

are all continuous functions of λ, and

$$\frac{\partial}{\partial\phi}\omega(\phi, \lambda) \text{ does not vanish for any } \phi \in \partial D \text{ and } \lambda \in [0, 1]. \tag{5.1.6}$$

Let now u_λ denote the harmonic map from D to $B(p, M)$ with boundary values $\omega(\cdot, \lambda)$ (the existence of u_λ follows from Lemma 4.5.4), and let $\lambda_n \in [0, 1]$ be a sequence converging to some $\lambda \in [0, 1]$. By Theorem 2.5.1, the Arzela–Ascoli theorem and the uniqueness Theorem 2.2.1, u_{λ_n} converges to the harmonic map u_λ in the $C^{1,\beta}$-topology, $0 < \beta < \alpha$. In particular,

$$p(\lambda) := \inf_{x \in D} |J(u_\lambda)(x)|$$

depends continuously on λ (here $J(u_\lambda)$ denotes the Jacobian of u_λ). We define $L := \{\lambda \in [0,1] : p(\lambda) > 0\}$. By Corollary 3.1.2, $0 \in L$ (u_0 is the conformal map k), and therefore L is not empty. Since we assumed (5.1.2) and (5.1.3) which implied (5.1.5) and (5.1.6) we can apply Theorem 2.7.3 to the extent that

$$p(\lambda) \geqslant p_0 > 0 \qquad \text{for } \lambda \in L. \tag{5.1.7}$$

Since $p(\lambda)$ depends continuously on λ, (5.1.7) implies $L = [0,1]$. Thus, u_1 is a local diffeomorphism and a diffeomorphism between the boundaries of D and $u_1(D)$, and consequently a global difffeomorphism by the homotopy lifting theorem.

Thus, the proof of Theorem 5.1.1. is complete, except for the approximation arguments.

So far, we have assumed that the boundary of the image is strictly convex, and, in addition, that the boundary values are a diffeomorphism of class C^2. We now have to prove the theorem also for the case that the boundary is only supposed to be convex and that the boundary values are only supposed to induce a homeomorphism of the boundaries.

We shall present only the first approximation argument. It is a modification of the corresponding one given by Heinz in [H3; Part II, pp. 178–183]. The reasoning for the second case can be taken over from [H3; Part I, pp. 351–352] in the case of $\partial\Psi(D) \in C^{2,\alpha}$.

Therefore, let us suppose that the boundary of the image $\Psi(D)$ is only convex, while the boundary values Ψ are still assumed to be a diffeomorphism of class C^2. Then we argue in the following way. Given a metric g_{ij} on the image with respect to which the boundary of $A := \Psi(D)$ is convex, there is a sequence $\{g^n_{ij}\}$ of metrics on A such that ∂A is even strictly convex with respect to g^n_{ij}, according to [H4, Section 4]. Moreover, $\{g^n_{ij}\}$ can be chosen to converge uniformly to g_{ij} on A together with their first and second derivatives, as $n \to \infty$. Keeping the boundary values Ψ fixed, we consider the map $u_n(x)$ which is harmonic in the metric g^n_{ij} and which solves the Dirichlet problem with boundary values Ψ. The existence of u_n is guaranteed by the arguments given above—at least for large values of n, when g^n_{ij} is so close to g_{ij} that the geometric conditions are satisfied.

By virtue of Theorem 2.7.2, on each disk $B(0,r)$, $r < 1$, there is an *a priori* bound of the fundamental determinant of $u_n(x)$ from below. Moreover, by virtue of Theorem 2.5.1, we can choose a subsequence of the functions $u_n(x)$ which converges uniformly on D together with the first derivatives to a map $u(x)$. In particular, the u_n converge to u strongly in $H^{1,2}$. Therefore, u is a weakly harmonic map w.r.t the metric g_{ij}, i.e. a weak solution of the corresponding Euler equations. Since u is also of class C^1, Theorem 2.5.1 implies that u is a classical solution, i.e. harmonic. Moreover, u is a local diffeomorphism in the interior, and since it is the uniform limit of diffeomorphisms, it is a diffeomorphism in the interior.

q.e.d.

Remarks: (1) Actually, using a further approximation argument, we do not even have to assume that the boundary values are homeomorphic. We need only that they are continuous and monotonic, i.e. a uniform limit of homeomorphisms. The corresponding harmonic solution of the Dirichlet problem still remains a diffeomorphism in the interior.

(2) In the case where both Ω and $\Psi(\Omega)$ are bounded, simply connected domains in the plane, the assertion of Theorem 5.1.1 was already obtained by Rado [R1], Kneser [Kn1], and Choquet [Cq]. Choquet also showed that the convexity of the image is necessary for Theorem 5.1.1 to hold. The reason is the following. Suppose there exist p, $q \in \partial\overline{\psi(\Omega)}$ for which the line connecting them is disjoint from $\overline{\psi(\Omega)}$. If the boundary values $\Psi(\partial\Omega)$ are concentrated near p and q, then by the mean value property of harmonic functions, some points of Ω will be mapped onto points between p and q not belonging to $\Psi(\Omega)$.

This is different to the case of conformal maps where convexity of the image is not necessary to guarantee that the solution is a diffeomorphism (cf. Corollary 3.1.2). Note that a conformal map is a solution of a free boundary value problem instead of a Dirichlet problem.

(3) The proof of Theorem 5.1.1. actually shows that when Ψ is a diffeomorphism between $\bar{\Omega}$ and its image, then u can be constructed to be isotopic (i.e. homotopic through diffeomorphisms) to Ψ. This is seen as follows.

First of all, in the proof of Theorem 3.1.1, one may restrict the class $\mathscr{D}$ to consist only of the diffeomorphisms in a given isotopy class, and the conformal diffeomorphism constructed in the proof then belongs to the same isotopy class. The same applies to Corollary 3.1.2 and Theorem 3.2.1.

We can therefore assume that the conformal map $k: D \to \Psi(D)$ utilized in the proof of Theorem 5.1.1 is isotopic to Ψ. Then also u is isotopic to Ψ, as it follows from the proof that u is isotopic to k. Of course, it is a topological result (cf. [Z]) that Ψ and u are automatically isotopic in the present situation, but our preceding observations allow us to give an analytic proof of such a result (cf. Lemma 6.1.1 below).

We proceed to the global result:

Theorem 5.1.2: *Suppose that Σ_1 and Σ_2 are compact surfaces without boundary, and that $h: \Sigma_1 \to \Sigma_2$ is a diffeomorphism. Then there exists a harmonic diffeomorphism $u: \Sigma_1 \to \Sigma_2$ isotopic to h. Furthermore, u is of least energy among all diffeomorphisms isotopic to h.*

The same assertion holds with 'isotopic' replaced by 'homotopic'. Actually, homotopic diffeomorphisms between surfaces are isotopic by Baer's Theorem (cf. [Z]). Theorems 5.1.2 and 2.2.3 provide an analytic proof of this result. Theorem 5.1.2 was proved by Jost and Schoen [JS]. (An earlier approach by Shibata [Sh] was erroneous; cf. however [Se] and [Hl1] for more recent results in this direction.)

Corollary 5.1.2: *Suppose that Σ_1 and Σ_2 are compact surfaces without boundary,* *and that* $\Psi:\Sigma_1\to\Sigma_2$ *is a covering map, i.e. a local diffeomorphism. Then there exists a harmonic covering map* $u:\Sigma_1\to\Sigma_2$*, homotopic to* Ψ.

Proof of Corollary 5.1.2: We pull the metric ds^2 of Σ_2 back via Ψ to obtain a surface Σ_2', diffeomorphic to Σ_1 and with metric $\Psi^* ds^2$. Then $\Psi:\Sigma_2'\to\Sigma_2$ is a local isometry. By Theorem 5.1.2, there is a harmonic diffeomorphism $u':\Sigma_1\to\Sigma_2'$, homotopic to the identity. $u:=\Psi\circ u'$ then is the desired harmonic covering map.

q.e.d.

Let us consider an attempt to prove Theorem 5.1.2. We want to minimize the energy in the class of all diffeomorphisms isotopic to h or some suitable closure thereof. One obtains a minimizing map u which is a weak $H^{1,2}$ as well as a uniform limit of diffeomorphisms, as in Chapter 3, for example, in the proof of Theorem 3.1.1. Lemma 1.2.4 implies that

$$\varphi\, dz^2 := \rho^2(u) u_z \bar{u}_z\, dz^2$$

(where $\rho^2(u)$ is the image metric) is a holomorphic quadratic differential.

We may assume that the functional determinant $J(h)$ of h is positive. Lemma 3.1.4 then implies

$$J(u) \geqslant 0 \qquad \text{almost everywhere}$$

i.e.

$$|u_{\bar{z}}| \leqslant |u_z| \qquad \text{almost everywhere.}$$

Hence

$$|u_{\bar{z}}| \leqslant \frac{1}{\rho^2(u)} |\varphi|^{1/2}.$$

Since the energy of u is controlled, as u is energy minimizing, the L^1-norm of φ is controlled, hence also the pointwise supremum of φ, as φ is holomorphic. Consequently

$$|u_{\bar{z}}| \leqslant \text{const.}$$

Since we also control the modulus of continuity of u, as in the proofs of Theorems 3.1.1 or 4.5.1, we obtain C^α-estimates for u from potential theory.

At this point, the difficulties arise. It is not clear how to obtain higher-order estimates for u; if u were of class C^2, the holomorphicity of φ would imply that u is harmonic (Lemma 1.3.4). Alternatively, one could try to show that u is energy minimizing on small disks w.r.t. to its own boundary values; this would imply that u is weakly harmonic and, being continuous, of class C^2 again; cf. Section 2.5. Since u is only known to be minimizing among limits of diffeomorphisms, it is, however, not clear whether we can perform arbitrary variations $u+\epsilon\varphi$. Actually, the minimizing property of u would follow, if u were known to be quasiconformal, by the result of [Hl1], but again quasiconformality is not clear.

The idea proposed to overcome these difficulties is to make local replacements in a minimizing sequence by local harmonic diffeomorphisms as constructed in Theorem 5.1.1., and to use the estimates of Section 2.7 for the Jacobian from below to show that a limit of such local replacements is again a harmonic diffeomorphism. By local uniqueness (Theorem 2.2.1), this does not increase the energy of the original minimizing sequence, and u itself locally has to coincide with a harmonic diffeomorphism.

Proof of Theorem 5.1.2 (following [JS]): If Σ_1 and Σ_2 are homeomorphic to S^2, then we can find a conformal (and hence harmonic) diffeomorphism homotopic to h by Theorem 3.1.1. The case where Σ_1 and Σ_2 are homeomorphic to the real projective plane is similarly handled by passing to two-sheeted coverings. Thus we can assume w.l.o.g. that $\pi_2(\Sigma_i)=0$ $(i=1,2)$. We define $\mathscr{D}_K$ to be the class of all diffeomorphisms $f:\Sigma_1 \to \Sigma_2$ isotopic to h with energy $E(f) \leqslant K$. Of course, K is chosen so large that $\mathscr{D}_K \neq \varnothing$. The argument of Lemma 3.3.6 implies the equicontinuity of $\mathscr{D}_K$, as $\pi_2(\Sigma_1)$ and $\pi_2(\Sigma_2)$ are trivial. We let $\bar{\mathscr{D}}_K$ be the class of all weak $H^{1,2}$-limits of sequences in $\mathscr{D}_K$. Since the energy is lower semicontinuous w.r.t. weak $H^{1,2}$ convergence, $E(f) \leqslant K$ for all $f \in \bar{\mathscr{D}}_K$. This in particular implies that $\bar{\mathscr{D}}_K$ is closed w.r.t. weak $H^{1,2}$-convergence. Also, using the weak compactness of energy-bounded sets in $H^{1,2}$ and in addition the equicontinuity of $\mathscr{D}_K$ and the Arzela–Ascoli theorem, each element in $\bar{\mathscr{D}}_K$ is the weak $H^{1,2}$ and uniform limit of a sequence of diffeomorphisms in $\mathscr{D}_K$. (These properties are not clear for the weak $H^{1,2}$-closure of the space of all diffeomorphisms in a given homotopy or isotopy class, and this is the reason for introducing the energy bound K.)

We now let $(f_n)_{n\in\mathbb{N}}$ be an energy-minimizing sequence in $\bar{\mathscr{D}}_K$:

$$E(f_n) \to \inf_{f\in\bar{\mathscr{D}}_K} E(f) \qquad \text{as } n\to\infty.$$

Using again the weak compactness of energy-bounded sets in $H^{1,2}$ and the equicontinuity of $\mathscr{D}_K$ and the Arzela–Ascoli theorem, after selection of a subsequence, $(f_n)_{n\in\mathbb{N}}$ converges weakly in $H^{1,2}$ and uniformly to some map u_0 in our isotopy class.

The lower semicontinuity of the energy implies

$$E(u_0) = \inf_{f\in\bar{\mathscr{D}}_K} E(f).$$

As observed above, there exists a sequence $(u_n)_{n\in\mathbb{N}} \subset \mathscr{D}_K$ of diffeomorphisms converging weakly in $H^{1,2}$ and uniformly to u_0. By approximation, we may also assume $u_n \in C^{1,\alpha}$ for all n and some fixed α $(0<\alpha<1)$.

The aim now is to show that u_0 is a harmonic diffeomorphism. We consider an arbitrary point $x_0 \in \Sigma_1$ and define

$$\mathring{B}_\sigma := B(u_0(x_0), \sigma)$$

i.e. the open disk in Σ_2 centred at $u_0(x_0)$ with radius σ. We restrict ourselves in the following to values of σ which are smaller than the injectivity radius of Σ_2 and smaller than $\pi/2\kappa$, where κ^2 is again an upper bound for the curvature of Σ_2. This will be crucial for the applications below of the local uniqueness Theorem 2.2.1. We define

$$\Omega_0 := u_0^{-1}(B_\sigma)$$
$$\Omega_n := u_n^{-1}(B_\sigma) \qquad n \in \mathbb{N}.$$

W.l.o.g., we can assume $x_0 \in \Omega_n$ for all n, since the u_n converge uniformly to u_0. As always D is the unit disk in the complex plane. We let

$$F_n: D \to \bar{\Omega}_n$$

be a conformal mapping which maps 0 to x_0 (cf. Theorem 3.2.1 and Corollary 3.2.1).

We note

$$E(F_n, D) = \operatorname{Area}(\Omega_n) \leqslant \operatorname{Area}(\Sigma_1)$$

so that the sequence (F_n) is energy bounded.

Since $\Gamma_n := \partial\Omega_n$ is a Jordan curve of class C^1 (because u_n is a diffeomorphism), F_n is a homeomorphism of D onto $\bar{\Omega}_n$, and therefore $u_n \circ F_n$ maps ∂D homeomorphically onto ∂B_σ. By Theorem 5.1.1. and Corollary 5.1.1, there exists a unique harmonic mapping $v_n: D \to B_\sigma$ which assumes the boundary values prescribed by $u_n \circ F_n$, and v_n minimizes energy in its homotopy class and is a diffeomorphism, and lies in the same isotopy class as $u_n \circ F_n$ (cf. Remark (3) above).

In particular,

$$E(v_n, D) \leqslant E(u_n \circ F_n, D) = E(u_n, \Omega_n) \leqslant K \tag{5.1.8}$$

by Lemma 1.3.2. Since the u_n converge uniformly to u_0, we can assume that $u_n \circ F_n(0)$ stays in an arbitrarily small neighbourhood of $u_0(x_0)$. Therefore, we can again apply the argument of Lemma 3.1.2 to show that the maps $u_n \circ F_n$ are equicontinuous on D. In particular, the boundary values of v_n, namely $u_n \circ F_n | \partial D$, are equicontinuous. By Theorems 2.5.1 and 2.2.1, we can therefore assume that the v_n converge uniformly on D to a map v_0 which is harmonic in the interior of D. Using Theorem 2.7.2, we see furthermore that v_0 is a diffeomorphism in the interior of D. We now define

$$\tilde{u}_n = \begin{cases} v_n \circ F_n^{-1} & \text{in } \Omega_n \\ u_n & \text{in } \Sigma_1 \backslash \Omega_n. \end{cases}$$

$\tilde{u}_n$ is a Lipschitz map and lies in $H^{1,2}$ and $E(\tilde{u}_n) \leqslant K$. Then, for each n, the functional determinant of $\tilde{u}_n$ is defined and bounded from below on $\Sigma_1 \backslash \Gamma_n$ by Corollary 2.7.1 recalling $u_n \in C^{1,\alpha}$. If u_n and $v_n \circ F_n^{-1}$ do not coincide on Ω_n, then u_n is not necessarily smooth on Γ_n, but in this case $E(\tilde{u}_n)$ is strictly less than K by uniqueness of the energy-minimizing map, and an approximation argument shows that u_n can be approximated in the $H^{1,2}$-norm by diffeomorphisms with energy bounded by K, and u_n lies therefore in $\bar{\mathscr{D}}_K$ in any case. As usual, using

Lemma 3.1.1, after selection of a subsequence, $(\tilde{u}_n)$ converges uniformly as well as weakly in $H^{1,2}$ to some $\tilde{u}_0 \in \bar{\mathscr{D}}_K$, and (F_n) converges uniformly on compact subsets of D to a conformal map F which maps the interior of D diffeomorphically onto some open set $\Omega \subset \Sigma_1$, and $F(0) = x_0$.

F is not necessarily smooth on ∂D, but that does not affect the following arguments.

$u_0 \circ F$ is the uniform limit of $u_n \circ F_n$ and thus extends continuously to D. Since $u_n \circ F_n$ and v_n coincide on ∂D, it follows that $u_0 \circ F$ and v_0 also coincide there, and, since v_0 is harmonic and therefore energy minimizing (by Theorem 2.2.1) in its homotopy class, then

$$E(v_0, D) \leqslant E(u_0 \circ F, D).$$

Since conformal maps preserve energy by Lemma 1.3.2, this implies

$$E(\tilde{u}_0, \Omega) \leqslant E(u_0, \Omega). \tag{5.1.9}$$

We are going to show that

$$E(\tilde{u}_0, \Sigma_1 \backslash \Omega) \leqslant E(u_0, \Sigma_1 \backslash \Omega). \tag{5.1.10}$$

Once (5.1.10) is shown, the proof is easily completed. Namely, by definition of u_0,

$$E(u_0, \Sigma_1) \leqslant E(\tilde{u}_0, \Sigma_1).$$

Comparing this with (5.1.9) and (5.1.10), we see

$$E(u_0, \Omega) = E(\tilde{u}_0, \Omega),$$

and hence also, since conformal maps are energy preserving,

$$E(u_0 \circ F, D) = E(v_0, D).$$

In particular, $u_0 \circ F$ is energy minimizing on D w.r.t its own boundary values. The local uniqueness of energy-minimizing maps (Theorem 2.2.1) then implies

$$u_0 \circ F = v_0 \qquad \text{on } D$$

since the boundary values on ∂D coincide. Thus $u_0 \circ F$ is a harmonic diffeomorphism on D, and so then is u_0 on Ω. Since Ω was a neighbourhood of an arbitrarily chosen point $x_0 \in \Sigma_1, u_0$ is then a diffeomorphism on Σ_1, as it is also isotopic to a diffeomorphism.

It remains to prove (5.1.10). It suffices to prove that u_0 and $\tilde{u}_0$ coincide almost everywhere on $\Sigma_1 \backslash \Omega$. We first show

$$\Sigma_1 \backslash \Omega \subset u_0^{-1}(\Sigma_2 \backslash B_\sigma). \tag{5.1.11}$$

We define

$$\rho_n(x) := d(u_n(x), u_0(x_0))$$
$$\rho_0(x) := d(u_0(x), u_0(x_0))$$

for $x \in \Sigma_1$. Let $x \in \Sigma_1 \backslash \Omega$. If

$$\rho_0(x) = \lim_{n \to \infty} \rho_n(x) \geqslant \sigma$$

then

$$x \in u_0^{-1}(\Sigma_2 \backslash B_\sigma).$$

Since the $\rho_n \circ u_n \circ F_n$ are equicontinuous and equal to σ on ∂D, $\rho_0(x) < \sigma$ implies that

$$d(F_0^{-1}(x), \partial D) \geqslant \delta > 0$$

for sufficiently large n.

Since the F_n converge uniformly to F on compact subsets of D, this would imply $x \in F(D) = \Omega$ which contradicts the assumption $x \in \Sigma_1 \backslash \Omega$. This proves (5.1.11).

We also have

$$u_0^{-1}(\Sigma_2 \backslash B_\sigma) = u_0^{-1}(\partial B_\sigma) \cup u_0^{-1}(\Sigma_2 \backslash \bar{B}_\sigma)$$

and since the sets $u_0^{-1}(\partial B_\sigma)$ cover a neighbourhood of x_0 and are disjoint, we can assume w.l.o.g. that the two-dimensional measure of $u_0^{-1}(\partial B_\sigma)$ vanishes for our chosen σ. If

$$x \in u_0^{-1}(\Sigma_2 \backslash \bar{B}_\sigma)$$

then

$$\lim_{n \to \infty} \rho_n(x) = \rho_0(x) > \sigma$$

and because of the equicontinuity of the functions ρ_n, there exists an open neighbourhood U of x such that $\rho_n | U > \sigma$ for sufficiently large n. This implies

$$\tilde{u}_0 = \lim_{n \to \infty} \tilde{u}_n = \lim_{n \to \infty} u_n = u_0 \qquad \text{on } U.$$

Therefore $u_0 = \tilde{u}_0$ almost everywhere on $u_0^{-1}(\Sigma_2 \backslash B_\sigma)$, and (5.1.10) now follows from (5.1.11). This completes the proof of Theorem 5.1.2.

q.e.d.

Remark: For a uniqueness result for harmonic diffeomorphism see [CH].

We also have the following result for the Dirichlet problem.

Theorem 5.1.3: *Let $\Omega \subset \Sigma_1$ be a two-dimensional domain with non-empty boundary $\partial\Omega$ consisting of Lipschitz curves, and let $\Psi: \bar{\Omega} \to \Sigma_2$ be a homeomorphism of $\bar{\Omega}$ onto its image $\Psi(\bar{\Omega})$ and suppose that the curves $\Psi(\partial\Omega)$ are of Lipschitz class and convex with respect to $\Psi(\Omega)$. Then there exists a harmonic diffeomorphism $u: \Omega \to \Psi(\Omega)$ which is homotopic to Ψ and satisfies $u = \Psi$ on $\partial\Omega$. Moreover, u is of least energy among all diffeomorphisms which are homotopic to Ψ and assume the same boundary values.*

This result is again taken from [JS]. The case of non-positive image curvature was solved in [SY1].

Proof: We assume first that $\partial\Omega$ and $\Psi(\partial\Omega)$ are of class $C^{2,\alpha}$ and that Ψ gives rise to a diffeomorphism between $\partial\Omega$ and $\Psi(\partial\Omega)$ and that $\Psi(\partial\Omega)$ is strictly convex with respect to $\Psi(\Omega)$.

In this case, the proof proceeds along the lines of the proof of Theorem 5.1.2 with an obvious change of the replacement argument at boundary points involving Theorem 2.7.3. The general case now follows by approximation arguments as in the proof of Theorem 5.1.1.

5.2. Local computations. Consequences for non-positively curved image metrics. Harmonic diffeomorphisms. Kneser's Theorem

We start with some computations that were given by Schoen and Yau [SY1] and in similar form by Sampson [Sa]. We let Σ_1 and Σ_2 be compact oriented surfaces with local conformal metrics.

$$\lambda^2(z)\,dz\,d\bar{z}$$

and

$$\rho^2(u)\,du\,d\bar{u}.$$

We recall the equation for a harmonic map $u: \Sigma_1 \to \Sigma_2$

$$u_{z\bar{z}} + \frac{2\rho_u}{\rho} u_z u_{\bar{z}} = 0. \tag{5.2.1}$$

We define

$$H := |\partial u|^2 := \frac{\rho^2}{\lambda^2} u_z \bar{u}_{\bar{z}} \tag{5.2.2}$$

and

$$L := |\bar{\partial} u|^2 := \frac{\rho^2}{\lambda^2} u_{\bar{z}} \bar{u}_z. \tag{5.2.3}$$

We denote the curvature of Σ_i by K_i, i.e.

$$K_1 = -\Delta \log \lambda = -\frac{4}{\lambda^2} \frac{\partial^2}{\partial z\,\partial \bar{z}} \log \lambda \tag{5.2.4}$$

and

$$K_2 = -\frac{4}{\rho^2} \frac{\partial^2}{\partial u\,\partial \bar{u}} \log \rho. \tag{5.2.5}$$

Lemma 5.2.1: *At points where H or L, respectively, is non-zero*

$$\Delta \log H = 2K_1 - 2K_2(H - L) \tag{5.2.6}$$

$$\Delta \log L = 2K_1 + 2K_2(H - L). \tag{5.2.7}$$

Proof: For a positive smooth function f on Σ_1,

$$\Delta \log f = \frac{1}{f} \Delta f - \frac{4}{\lambda^2} \frac{1}{f^2} f_z f_{\bar{z}}. \tag{5.2.8}$$

Also, from (5.2.4),

$$\Delta \log \frac{1}{\lambda^2} = 2K_1. \tag{5.2.9}$$

Then

$$\begin{aligned}
\Delta(\rho^2 u_z \bar{u}_{\bar{z}}) &= \frac{4}{\lambda^2}\frac{\partial}{\partial \bar{z}}\left(\rho^2\left(u_{zz} + \frac{2\rho_u}{\rho}u_z u_z\right)\bar{u}_{\bar{z}}\right) \\
&= \frac{4}{\lambda^2}\rho^2\left\{\frac{\partial}{\partial \bar{z}}\left(u_{zz} + \frac{2\rho_u}{\rho}u_z u_z\right) + \frac{2\rho_{\bar{u}}}{\rho}\left(u_{zz} + \frac{2\rho_u}{\rho}u_z u_z\right)\bar{u}_{\bar{z}}\right\}\bar{u}_{\bar{z}} \\
&\quad + \frac{4}{\lambda^2}\rho^2\left(u_{zz} + \frac{2\rho_u}{\rho}u_z u_z\right)\left(\bar{u}_{\bar{z}\bar{z}} + \frac{2\rho_{\bar{u}}}{\rho}\bar{u}_{\bar{z}}\bar{u}_{\bar{z}}\right).
\end{aligned} \tag{5.2.10}$$

Differentiating (5.2.1), i.e.

$$\frac{\partial}{\partial z}\left(u_{z\bar{z}} + \frac{2\rho_u}{\rho}u_z u_{\bar{z}}\right) = 0$$

and using (cf. (5.2.5) and (5.2.8))

$$K_2 = -\frac{4}{\rho^2}\left(\frac{\rho_{u\bar{u}}}{\rho} - \frac{\rho_u \rho_{\bar{u}}}{\rho^2}\right)$$

the first term on the right-hand side of (5.2.10) yields a commutator term, and we find

$$\Delta(\rho^2 u_z u_{\bar{z}}) = -2K_2 H(H-L) + \frac{4}{\lambda^2}\frac{\partial}{\partial z}(\rho^2 u_z \bar{u}_{\bar{z}})\frac{\partial}{\partial \bar{z}}(\rho^2 u_z \bar{u}_{\bar{z}})\frac{1}{\rho^2 u_z \bar{u}_{\bar{z}}}. \tag{5.2.11}$$

Equations (5.2.8), (5.2.9) and (5.2.11) imply (5.2.6). Equation (5.2.7) can be computed in the same way, or directly deduced from (5.2.6), as $|\bar{\partial} u|^2 = |\partial \bar{u}|^2$ and complex conjugation in the image can be considered as a change of orientation.

q.e.d.

It may seem preferable to carry out the computations in the proof of Lemma 5.2.1 in intrinsic notation. One then defines ∇ as the covariant derivative in $u^{-1}T\Sigma_2$, e.g.

$$\nabla_{\partial/\partial z} u_z = u_{zz} + \frac{2\rho_u}{\rho}u_z u_z.$$

Equation (5.2.1) then becomes

$$\nabla_{\partial/\partial z} u_{\bar{z}} = 0. \tag{5.2.12}$$

Also

$$\rho^2 u_z \bar{u}_{\bar{z}} = \langle u_z, \bar{u}_{\bar{z}} \rangle_{u^{-1}T\Sigma_2}$$

and one computes

$$\Delta(\rho^2 u_z \bar{u}_{\bar{z}}) = \frac{4}{\lambda^2}\frac{\partial}{\partial \bar{z}}\langle \nabla_{\partial/\partial z} u_z, \bar{u}_{\bar{z}}\rangle \qquad \text{(using (5.2.12))}$$

$$= \frac{4}{\lambda^2}\langle \nabla_{\partial/\partial \bar{z}}\nabla_{\partial/\partial z} u_z, \bar{u}_{\bar{z}}\rangle + \frac{4}{\lambda^2}\langle \nabla_{\partial/\partial z} u_z, \nabla_{\partial/\partial \bar{z}} \bar{u}_{\bar{z}}\rangle$$

$$= \frac{4}{\lambda^2}\left\langle R\left(u_*\left(\frac{\partial}{\partial \bar{z}}\right), u_*\left(\frac{\partial}{\partial z}\right)\right)u_z, \bar{u}_{\bar{z}}\right\rangle + \frac{4}{\lambda^2}\langle \nabla_{\partial/\partial z} u_z, \nabla_{\partial/\partial \bar{z}} \bar{u}_{\bar{z}}\rangle$$

$$= -2K_2 H(H-L) + \frac{4}{\lambda^2}\frac{\partial}{\partial z}\langle u_z, \bar{u}_{\bar{z}}\rangle \frac{\partial}{\partial \bar{z}}\langle u_z, \bar{u}_{\bar{z}}\rangle \frac{1}{\rho^2 u_z \bar{u}_{\bar{z}}}$$

(in which R is the curvature tensor of Σ_2 and u_* is the differential of u) and combining this with (5.2.8) and (5.2.9), (5.2.6) again follows.

The next result is similar to Lemma 2.7.1.

Lemma 5.2.2: *Suppose* $f \in C^1(\Sigma_1, \mathbb{C})$,

$$f_{\bar{z}} = f\omega \qquad \textit{with } \omega \in C^1 \tag{5.2.13}$$

and the zeros of f *are isolated. Then locally*

$$f = e^{-g} h \tag{5.2.14}$$

where $g \in C^1$ *and* h *is holomorphic.*

Proof: We can find $g \in C^1$ with $g_{\bar{z}} = \omega$ (locally) and $h := f e^{-g}$ is then holomorphic.

q.e.d.

Corollary 5.2.1: *Unless* $H \equiv 0$, *the zeros of* H *are isolated, and near each zero* z_i *of* H

$$H = a_i |z - z_i|^{n_i} + o(|z - z_i|^{n_i}) \tag{5.2.15}$$

for some $a_i > 0$ *and some* $n_i \in \mathbb{N}$.

Likewise, unless $L \equiv 0$, *then near each zero* z_j *of* L

$$L = b_j |z - z_j|^{m_j} + o(|z - z_j|^{m_j}) \tag{5.2.16}$$

for some $b_j > 0$ *and some* $m_j \in \mathbb{N}$.

Proof: We recall (Lemma 1.3.3) that

$$\varphi \, dz^2 := \rho^2(u) u_z \bar{u}_z \, dz^2$$

is holomorphic

$$HL = \frac{1}{\lambda^4}\varphi\bar{\varphi}. \tag{5.2.17}$$

Thus, unless $H \equiv 0$ or $L \equiv 0$, the zeros of H and L are isolated. If on the other hand, for example, $L \equiv 0$, then u is conformal, and the zeros of $H = (\rho^2/\lambda^2)u_z\bar{u}_{\bar{z}}$ are likewise isolated.

Moreover, putting $f = u_z$, from (5.2.1)

$$f_{\bar{z}} = f\omega \qquad \text{with} \qquad \omega = \frac{2\rho_u}{\rho} u_{\bar{z}}$$

and the local expansion (5.2.15) then follows from Lemma 5.2.2. Expansion (5.2.16) is proved in the same way.

q.e.d.

Lemma 5.2.3: *If $H \not\equiv 0$,*

$$2\int_{\Sigma_1} K_1 - 2\int_{\Sigma_2} K_2(H - L) = -\sum n_i \tag{5.2.18}$$

where n_i is defined in Corollary 5.2.1, and similarly, if $L \not\equiv 0$,

$$2\int_{\Sigma_1} K_1 + 2\int_{\Sigma_2} K_2(H - L) = -\sum m_j. \tag{5.2.19}$$

Proof: This follows from Lemma 5.2.1 and Corollary 5.2.1.

q.e.d.

As a consequence, we note the following result of Eells and Wood [EW].

Theorem 5.2.1: *Let Σ_1 and Σ_2 be closed orientable surfaces with conformal metrics, and let $u: \Sigma_1 \to \Sigma_2$ be a harmonic map of degree* deg(u). *If*

$$\chi(\Sigma_1) + |\deg(u)||\chi(\Sigma_2)| > 0 \tag{5.2.20}$$

(χ denotes the Euler characteristic), then u is holomorphic or antiholomorphic.

Proof: From (5.2.18) we obtain

$$\chi(\Sigma_1) - \deg(u)\chi(\Sigma_2) \leqslant 0 \qquad \text{unless } H \equiv 0 \tag{5.2.21}$$

and from (5.2.19)

$$\chi(\Sigma_1) + \deg(u)\chi(\Sigma_2) \leqslant 0 \qquad \text{unless } L \equiv 0 \tag{5.2.22}$$

and $H \equiv 0$ ($L \equiv 0$) means that u is antiholomorphic (holomorphic).

q.e.d.

As a corollary, Eells and Wood [EW] obtained an analytic proof of the following topological result of Kneser [Kn2].

Corollary 5.2.2: *Let Σ_1, Σ_2 be closed orientable surfaces, $\chi(\Sigma_2) < 0$. Then for* *any continuous* $g: \Sigma_1 \to \Sigma_2$ *of degree* $\deg(g) \neq 0$

$$|\deg(g)|\chi(\Sigma_2) \geqslant \chi(\Sigma_1). \tag{5.2.23}$$

Proof: Let $u: \Sigma_1 \to \Sigma_2$ be homotopic to g and harmonic w.r.t. some metrics (cf. Theorem 4.5.1). By Theorem 5.2.1, u is $\pm$ holomorphic if $|\deg(g)|\chi(\Sigma_2) < \chi(\Sigma_1)$. This contradicts the Riemann–Hurwitz formula which says

$$|\deg(u)|\chi(\Sigma_2) = \chi(\Sigma_1) + r \qquad r \geqslant 0$$

if u is $\pm$ holomorphic, $u \neq$ const. Thus, (5.2.23) has to hold.

q.e.d.

Theorem 5.2.1 also implies Proposition 4.5.3(a), as well as

Corollary 5.2.3 ([EW]): *Let* Σ *be a closed orientable surface of genus* $p \geqslant 2$. *Then there exists a harmonic map* $u: \Sigma \to S^2$ *of degree* p *if and only if* Σ *is not hyperelliptic.*

Proof: By Theorem 5.2.1, such a harmonic map has to be holomorphic, and a holomorphic map $u: \Sigma \to S^2$ of degree p exists precisely if Σ is not hyperelliptic.

q.e.d.

Another consequence of Lemma 5.2.3 is:

Corollary 5.2.4: *Let* $u: \Sigma_1 \to \Sigma_2$ *be a harmonic map between closed orientable surfaces. If*

$$\chi(\Sigma_1) - \deg(u)\chi(\Sigma_2) = 0 \qquad \textit{and} \qquad \deg(u) > 0 \tag{5.2.24}$$

then $H > 0$ *on* Σ_1.

Likewise, if

$$\chi(\Sigma_1) + \deg(u)\chi(\Sigma_2) = 0 \qquad \textit{and} \qquad \deg(u) < 0 \tag{5.2.25}$$

then $L > 0$ *on* Σ_1.

Proof: This follows directly from Lemma 5.2.3, as under assumption (5.2.24), for example, either $H \equiv 0$, contradicting, however, $\deg(u) > 0$, or H has no zeros, as a consequence of (5.2.18).

q.e.d.

We proceed with:

Lemma 5.2.4: *Let* $u: \Sigma_1 \to \Sigma_2$ *be a harmonic map between closed orientable surfaces, and suppose* (5.2.24) *holds. If* $K_2 \leqslant 0$, *then the functional determinant* $J(u) = H - L$ *satisfies*

$$J(u) \geqslant 0$$

on Σ_1.

Proof: Since $H>0$ by Corollary 5.2.3, we have $L>0$ in the region where $J(u)\leqslant 0$. Therefore, from Lemma 5.2.1

$$\Delta \log(H/L) = -4K_2(H-L) \leqslant 0$$

whenever $J(u)\leqslant 0$. Therefore, $\log(H/L)$ is superharmonic where $J(u)\leqslant 0$, and hence cannot have an interior minimum there. Thus, no point with $J(u)(z)<0$ exists.

q.e.d.

As a consequence, we obtain the following special case of Theorem 5.1.2 originally due to Schoen and Yau [SY1] and Sampson [Sa] by similar arguments as used here:

Theorem 5.2.2: *Let $u\colon \Sigma_1 \to \Sigma_2$ be a harmonic map between closed oriented surfaces of the same genus with* $\deg(u) = \pm 1$. *Suppose $K_2 \leqslant 0$ (K_2 = curvature of Σ_2). Then u is a diffeomorphism.*

Proof: W.l.o.g. $\deg(u)=1$. By Lemma 5.2.4, $J(u)\geqslant 0$ on Σ and by Corollary 5.2.3, $H>0$ on Σ_1.

Suppose $J(u)(z_0)=0$. Then $H(z_0)=L(z_0)>0$.

From Lemma 5.2.1, in a neighbourhood V of z_0,

$$\Delta \log(H/L) = -4K_2 J(u).$$

Since $J(u)\geqslant 0$, there exist constants c_1, c_2 with (noting $K_2\leqslant 0$)

$$\begin{aligned} -4K_2 J(u) &\leqslant c_1(H/L-1) \\ &\leqslant c_2 \log(H/L) \qquad \text{in } V. \end{aligned}$$

Therefore,

$$\Delta \log(H/L) \leqslant c_2 \log(H/L) \qquad \text{in } V$$

and the strong maximum principle implies that $\log(H/L)$ cannot assume a non-positive minimum in the interior of V, unless $\log(H/L)\equiv$ const. It follows that $J(u)(z_0)=0$ implies $\log(H/L)=0$ in a neighbourhood of z_0. Since the set where $\log(H/L)=0$ is closed, we conclude $J(u)\equiv 0$, i.e. $\deg(u)=0$, a contradiction. Therefore $J(u)>0$, and since $\deg(u)=1$, u is a diffeomorphism.

q.e.d.

Corollary 5.2.5: *Let Σ_1, Σ_2 be closed oriented Riemannian surfaces of the same genus, and let Σ_2 be equipped with a metric of negative curvature. Then any $g\colon \Sigma_1\to\Sigma_2$ of degree ± 1 is homotopic to a unique harmonic map u, and u is a diffeomorphism.*

Proof: This follows from Theorems 4.5.1, 2.2.3, and 5.2.2.

q.e.d.

The result of Corollary 5.2.4 actually already holds if Σ_2 has a metric of non-positive curvature, provided the genus is at least 2. This follows from a more careful analysis of the uniqueness statement of Theorem 2.2.3. Still another proof of Theorem 5.2.2 can be given based on the estimate of Theorem 2.7.1 via a continuity argument. For this, one chooses a smooth family of surfaces, $(\Sigma^t)_{t\in[0,1]}$, where Σ^0 is conformally equivalent to Σ_1, Σ^1 is isometric to Σ_2, and all have non-positive sectional curvature. We choose $u_t\colon \Sigma_1 \to \Sigma^t$ as a harmonic map in the given homotopy class. By the uniqueness Theorem 2.2.3 and the *a priori* estimates of Theorem 2.5.1, we can use Theorem 2.7.1 to show that the functional determinant of u_t, since varying smoothly with t, stays bounded away from zero (note that u_0 is conformal). We conclude $J(u_1) > 0$ as desired (cf. the proof of Theorem 5.1.1 for a similar argument).

5.3. Miscellaneous results about harmonic branched coverings and harmonic diffeomorphisms

One may also ask under what conditions is a harmonic map between closed surfaces a branched covering.

First of all, not every continuous map between closed surfaces is homotopic to a branched covering (cf. e.g. [Ed]), and thus a harmonic map in such a homotopy class has to exhibit other geometric singularities besides branch points (the possible such singularities are listed in [Wo]). But even a harmonic map which is homotopic to a branched covering need not be a branched covering itself. Let us exhibit the following example of Eells and Lemaire [EL3].

In this section, Σ_p will always denote a closed orientable surface of genus p equipped with a hyperbolic metric, and T will always be a flat torus.

Let $k\colon \Sigma_3 \to \Sigma_2$ be an (unbranched) holomorphic covering of degree 2, and let $g\colon \Sigma_2 \to T$ be harmonic of degree 1; for topological reasons, g cannot be a branched covering.

Then $g \circ k\colon \Sigma_3 \to T$ is harmonic (cf. Lemma 1.3.2) of degree 2, homotopic to a branched cover (cf. [Ed]), but not a branched cover itself. By 'uniqueness' (Theorem 2.2.3), $g \circ k$ also is not homotopic to a *harmonic* branched covering.

This example is explained by:

Theorem 5.3.1: *Let $h\colon \Sigma_p \to T$ be harmonic (Σ_p = closed oriented surface of genus p, T = flat torus).*

Then h is a branched cover if and only if the conformal structure of T can be deformed into a conformal structure T' (still equipped with a flat metric), so that there exists a $\pm$ holomorphic map $h'\colon \Sigma_p \to T'$ which is homotopic to h.

Proof: '$\Leftarrow$' By Theorem 2.2.3 and Lemma 1.3.2, $h = g \circ h'$, where $g\colon T' \to T$ is a harmonic diffeomorphism.

'$\Rightarrow$' Lemma 5.2.3 implies that h_z and $h_{\bar z}$ have the same number of zeros (unless one of them vanishes identically). Since h is a branched cover, $|h_z|^2 - |h_{\bar z}|^2$ does not change sign, and

$$|h_z(z_0)|^2 - |h_{\bar{z}}(z_0)|^2 = 0 \qquad \text{(for some } z_0 \in \Sigma_p)$$

implies

$$h_z(z_0) = h_{\bar{z}}(z_0) = 0.$$

Therefore,

$$|h_z|^2/|h_{\bar{z}}|^2 \text{ is different from 0 and } \infty.$$

From Lemma 5.2.1, we can thus conclude

$$\Delta \log(|h_z|^2/|h_{\bar{z}}|^2) = 0 \qquad \text{on } \Sigma_p$$

and hence $(|h_z|^2/|h_{\bar{z}}|^2) \equiv \text{const}$.

Therefore, we can find a linear transformation $g': T \to T'$ so that $h' := g' \circ h$ is $\pm$ holomorphic.

q.e.d.

Theorem 5.3.2 (Schiffer and Spencer [SS]): *Let Σ be a compact Riemann surface of genus p, and let B be some disk contained in Σ. Let Σ' be another Riemann surface homeomorphic to Σ. Then Σ' is conformally equivalent to a Riemann surface obtained from Σ by deleting the disk B and replacing it by a suitable other disk.*

Proof: Let $h_0: \Sigma \to \Sigma'$ be a harmonic diffeomorphism w.r.t. some (e.g. hyperbolic in case $p > 1$) metric on Σ' (cf. Theorem 5.2.2 or 5.1.2). Let $B_0 := h_0(B)$.

We obtain a surface Σ_1 by deleting B from Σ and inserting B_0 in such a way that each point in ∂B_0 is identified with its pre-image under h_0.

By Theorem 3.3.1, Σ_1 acquires the structure of a Riemann surface. We take the map $k_1: \Sigma_1 \to \Sigma'$ coinciding with h_0 on $\Sigma \backslash B$ and being conformal on the inserted disk. Then

$$E(k_1) \leqslant E(h_0)$$

by Lemma 1.3.1.

We let $h_1: \Sigma_1 \to \Sigma'$ be the harmonic diffeomorphism homotopic to k_1. Then

$$E(h_1) \leqslant E(k_1).$$

We put $B_1 := h_1(B_0)$, obtain a surface Σ_2 by deleting B_0 from Σ_1 and inserting B_1 in such a way that each point in ∂B_1 is identified with its pre-image under h_1, and construct maps k_2 and h_2 as before. We iterate this construction, obtaining Riemann surfaces Σ_n and harmonic diffeomorphisms

$$h_n: \Sigma_n \to \Sigma'$$

with

$$E(h_n) \leqslant E(h_{n-1}) \qquad \text{for all } n \geqslant 1. \tag{5.3.1}$$

From the argument of Lemma 3.3.4 (or easier constructions if $p \leqslant 1$), we see that $(\Sigma_n)_{n \in \mathbb{N}}$ is pre-compact. Thus, after selection of a subsequence, it converges to some surface $\tilde{\Sigma}$. Likewise, the h_n then converge to a harmonic diffeomorphism

$$\tilde{h}: \tilde{\Sigma} \to \Sigma'$$

and

$$E(h_n) \to E(\tilde{h}) = \inf E(h_n) \qquad \text{by (5.3.1).} \tag{5.3.2}$$

This implies that $h_n|_{B_{n-1}}$, where B_{n-1} are the inserted disks, converges to a conformal map, as otherwise one could further decrease energy by the above replacement procedure, thus violating (5.3.2). Thus $\tilde{h}$ is conformal on some disk $\tilde{B} \subset \tilde{\Sigma}$. Therefore, the associated holomorphic quadratic differential of $\tilde{h}$ vanishes on $\tilde{B}$ (Lemma 1.3.3). It then vanishes identically on $\tilde{\Sigma}$ because it is holomorphic. Consequently, $\tilde{h}$ is conformal everywhere.

q.e.d.

The following result was obtained in [Bö] by a different method:

Corollary 5.3.1: *Let Σ be a compact Riemann surface of genus p, Σ' another Riemann surface homeomorphic to Σ, $\Sigma'_0 \subset \Sigma'$ an open Riemann surface with $\Sigma' \setminus \Sigma'_0$ containing an open disk. Then Σ'_0 is conformally equivalent to a subset of Σ.*

Proof: Let B be a disk contained in $\Sigma' \setminus \Sigma'_0$. Theorem 5.3.2 (with the roles of Σ and Σ' interchanged) implies that $\Sigma' \setminus B$ is conformally equivalent to a subset of Σ. Since $\Sigma'_0 \subset \Sigma' \setminus B$, the claim follows.

q.e.d.

6 HARMONIC MAPS AND TEICHMÜLLER SPACES

6.1. The basic definitions

We let Σ be a closed oriented surface of genus p. The diffeomorphisms of Σ form a group, denoted by

$$\mathscr{D} = \mathrm{Diff}(\Sigma).$$

Similarly, the group of orientation-preserving diffeomorphisms is denoted by $\mathscr{D}_+$, and $\mathscr{D}_0$ is the group of diffeomorphisms homotopic to the identity.

Lemma 6.1.1: *Suppose $h_0, h_1 \in \mathscr{D}$ are homotopic. Then they are also isotopic, i.e. there exists a homotopy h_t, $t \in [0, 1]$, for which each h_t is a diffeomorphism.*

Proof: We put any hyperbolic metric g on Σ. By Theorem 5.1.2, we can minimize the energy in the isotopy class of h_i $(i = 0, 1)$ (of course, the energy is defined w.r.t. the metric g on domain and image), getting a harmonic diffeomorphism u_i isotopic to h_i. Since h_0 and h_1 are homotopic, $u_0 = u_1$ by uniqueness (Theorem 2.2.3), and the claim follows.

q.e.d.

Corollary 6.1.1: $\pi_0(\mathscr{D}) = \mathscr{D}/\mathscr{D}_0$, $\pi_0(\mathscr{D}_+) = \mathscr{D}_+/\mathscr{D}_0$.

Proof: Lemma 6.1.1 implies that $\mathscr{D}_0$ is the connected component of the identity in $\mathscr{D}$.

q.e.d.

Definition 6.1.1: A conformal structure on Σ is an atlas of coordinate charts covering Σ with conformal transition functions.

We saw in Chapter 3 (Theorem 3.3.1) that each L^∞-metric on Σ induces a conformal structure. We also recall:

Lemma 6.1.2: *For each conformal structure on Σ, there exists a unique hyperbolic metric on Σ. Here, 'hyperbolic' means that the Gauss curvature is identically -1.*

Proof: The existence is contained in Theorem 3.3.1. For uniqueness, we note that a conformal map h (without branch points) between hyperbolic metrics is an isometry. This can, for example, be seen as follows.

In conformal coordinates, by (5.2.6),

$$\Delta \log|h_z|^2 = -2 + 2|h_z|^2$$

and the maximum principle implies that

$$\max |h_z|^2 \leqslant 1$$
$$\min |h_z|^2 \geqslant 1.$$

Hence $|h_z| \equiv 1$. Thus, the identity map of Σ is an isometry between any two conformally equivalent hyperbolic metrics, and hence they coincide.

q.e.d.

In the following, we shall frequently identify a conformal structure with the corresponding hyperbolic metric.

Of course, we do not want to distinguish between equivalent metrics or conformal structures. Here, two metric structures are considered equivalent if there exists an (orientation-preserving) isometric map between them, and two conformal structures are equivalent if there exists an (orientation-preserving)[1] conformal diffeomorphism between them. In other words, if g is a conformal structure (or metric) on Σ, and $v: \Sigma \to \Sigma$ a diffeomorphism, then

$$v: (\Sigma, v^*g) \to (\Sigma, g)$$

is conformal (respectively isometric) and for this reason, we do not want to consider the structure v^*g as different from g.

Definition 6.1.2: We let M_p denote the space of equivalence classes of conformal structures (or hyperbolic metrics) on Σ, and call it the moduli space of surfaces of genus p.

If $\mathcal{M}_p$ denotes the space of hyperbolic metrics on Σ, then

$$M_p = \mathcal{M}_p / \mathcal{D}_+. \tag{6.1.1}$$

Unfortunately, the topology of M_p is rather complicated. It does not have a manifold structure; singularities arise from conformal structures that admit non-trivial automorphisms (conformal self-maps).

[1] An orientation-reversing equivalence of conformal structures will be called anticonformal. In the present context, anticonformal diffeomorphisms are not considered as inducing the same structure.

For this reason, Teichmüller introduced a slightly weaker equivalence relation, where two conformal structures (Σ, g_1) and (Σ, g_2) are considered to be equivalent, if there exists a conformal diffeomorphism between them which is homotopic to the identity map of Σ. In other words, we now look at triplets

$$(\Sigma, g, f)$$

where g is a conformal structure on Σ and $f: \Sigma \to \Sigma$ is a diffeomorphism, and consider (Σ, g_0, f_0) and (Σ, g_1, f_1) as equivalent, if there exists a conformal map k such that

$$\begin{array}{ccc} \Sigma & \xrightarrow{f_0} & (\Sigma, g_0) \\ & & \big\downarrow k \\ \Sigma & \xrightarrow{f_1} & (\Sigma, g_1) \end{array}$$

commutes up to homotopy.

Definition 6.1.3: The space of these equivalence classes is denoted by T_p and called the Teichmüller space of surfaces of genus p.

We also note

$$T_p = \mathscr{M}_p / \mathscr{D}_0 \tag{6.1.2}$$

and

$$M_p = T_p / \pi_0(\mathscr{D}_+) \tag{6.1.3}$$

We could use these relations to define a topological structure on T_p and M_p and try to develop the theory of T_p from this relation, following Earle and Eells [EE] and Fischer and Tromba (cf. [Tr 4]). We shall use a different approach. It is not difficult to check, however, that the topological and differentiable structure introduced below on T_p coincides with that arising from (6.1.2). Our approach will depend on the existence and uniqueness of harmonic diffeomorphisms (Corollary 5.2.5) and on computations determining the effect of infinitesimal variations of the domain and image metric on the energy of the corresponding harmonic map. Variations of the image metric will be of particular importance. The necessary formulae, displayed in Section 6.2, are due to Wolf [Wf].

6.2. The topological and differentiable structure of T_p. Teichmüller's Theorem

Our approach will be based on the existence and uniqueness results for harmonic diffeomorphisms (Corollary 5.2.5) recalled here as:

Lemma 6.2.1: *Given any conformal structure* (Σ, g), *for any other hyperbolic* *structure* (Σ, γ), *there exists a unique harmonic map*

$$u(g, \gamma): (\Sigma, g) \to (\Sigma, \gamma)$$

homotopic to the identity of Σ.

$u(g, \gamma)$ *is a diffeomorphism.* $\varphi(u) := \rho^2 u_z \bar{u}_z \, dz^2$ *is holomorphic, where* z *is a conformal coordinate on* (Σ, g), *and the hyperbolic metric* γ *is locally represented as* $\rho^2 \, du \, d\bar{u}$.

$$\varphi \equiv 0 \Leftrightarrow u \text{ is conformal} \Leftrightarrow g = \gamma.$$

We get an induced map $q(g)$: $T_p \to Q(g)$ where $Q(g)$ is the space of holomorphic quadratic differentials on (Σ, g).

Theorem 6.2.1: $q(g)$ *is bijective* (*for any* $g \in T_p$).

Remark 6.2.1: Theorem 6.2.1 was proved by Wolf [Wf] under the assumption, however, that the dimension of T_p is already known. Our proof will not need this assumption, but will rather determine $\dim T_p$ as a corollary; cf. Theorem 6.2.2.

Before proving Theorem 6.2.1, we need to develop some infinitesimal computations for the variation of a harmonic map. These are due to Wolf and will play an essential role in our approach to the different structures of T_p (topological, differentiable, complex, Riemannian, and Kählerian).

Let $u: (\Sigma, g) \to (\Sigma, \gamma)$ be harmonic, z a conformal coordinate on (Σ, g), γ represented as $\rho^2 \, du \, d\bar{u}$. Then

$$u^*(\rho^2 \, du \, d\bar{u}) = \rho^2 u_z \bar{u}_z \, dz^2 + \rho^2 (u_z \bar{u}_{\bar{z}} + \bar{u}_z u_{\bar{z}}) \, dz \, d\bar{z} + \rho^2 u_{\bar{z}} \bar{u}_{\bar{z}} \, d\bar{z}^2.$$

We now assume that $\psi \, dz^2$, $\varphi_1 \, dz^2$, $\varphi_2 \, dz^2 \in Q(g)$ and that $\gamma(t_1, t_2)$ is a family in T_p so that with $t = (t_1, t_2)$ for the corresponding harmonic maps $u^t: (\Sigma, g) \to (\Sigma, \gamma^t)$ and metrics $\rho^2 \, du^t \, d\bar{u}^t$,

$$\rho^2 \, du^t \, d\bar{u}^t = \psi + t_1 \varphi_1 + t_2 \varphi_2.$$

We let g be represented by $\lambda^2 \, dz \, d\bar{z}$, and define

$$H(t) := |u_z^t|^2 = \frac{\rho^2(u^t)}{\lambda^2(z)} u_z^t \bar{u}_{\bar{z}}^t$$

$$L(t) := |\bar{u}_z^t|^2 = \frac{\rho^2(u^t)}{\lambda^2(z)} u_{\bar{z}}^t \bar{u}_z^t.$$

In this notation

$$\begin{aligned} \rho^2 \, du^t \, d\bar{u}^t = {} & (\psi + t_1 \varphi_1 + t_2 \varphi_2) \, dz^2 + \lambda^2 (H(t) + L(t)) \, dz \, d\bar{z} \\ & + (\bar{\psi} + t_1 \bar{\varphi}_1 + t_2 \bar{\varphi}_2) \, d\bar{z}^2. \end{aligned} \tag{6.2.1}$$

We want to calculate the derivatives of $H(t)$ and $L(t)$ at $t = 0$. For certain reasons, it is more convenient to work with harmonic Beltrami differentials instead of holomorphic quadratic differentials.

Thus, we get

$$\mu := \frac{\bar{\psi}}{\lambda^2} \qquad \alpha = \frac{\bar{\varphi}_1}{\lambda^2} \qquad \beta = \frac{\bar{\varphi}_2}{\lambda^2}$$

for $\psi, \varphi_1, \varphi_2 \in Q(g)$.

These differentials then satisfy

$$\mu_z + \frac{2\lambda_z}{\lambda}\mu = 0 \tag{6.2.2}$$

because $\bar{\psi}_z = 0$.

Solutions of (6.2.2) are called harmonic Beltrami differentials on (Σ, g), and the space of these differentials is denoted by $H(g)$. Of course, $Q(g)$ and $H(g)$ are dual to each other, and the duality is given by the metric λ^2. Then

$$H(t)L(t) = (\mu + t_1\alpha + t_2\beta)(\bar{\mu} + t_1\bar{\alpha} + t_2\bar{\beta}) \tag{6.2.3}$$

$$\Delta \log H(t) = -2 + 2(H(t) - L(t)) \tag{6.2.4}$$

(cf. (5.2.6) and note $H(t) \neq 0$ by (5.2.24)).

We then denote differentiation w.r.t. α, $\bar{\alpha}$, etc. by corresponding subscripts and obtain at $t = 0$

$$H_\alpha L + HL_\alpha = \mu\bar{\alpha} + \bar{\mu}\alpha \tag{6.2.5}$$

$$\Delta\frac{H_\alpha}{H} = 2(H_\alpha - L_\alpha) \tag{6.2.6}$$

$$H_{\alpha\bar{\beta}}L + H_\alpha L_{\bar{\beta}} + H_{\bar{\beta}}L_\alpha + HL_{\alpha\bar{\beta}} = \alpha\bar{\beta} \tag{6.2.7}$$

$$\Delta\left[\frac{H_{\alpha\bar{\beta}}}{H} - \frac{H_\alpha H_{\bar{\beta}}}{H^2}\right] = 2(H_{\alpha\bar{\beta}} - L_{\alpha\bar{\beta}}) \tag{6.2.8}$$

and similarly, for three- and four-parameter variations,

$$H_{\alpha\bar{\beta}\gamma}L + H_{\alpha\bar{\beta}}L_\gamma + H_{\alpha\gamma}L_{\bar{\beta}} + H_{\bar{\beta}\gamma}L_\alpha + H_\alpha L_{\bar{\beta}\gamma} + H_{\bar{\beta}}L_{\alpha\gamma} + H_\gamma L_{\alpha\bar{\beta}} + HL_{\alpha\bar{\beta}\gamma} = 0 \tag{6.2.9}$$

$$\Delta\left[\frac{H_{\alpha\bar{\beta}\gamma}}{H} \pm \cdots\right] = 2(H_{\alpha\bar{\beta}\gamma} - L_{\alpha\bar{\beta}\gamma}) \tag{6.2.10}$$

$$H_{\alpha\bar{\beta}\gamma\bar{\delta}}L \pm \cdots + HL_{\alpha\bar{\beta}\gamma\bar{\delta}} = 0 \tag{6.2.11}$$

$$\Delta\left[\frac{H_{\alpha\bar{\beta}\gamma\bar{\delta}}}{H} \pm \cdots\right] = 2(H_{\alpha\bar{\beta}\gamma\bar{\delta}} - L_{\alpha\bar{\beta}\gamma\bar{\delta}}). \tag{6.2.12}$$

Since $H - L$ is the functional determinant of u^t, we obtain

$$\int (H(t) - L(t))\lambda^2(z)\,dz\,d\bar{z} = \text{const.} \qquad \text{(independent of } t\text{)} \tag{6.2.13}$$

and hence

$$\int H_\alpha = \int L_\alpha \qquad \int H_{\alpha\bar{\beta}} = \int L_{\alpha\bar{\beta}} \quad \text{etc.} \tag{6.2.14}$$

(this, of course, also follows from (6.2.6), (6.2.8)). Furthermore

$$E(t) := E(u^t) = \int (H(t) + L(t))\lambda^2 \, dz \, d\bar{z} \tag{6.2.15}$$

is the energy of u^t, hence

$$E_\alpha = \int H_\alpha + L_\alpha = 2 \int L_\alpha \qquad \text{by (6.2.14)} \tag{6.2.16}$$

$$E_{\alpha\bar{\beta}} = \int H_{\alpha\bar{\beta}} + L_{\alpha\bar{\beta}} = 2 \int L_{\alpha\bar{\beta}} \tag{6.2.17}$$

and so on.

Of special importance is the case $\psi = 0$. If $\psi = 0$, then $g = \gamma(0)$ and u is hence conformal and thus

$$L(0) \equiv 0 \qquad H(0) \equiv 1. \tag{6.2.18}$$

We then obtain from (6.2.5) (always at $t = 0$) that

$$L_\alpha \equiv 0 \tag{6.2.19}$$

from (6.2.6) that

$$H_\alpha \equiv 0 \tag{6.2.20}$$

from (6.2.7), (6.2.19) that

$$L_{\alpha\bar{\beta}} \equiv \alpha\bar{\beta} \tag{6.2.21}$$

and from (6.2.8), (6.2.20), (6.2.21) that

$$H_{\alpha\bar{\beta}} = -2(\Delta - 2)^{-1}(\alpha\bar{\beta}). \tag{6.2.22}$$

Furthermore

$$L_{\alpha\bar{\beta}\gamma} \equiv 0 \equiv H_{\alpha\bar{\beta}\gamma} \tag{6.2.23}$$

and

$$L_{\alpha\bar{\beta}\gamma\bar{\delta}} = -H_{\alpha\bar{\beta}}L_{\gamma\bar{\delta}} - H_{\gamma\bar{\delta}}L_{\alpha\bar{\beta}} - H_{\alpha\bar{\delta}}L_{\gamma\bar{\beta}} - H_{\gamma\bar{\beta}}L_{\alpha\bar{\delta}}. \tag{6.2.24}$$

Also, from (6.2.15) and (6.2.19),

$$E_\alpha = 0 \tag{6.2.25}$$

and from (6.2.16), (6.2.21),

$$E_{\alpha\bar{\beta}} = 2 \int \alpha\bar{\beta}\lambda^2 \, dz \, d\bar{z}. \tag{6.2.26}$$

For later purposes, we also observe:

Lemma 6.2.2: *Assume that $\mu \in H(g)$ is a critical point for E, i.e. $E_\alpha = 0$ for all variations $\mu + t\alpha$, $\alpha \in H(g)$. Then $\mu \equiv 0$.*

Proof: We choose $\alpha = \mu$. From (6.2.15) and (6.2.14) we obtain

$$\int H_\alpha \lambda^2 \, dz \, d\bar{z} = 0. \tag{6.2.27}$$

If H_α is not identically zero,

$$H_\alpha(z_0) := \min H_\alpha < 0.$$

Then

$$\Delta H_\alpha(z_0) > 0.$$

From (6.2.6), using $H > 0$,

$$L_\alpha(z_0) < 0$$

contradicting (6.2.5) (note that $\alpha = \mu$ in the present case). Hence

$$H_\alpha \equiv 0,$$

and from (6.2.6) we obtain

$$L_\alpha \equiv 0$$

and hence from (6.2.5) again

$$\mu \equiv 0$$

(and thus also $L \equiv 0$, since $H > 0$)

q.e.d.

Proof of Theorem 6.2.1: We first show that $q(g)$ is injective. This is due to Sampson [Sa]. Let

$$u^i : (\Sigma, g) \to (\Sigma, \gamma^i) \qquad (i = 1, 2)$$

be harmonic and leading to the same quadratic differential, i.e. $\psi_1 = \psi_2$ (with $H^i = |u_z^i|^2$, $L^i = |u_{\bar{z}}^i|^2$, $\psi_i = \rho^2 u_z^i \bar{u}_z^i$). Consequently (cf. (6.2.2))

$$H^1 L^1 = H^2 L^2. \tag{6.2.28}$$

From (6.2.3), using $H^i > 0$,

$$\Delta \log(H^1/H^2) = 2(H^1 - H^2 + L^2 - L^1). \tag{6.2.29}$$

Equations (6.2.28), (6.2.29) and the maximum principle imply

$$\max(H^1/H^2) \leqslant 1 \leqslant \min(H^1/H^2).$$

Hence $H^1 = H^2$ and by (6.2.28) also $L^1 = L^2$. Thus (cf. (6.2.1))

$$\rho^2 \, du^1 \, d\bar{u}^1 = \rho^2 \, du^2 \, d\bar{u}^2$$

i.e. $\gamma^1 = \gamma^2$, proving the injectivity of $q(g)$.

We point out that the preceding argument uses the fact that the image curvature is constant in an essential way.

To show that $q(g)$ is surjective, we proceed as follows. Let $\varphi \in Q(g)$. We look at the path $t\varphi$, $t \in [0, 1]$, in $Q(g)$ and try to find corresponding metrics $\gamma(t)$ and harmonic maps

$$u^t\colon(\Sigma, g)\to(\Sigma, \gamma^t)$$

with

$$\rho^2 u_z^t \bar{u}_z^t = t\varphi. \tag{6.2.30}$$

Suppose this is possible for some $\tau\in[0,1]$. Then, at τ

$$H(\tau)\geqslant 1 \tag{6.2.31}$$

from (6.2.3) and the minimum principle, since $H(z)\neq 0$ (cf. (5.2.24)). Moreover, at τ

$$H(\tau)L(\tau)=\tau^2\frac{\varphi\bar{\varphi}}{\lambda^2}. \tag{6.2.32}$$

We now take real instead of complex derivatives and denote a derivative in the direction $t\varphi$ by an overdot:

$$\dot{H}(\tau)L(\tau)+H(\tau)\dot{L}(\tau)=2\tau|\varphi\,\mathrm{d}z^2|^2 \qquad \text{from (6.2.5)} \tag{6.2.33}$$

and hence, from (6.2.6), (6.2.32), (6.2.33),

$$\Delta\frac{\dot{H}(\tau)}{H(\tau)}=2\left(\frac{\dot{H}(\tau)}{H(\tau)}\left(H(\tau)+\frac{\tau^2|\varphi\,\mathrm{d}z^2|^2}{H(\tau)}\right)-\frac{2\tau^2|\varphi\,\mathrm{d}z^2|^2}{H(\tau)}\right). \tag{6.2.34}$$

Equations (6.2.31), (6.2.34) and the maximum principle imply

$$-2\sup|\varphi\,\mathrm{d}z^2|^2\leqslant\frac{\dot{H}(\tau)}{H(\tau)}\leqslant 2\sup|\varphi\,\mathrm{d}z^2|^2$$

i.e.

$$|\dot{H}(\tau)|\leqslant cH(\tau). \tag{6.2.35}$$

From (6.2.35), one easily shows that the set of those $\tau\in[0,1]$ for which a solution $H(\tau)$ of (6.2.34) exists, is open and closed and hence coincides with the whole interval $[0,1]$, since it contains $\tau=0$. Equation (6.2.32) then determines $L(\tau)$.

The desired metric is then

$$\tau\varphi\,\mathrm{d}z^2+\lambda^2(H(\tau)+L(\tau))\,\mathrm{d}z\,\mathrm{d}\bar{z}+\tau\bar{\varphi}\,\mathrm{d}\bar{z}^2.$$

One may check that it has constant negative curvature.

q.e.d.

We use $q(g)$ to define on T_p the structure of a differentiable manifold. In order to make this structure canonical, i.e. independent of the choice of g, we have to show that the transition maps

$$q(g_2)\circ q(g_1)^{-1}$$

are (continuous and) differentiable.

This is demonstrated as follows. Let φ, $\psi\in Q(g_1)$. We look at

$$q(g_2)\circ q(g_1)^{-1}(\psi+t\varphi), \qquad t\in\mathbb{R}$$

in a neighbourhood of $t = 0$. Let $\lambda^2\,dz\,d\bar{z}$ be a conformal metric on (Σ, g_1). Then

$$q(g_1)^{-1}(\psi + t\varphi) = (\psi + t\varphi)\,dz^2 + \lambda^2(H(t) + L(t))\,dz\,d\bar{z} + (\bar{\psi} + t\bar{\varphi})\,d\bar{z}^2$$

(where $H(t) + L(t)$ is the energy density of the harmonic map as before) describes a differentiable family of hyperbolic metrics. The harmonic maps

$$u^t: (\Sigma, g_2) \to (\Sigma, q(g_1)^{-1}(\psi + t\varphi))$$

depend differentiably on the image metric (cf. Theorem 2.5.6), and hence on t. Therefore the holomorphic quadratic differential defined by u^t on (Σ, g_2) also depends differentiably on t. This shows that the composition map is differentiable; and hence that the differentiable structure defined on T_p is canonical. The same argument shows that the composition map is actually of class C^∞.

Remark 6.2.2: One can even show that the composition maps are real analytic.

As a consequence of Theorem 6.2.1 we therefore obtain:

Teichmüller's Theorem (6.2.2): *T_p is diffeomorphic to $\mathbb{R}^{6p-6}$.*

Proof: $Q(g)$ is a real vector space of dimension $6p - 6$ by the Riemann–Roch Theorem.

q.e.d.

Remark 6.2.3: By the preceding considerations, our differentiable structure is canonical since it is independent of g. In order to show that the above result is actually the same as in Teichmüller's original theorem, one needs to verify that our differentiable structure is the same as Teichmüller's.

Actually, comparing other procedures of introducing a differentiable structure on T_p (cf. e.g. [A1], [Tr4]), it is not too difficult to see that they all agree with ours.

We also point out that Theorem 6.2.2 was already proved by Fricke–Klein [FK], long before Teichmüller's work [Tm1]. The connection of T_p with holomorphic quadratic differentials, however, was only recognized by Teichmüller.

By Theorem 6.2.1, any $\varphi \in Q(g)$ represents a hyperbolic metric $m(\varphi)$. We put

$$E(\varphi) := E(u(g, m(\varphi)))$$

in other words $E(\varphi)$ is the energy of the harmonic map (homotopic to the identity of Σ) from g into $m(\varphi)$.

Lemma 6.2.3: *$E(\varphi)$ is a proper function, i.e.*

$$E_c := \{\varphi \in Q(g) : E(\varphi) \leq c\}$$

is compact for any $c \in \mathbb{R}$.

Proof: Let $c \in \mathbb{R}$, $\varphi \in E_c$. Abbreviating, we write

$$u = u(g, m(\varphi)) \qquad H = |u_z|^2 \qquad L = |\bar{u}_z|^2 \qquad \text{etc.}$$

Recalling

$$H - L \geqslant 0 \text{ since } u \text{ is a diffeomorphism,}$$

we obtain from (6.2.3)

$$\Delta \log H \geqslant -2 \tag{6.2.36}$$

Let $0 < \rho_0 < \frac{1}{2}$ injectivity radius of (Σ, g)). For $\rho_0 \leqslant \rho \leqslant 2\rho_0$, we obtain from (6.2.36) by Green's representation formula, noting $H \geqslant 1$ (6.2.31)

$$\left| \log H(z) - \frac{1}{\rho} \int_{\partial B(z,\rho)} \log H(z) \right| \leqslant 2 \int_{B(z,\rho)} \left| \log \frac{\operatorname{dist}(\zeta, z)}{\rho} \right| d\zeta.$$

We choose ρ in such a way that

$$\int_{\partial B(z,\rho)} \log H(\zeta) \leqslant \frac{2}{\rho_0} \int_{B(z,\rho_0)} \log H(\zeta).$$

Since $\log H \leqslant H$, as $H \geqslant 1$, and

$$\int H \leqslant E(u) \leqslant c \qquad \text{by assumption,}$$

we infer that

$$\log H \leqslant k \qquad \text{where } k \text{ depends on } c.$$

Since $L \leqslant H$, we obtain

$$H + L \leqslant K$$

where K again depends on c.
Hence also

$$|\varphi|^2 = HL \leqslant K^2$$

proving the claim.

q.e.d.

Remark 6.2.4: $E(\varphi)$ by Lemma 6.2.3 provides an exhaustion function of Teichmüller space, which has precisely one critical point (Lemma 6.2.2), and at this point the Hessian is strictly positive definite (6.2.26). This is related to Tromba's argument deducing the result that T_p is a cell from the existence of such an exhaustion function via elementary Morse theory (cf. [Tr2]). He considered the energy, however, as a function of the domain metric, thereby making the proof of properness somewhat more difficult; the proof depends on Lemma 3.3.4.

6.3. The complex structure

We first define an almost-complex structure as follows. We identify the cotangent space of T_p at (Σ, γ) with $Q(\gamma)$, the space of holomorphic quadratic differentials on (Σ, γ) (The tangent space is given by $H(\gamma)$, the space of harmonic Beltrami differentials on (Σ, γ). The relation between these spaces is an example of Serre duality.). We define

$$J: Q(\gamma) \to Q(\gamma)$$

as multiplication by i $(=\sqrt{-1})$ in $Q(\gamma)$. Clearly

$$J^2 = -\,\mathrm{id}.$$

Lemma 6.3.1:

$$N(J)(X, Y) := [JX, JY] - J[JX, Y] - J[X, JY] - [X, Y] = 0$$

for any smooth vector fields X and Y on T_p.

Proof: Let $\gamma \in T_p$ and choose the chart centred at γ. Let $X(\gamma) = \phi \in Q(\gamma)$, $Y(\gamma) = \psi \in Q(\gamma)$ be the corresponding holomorphic quadratic differentials. Because of the tensorial character of $N(J)(X, Y)$, we can assume that in this chart, X and Y are constant, i.e.

$$X(\delta) = \phi \in Q(\gamma)$$
$$Y(\delta) = \psi \in Q(\gamma)$$

for all $\delta \in T_p$. (In other words, we have to show that if X and Y commute ($[X, Y] = 0$), then any two of the vector fields X, JX, Y, JY commute as well). We identify T_p with $Q(\gamma)$ via the chart centred at γ. We denote by J_δ the almost-complex structure at δ.

We shall only show that JX and Y commute, the proof for the other cases being the same. Let f be a smooth function in a neighbourhood of $0 \in Q(\gamma)$. We have to show that

$$\begin{aligned} 0 &= [JX, Y] f(0) \\ &= \lim_{s,t \to 0} \frac{1}{st} \{ f(sJ_0\phi + t\psi) - f(t\psi) - f(sJ_0\phi) + f(0) \\ &\quad - (f(t\psi + J_{t\psi}(s\phi)) - f(sJ_0\phi) - f(t\psi) + f(0)) \} \\ &= \lim_{s,t \to 0} \frac{1}{st} \{ f(sJ_0\phi + t\psi) - f(t\psi + J_{t\psi}(s\phi)) \}. \end{aligned}$$

It hence suffices to show that

$$\lim_{s,t \to 0} \frac{1}{st} J_{t\psi}(s\phi) = 0. \tag{6.3.1}$$

For $\omega \in Q(\gamma)$, we let $m(\omega)$ be the corresponding hyperbolic metric (via the correspondence established in Theorem 6.2.1). As in the proof of Theorem 6.2.1, we look at the harmonic map $\hat{u}$ from $m(t\psi)$ into $m(t\psi + s\phi)$ and expand the metric $m(t\psi + s\phi)$ in terms of this map. Writing $m(t\psi)$ as $\rho^2\,du^{t,0}\,d\bar{u}^{t,0}$, this gives

$$m(t\psi + s\phi) \sim s\tilde{\varphi}^t(du^{t,0})^2 + \rho^2(1 + \tfrac{1}{2}s^2(\hat{H}'' + \hat{L}'') + o(s^2))\,du^{t,0}\,d\bar{u}^{t,0} + s\bar{\tilde{\varphi}}^t(d\bar{u}^{t,0})^2$$

where $\hat{H} + \hat{L}$ is the energy density of the harmonic map as usual and a prime denotes a derivative w.r.t. s., and where $\tilde{\varphi}^t$ is holomorphic w.r.t. the conformal structure $du^{t,0}\,d\bar{u}^{t,0}$.

Now $m(t\varphi + J_{t\varphi}(s\phi))$ is obtained from this expression by simply replacing $\tilde{\varphi}$ by $i\tilde{\varphi}$ and calculating for the middle term the energy density $\hat{\hat{H}} + \hat{\hat{L}}$ of the harmonic map from $m(t\psi)$ into $m(t\psi + J_{t\psi}(s, \phi))$, i.e.

$$\begin{aligned} m(t\psi + J_{t\psi}(s\phi)) \sim{} & (is\tilde{\varphi}^t + o(s^2))(du^{t,0})^2 \\ & + \rho^2(1 + \tfrac{1}{2}s^2(\hat{\hat{H}}'' + \hat{\hat{L}}'') + o(s^2))\,du^{t,0}\,d\bar{u}^{t,0} \\ & + (-is\bar{\tilde{\varphi}}^t + o(s^2))(d\bar{u}^{t,0})^2. \end{aligned}$$

The important point is that the middle term has no linear term in s. We then let $u^{t+s,t}$ and $u^{t+s,0}$ be the harmonic maps from $m(t\psi)$ and $m(0)$ respectively into $m(t\psi + s\phi)$ and $v := (u^{t+s,t})^{-1} \circ u^{t+s,0}$.

We then pull the preceding expression for $m(t\psi + J_{t\psi}(s\phi))$ back under v from $m(t\psi)$ to $m(0)$. If we neglect terms that grow quadratically in s or t, we get the same expression as when pulling back $m(t\psi + s\phi)$ under v, except that ϕ is replaced by $i\phi$, i.e. we obtain, putting $\phi = \varphi_0\,dz^2$, $\psi = \psi_0\,dz^2$,

$$\begin{aligned} m(t\psi + J_{t\psi}(s\phi)) \sim{} & (is\varphi_0 + t\psi_0)\,dz^2 + \lambda^2(1 + 2is(\varphi_0 v_z v_{\bar{z}} - \bar{\varphi}_0 \bar{v}_{\bar{z}} \bar{v}_z))\,dz\,d\bar{z} \\ & + (-is\bar{\varphi}_0 + t\bar{\psi}_0)\,d\bar{z}^2 + o(s^2) + o(t^2). \end{aligned}$$

Of course

$$\begin{aligned} m(t\psi + s\phi) \sim{} & (s\varphi_0 + t\psi_0)\,dz^2 + \lambda^2(1 + 2s(\varphi_0 v_z v_{\bar{z}} + \bar{\varphi}_0 \bar{v}_{\bar{z}} \bar{v}_z))\,dz\,d\bar{z} \\ & + (s\bar{\varphi}_0 + t\bar{\psi}_0)\,d\bar{z}^2 + o(s^2) + o(t^2) \end{aligned}$$

Hence, cf. (6.2.17),

$$\begin{aligned} m(t\psi + J_0(s\phi)) &= m(t\psi + is\phi) \\ &\sim (is\varphi_0 + t\psi_0)\,dz^2 + \lambda^2(1 + 2is(\varphi_0 v_z v_{\bar{z}} - \bar{\varphi}_0 \bar{v}_{\bar{z}} \bar{v}_z))\,dz\,d\bar{z} \\ &\quad + (-is\bar{\varphi}_0 + t\bar{\psi}_0)\,d\bar{z}^2 + o(s^2) + o(t^2). \end{aligned}$$

We conclude that at $s = t = 0$, the metrics $m(t\psi + J_{t\psi}(s\phi))$ and $m(t\psi + sJ_0\phi)$ coincide together with their $\partial/\partial s$, $\partial/\partial t$, and $\partial^2/\partial s\partial t$ derivatives. Since the image metric uniquely determines the harmonic map from $m(0)$ into this metric, we conclude that the harmonic maps from $m(0)$ into $m(t\psi + sJ_0\phi)$ and $m(t\psi + J_{t\psi}(s\phi))$ also coincide at $s = t = 0$ together with their $\partial/\partial s$, $\partial/\partial t$, and $\partial^2/\partial s\partial t$ derivatives. Hence the same holds for the associated holomorphic quadratic differentials, and we conclude

$$0 = \frac{\partial^2}{\partial s \partial t}(t\psi + sJ_0\phi)|_{s=t=0}$$

$$= \frac{\partial^2}{\partial s \partial t}(t\psi + J_{t\psi}(s\phi))|_{s=t=0},$$

i.e. (6.3.1).

q.e.d.

As a consequence of Lemma 6.3.1, the fact that our differentiable structure is of class C^∞ (cf. Section 6.2), and the Theorem of Newlander and Nirenberg [NN], we obtain:

Theorem 6.3.1: *The almost complex structure J defined on T_p is integrable. Hence T_p acquires the structure of a complex manifold.*

Remark 6.3.1: A complex structure on Teichmüller space was first introduced by Teichmüller [Tm2]. More systematically, it was first investigated by Bers (cf. [Be1, [Be2], [A1]). Again, one can check that the present complex structure agrees with that of Bers.

Remark 6.3.2: Using Remark 6.2.2, one can actually avoid appealing to the Newlander–Nirenberg Theorem. Namely, in the real analytic case, the integrability of an almost-complex structure J with vanishing Nijenhuis tensor $N(J)$ can be reduced to the classical Theorem of Frobenius.

Remark 6.3.3: It might be worth pointing out that the transition functions $q(g_2)\circ q(g_1)^{-1}$ are not holomorphic.

Our complex structure on T_p induces a complex structure on $T_p \times H$ with the following property. If $g \in T_p$ and (Σ, g) is represented by H/Γ_g where Γ_g is a discrete subgroup of Aut(H), then of course multiplication by i on $Q(g) \times H$ remains invariant under the operation of Γ_g. Therefore we obtain:

Theorem 6.3.2: *The complex structure on T_p induces a complex structure on the Teichmüller curve, the fibre bundle over T_p where the fibre over $g \in T_p$ is conformally equivalent to the surface (Σ, g) (topologically, in the chart $Q(g)$, the fibre over γ can be identified with the fibre over g via the harmonic map $u(g, \gamma): (\Sigma, g) \to (\Sigma, \gamma)$).*

Such a complex structure is essentially unique, and this observation probably yields the easiest way to check the assertion of Remark 6.3.1. In any case, it is clear from our construction that the complex structure introduced above does not depend on the choice of a conformal structure g. From these observations, it follows that the complex structure is natural and canonical.

6.4. The energy as a function of the domain metric

In Section 6.2, we presented the computations of Wolf describing the effect of infinitesimal variations of the image metric. For the next section, where we discuss the Weil–Petersson metric and its properties, we also need to consider the effect of infinitesimal variations of the domain metric. The formulae of the present section, describing this effect, are mostly due to Tromba (cf. [Tr5], where they were, however, obtained in a somewhat different way than here). The advantage of our method will show itself clearly in the next section.

We let g and γ be hyperbolic structures on Σ, and $u:(\Sigma, g)\to(\Sigma,\gamma)$ be a (differentiable) map.

We let $g=\lambda^2\,dz\,d\bar{z}$ and $\gamma=\rho^2\,du\,d\bar{u}$,

$$E(u,g,\gamma)=\int_\Sigma \rho^2(u_z\bar{u}_{\bar{z}}+u_{\bar{z}}\bar{u}_z)\,dz\,d\bar{z}$$

be the energy of u.

We also let $u(g,\gamma)$ be the harmonic map from (Σ, g) into (Σ,γ) homotopic to the identity of Σ. $u(g,\gamma)$ is uniquely determined and a diffeomorphism (cf. Lemma 6.2.1).

We represent (Σ, g) as a fundamental region H/T in the upper half-plane H, where we let $\pi_1(\Sigma)$ operate as a group T of automorphisms of H; cf. Theorem 3.3.1. We shall start with some computations already used in Section 3.3.

We let $\zeta_t\colon H\to H$ be a family of diffeomorphisms depending differentiably on t, with $\zeta_0=\mathrm{id}$. For every t, we assume for $\Gamma\in\mathrm{T}$

$$\zeta_t\circ\Gamma=\Gamma^t\circ\zeta_t \tag{6.4.1}$$

where $\Gamma^t\in\mathrm{T}^t$ and T^t is isomorphic to T (the isomorphism is understood to be between abstract groups, not necessarily between subgroups of $\mathrm{Aut}(H)=\mathrm{PSL}(2,\mathbb{R})$).

We put

$$\mu:=\left.\frac{\partial\zeta_t}{\partial t}\right|_{t=0}$$

and

$$\varphi:=\rho^2(u)u_z\bar{u}_z$$

and

$$(\Sigma, g_t)=H/\mathrm{T}^t.$$

We compute, with $\zeta=\zeta_t$,

$$E(u\circ\zeta_t^{-1},g_t,\gamma)=\int_{H/\mathrm{T}}\rho^2\{(u_z\bar{u}_{\bar{z}}+\bar{u}_z u_{\bar{z}})(\zeta_z\bar{\zeta}_{\bar{z}}+\bar{\zeta}_z\zeta_{\bar{z}})$$

$$-2u_z\bar{u}_z\zeta_{\bar{z}}\bar{\zeta}_{\bar{z}}-2\bar{u}_{\bar{z}}u_{\bar{z}}\zeta_z\bar{\zeta}_z)\}\frac{dz\,d\bar{z}}{\zeta_z\bar{\zeta}_{\bar{z}}-\zeta_{\bar{z}}\bar{\zeta}_z} \tag{6.4.2}$$

(a subscript z or $\bar{z}$ denotes, of course, a corresponding derivative) and

$$\frac{\mathrm{d}}{\mathrm{d}t} E(u \circ \zeta_t^{-1}, g_t, \gamma)|_{t=0}$$

$$= \int_{H/\mathrm{T}} \rho^2 \Big\{ ((u_z \bar{u}_{\bar{z}} + \bar{u}_z u_{\bar{z}})(\mu_z \bar{\zeta}_{\bar{z}} + \zeta_z \bar{\mu}_{\bar{z}} + \bar{\mu}_z \zeta_{\bar{z}} + \bar{\zeta}_z \mu_{\bar{z}})$$

$$- 2u_z \bar{u}_z (\mu_{\bar{z}} \bar{\zeta}_{\bar{z}} + \zeta_{\bar{z}} \bar{\mu}_{\bar{z}}) - 2u_{\bar{z}} \bar{u}_{\bar{z}} (\mu_z \bar{\zeta}_z + \zeta_z \bar{\mu}_z)) \frac{1}{\zeta_z \bar{\zeta}_{\bar{z}} - \zeta_{\bar{z}} \bar{\zeta}_z}$$

$$- (\mu_z \bar{\zeta}_{\bar{z}} + \zeta_z \bar{\mu}_{\bar{z}} - \mu_{\bar{z}} \bar{\zeta}_z - \zeta_{\bar{z}} \bar{\mu}_z)((u_z \bar{u}_{\bar{z}} + \bar{u}_z u_{\bar{z}})(\zeta_z \bar{\zeta}_{\bar{z}} + \bar{\zeta}_z \zeta_{\bar{z}})$$

$$- 2u_z u_{\bar{z}} \zeta_{\bar{z}} \bar{\zeta}_{\bar{z}} - 2u_{\bar{z}} \bar{u}_{\bar{z}} \zeta_z \bar{\zeta}_z) \frac{1}{(\zeta_z \bar{\zeta}_{\bar{z}} - \zeta_{\bar{z}} \bar{\zeta}_z)^2} \Big\} \mathrm{d}z \, \mathrm{d}\bar{z}$$

$$= - \int (\rho^2 u_z \bar{u}_z \mu_{\bar{z}} + \rho^2 u_{\bar{z}} \bar{u}_{\bar{z}} \bar{\mu}_z) \, \mathrm{d}z \, \mathrm{d}\bar{z},$$

since at $t = 0, \zeta = z$,

$$= -2 \operatorname{Re} \int \varphi \mu_{\bar{z}} \, \mathrm{d}z \, \mathrm{d}\bar{z}. \tag{6.4.3}$$

We first look at those μ which vanish in H/T outside the neighbourhood of a point and put on H/T

$$\zeta_t = z + t\mu$$

and

$$\zeta_t(\Gamma z) = \Gamma \zeta_t(z) \qquad \text{for } \Gamma \in \mathrm{T}.$$

Since $u(g, \gamma)$ is harmonic, we get, for those μ, (note $g_t = g$ here)

$$0 = \frac{\mathrm{d}}{\mathrm{d}t} E(u(g, \gamma) \circ \zeta_t^{-1}, g, \gamma)|_{t=0}$$

$$= -2 \operatorname{Re} \int \varphi \mu_{\bar{z}} \, \mathrm{d}z \, \mathrm{d}\bar{z}.$$

Since the same holds for $\mathrm{i}\mu$ instead of μ, we get

$$\varphi_{\bar{z}} = 0.$$

Since φ transform via

$$\varphi = (\varphi \circ \Gamma)(\Gamma_z)^2 \qquad \text{for } \Gamma \in \mathrm{T}$$

φ is a holomorphic quadratic differential on H/T.

We can also interpret these calculations in the following way. If u is any diffeomorphism, not necessarily harmonic, we can represent infinitesimal variations of u by some μ as above, and

$$\frac{\partial}{\partial u}E(u,g,\gamma)\mu=\frac{\mathrm{d}}{\mathrm{d}t}E(u\circ\zeta_t^{-1},g,\gamma)|_{t=0}$$

$$=-2\,\mathrm{Re}\int\rho^2u_z\bar{u}_z\mu_{\bar{z}}\,\mathrm{d}z\,\mathrm{d}\bar{z}$$

and hence

$$\frac{\partial}{\partial u}E(u,g,\gamma)=0\Leftrightarrow\rho^2u_z\bar{u}_z\mathrm{d}z^2 \text{ is a holomorphic quadratic differential.} \tag{6.4.4}$$

Moreover, we saw in Section 6.2, that a holomorphic quadratic differential ψ on $H/\mathrm{T}=(\Sigma,g)$ induces a variation g_t of the hyperbolic structure g via

$$t\psi\,\mathrm{d}z^2+\frac{1}{y^2}(H(t)+L(t))\,\mathrm{d}z\,\mathrm{d}\bar{z}+t\bar{\psi}\,\mathrm{d}\bar{z}^2.$$

We put

$$\alpha(z):=\bar{\psi}(z)\cdot y^2$$

which transforms via

$$(\alpha\circ\Gamma)\bar{\Gamma}_{\bar{z}}=\alpha\Gamma_z \qquad \text{for } \Gamma\in\mathrm{T}. \tag{6.4.5}$$

If $|t|$ is small enough, $|t\alpha|<1$ on H/T.

Then the metric

$$\frac{1}{y^2}|\mathrm{d}z+t\alpha\,\mathrm{d}\bar{z}|^2=\frac{1}{y^2}(t\bar{\alpha}\,\mathrm{d}z^2+(1+t^2\alpha\bar{\alpha})\,\mathrm{d}z\,\mathrm{d}\bar{z}+t\alpha\,\mathrm{d}z^2)$$

agrees with g_t to first order, since

$$\frac{\mathrm{d}}{\mathrm{d}t}(H(t)+L(t))|_{t=0}=0 \qquad \text{(cf. (6.2.19), (6.2.20))}.$$

By Theorem 3.2.1, there exists a conformal map

$$\zeta_t\colon\left(H,\frac{1}{y^2}\mathrm{d}z\,\mathrm{d}\bar{z}\right)\to\left(H,\frac{1}{y^2}|\mathrm{d}z+t\alpha\,\mathrm{d}\bar{z}|^2\right)$$

normalized in such a way that

$$\zeta_0(z)=z. \tag{6.4.6}$$

We have

$$\zeta_{t,\bar{z}}=t\alpha\zeta_{t,z}, \tag{6.4.7}$$

and hence from (6.4.6)

$$\frac{\partial}{\partial t}\zeta_{t,\bar{z}}|_{t=0}=\alpha$$

or, in the above notations, $\mu_{\bar{z}} = \alpha$. Equation (6.4.5) implies that $\zeta_t \circ \Gamma$ also solves (6.4.7) for any $\Gamma \in \mathrm{T}$. Hence ζ_t and $\zeta_t \circ \Gamma$ differ only by a conformal automorphism of H (cf. Lemma 6.1.2) and therefore

$$\zeta_t \circ \Gamma = \Gamma^t \circ \zeta_t$$

where $\Gamma^t \in \mathrm{T}^t$, and T^t is isomorphic to T; i.e. (6.4.1) holds.

This yields the following interpretation:

$$\frac{\mathrm{d}}{\mathrm{d}g} E(u(g,\gamma), g, \gamma)\psi = \frac{\mathrm{d}}{\mathrm{d}t} E(u(g,\gamma) \circ \zeta_t^{-1}, g_t, \gamma)$$

$$= -2\,\mathrm{Re} \int \varphi \bar{\psi} y^2 \,\mathrm{d}z\,\mathrm{d}\bar{z} \text{ at } t = 0. \qquad (6.4.8)$$

(Note that we can neglect the variation $u(g_t, \gamma)$ of the harmonic map $u(g,\gamma)$ because of (6.4.4)).

In particular, g is a critical point of E, if and only if

$$\begin{aligned} &\varphi \equiv 0 \\ \Leftrightarrow\quad & u \text{ is conformal} \\ \Leftrightarrow\quad & g = \gamma. \end{aligned}$$

We now want to study second variations at a critical point g of E. First of all, we wânt to take the derivative w.r.t. variations of u of

$$\frac{\mathrm{d}}{\mathrm{d}g} E(u, g, \gamma)\psi = -2 \int \mathrm{Re}\, \varphi \bar{\psi} y^2 \,\mathrm{d}z\,\mathrm{d}\bar{z}$$

where $u = u(g,\gamma)$, $g = \gamma$, $\varphi = \rho^2(u) u_z \bar{u}_z$. We consider variations $u \circ \zeta_t$ by a smooth family of diffeomeorphisms as before. We get corresponding variations

$$\varphi_t = \rho^2(u \circ \zeta_t)(u \circ \zeta_t)_z \overline{(u \circ \zeta_t)_z}$$

and since u is conformal, putting $\dot{u} := (\mathrm{d}/\mathrm{d}t) u \circ \zeta_t|_{t=0}$,

$$\frac{\mathrm{d}}{\mathrm{d}t} \varphi_t|_{t=0} = \rho^2(u) u_z \dot{\bar{u}}_z.$$

Hence

$$\frac{\mathrm{d}}{\mathrm{d}t} \frac{\partial}{\partial g} E(u + th, g, \gamma)\psi|_{t=0} = -2 \int \mathrm{Re}\, \rho^2 u_z \dot{\bar{u}}_z \bar{\psi} y^2 \,\mathrm{d}z\,\mathrm{d}\bar{z}.$$

Since $g = \gamma$ and u is conformal, $\rho^2(u(z)) = 1/y^2, u_z = 1$, hence

$$\frac{\mathrm{d}}{\mathrm{d}t} \frac{\partial}{\partial g} E(u + th, g, \gamma)\psi|_{t=0} = 2 \int \dot{u} \bar{\psi}_z \,\mathrm{d}z\,\mathrm{d}\bar{z} = 0$$

since ψ is holomorphic.

This shows that the mixed derivatives

$$\frac{\partial}{\partial u}\frac{\partial}{\partial g}E(u(g,\gamma),g,\gamma)=0 \tag{6.4.9}$$

at $g=\gamma$.

A variation ψ of g induces a variation u_ψ of u.

Since $u(g,\gamma)$ is harmonic for every g,

$$\frac{\partial}{\partial u}E(u(g,\gamma),g,\gamma)=0 \qquad \text{for all } g. \tag{6.4.10}$$

Hence

$$\begin{aligned}0&=\frac{\mathrm{d}}{\mathrm{d}g}\frac{\partial}{\partial u}E(u(g,\gamma),g,\gamma)\psi\\&=\frac{\partial^2}{\partial u^2}E(u(g,\gamma),g,\gamma)u_\psi\cdot\psi+\frac{\partial}{\partial u}\frac{\partial}{\partial g}E(u(g,\gamma),g,\gamma)\psi\\&=\frac{\partial^2}{\partial u^2}E(u(g,\gamma),g,\gamma)u_\psi\cdot\psi \qquad \text{(by (6.4.9))}.\end{aligned} \tag{6.4.11}$$

This together with (6.4.9) in turn implies (again at $g=\gamma$)

$$\frac{\mathrm{d}^2}{\mathrm{d}g^2}E(u(g,\gamma),g,\gamma)(\psi_1,\psi_2)=\frac{\partial^2}{\partial g^2}E(u(g,\gamma),g,\gamma)(\psi_1,\psi_2)$$

i.e. we can again neglect the induced variations of u. Actually, since the second variation of E w.r.t. u is strictly positive definite, (cf. (1.2.29) and note that the image has negative curvature), we conclude from (6.4.11)

$$u_\psi=0 \tag{6.4.12}$$

i.e. the variation of $u(g,\gamma)$ induced by a variation of g at $g=\gamma$ vanishes.

We now compute

$$\frac{\partial^2}{\partial g^2}E(u(g,\gamma),g,\gamma)(\psi_1,\psi_2) \qquad \text{at } g=\gamma.$$

As before, we put

$$\alpha_i(z):=\bar{\psi}_i(z)\cdot y^2 \qquad (i=1,2),$$

take the metric

$$\begin{aligned}\frac{1}{y^2}|\mathrm{d}z+(s\alpha_1+t\alpha_2)\,\mathrm{d}\bar{z}|^2&=\frac{1}{y^2}((s\bar{\alpha}_1+t\bar{\alpha}_2)\,\mathrm{d}z^2+(1+s^2\alpha_1\bar{\alpha}_1\\&\quad+st(\alpha_1\bar{\alpha}_2+\alpha_2\bar{\alpha}_1)+t^2\alpha_2\bar{\alpha}_2)\,\mathrm{d}z\,\mathrm{d}\bar{z}+(s\alpha_1+t\alpha_2)\,\mathrm{d}\bar{z}^2)\end{aligned}$$

and compare it with

$$(s\psi_1 + t\psi_2)\,\mathrm{d}z^2 + \frac{1}{y^2}(H(s,t) + L(s,t))\,\mathrm{d}z\,\mathrm{d}\bar{z} + (s\bar{\psi}_1 + t\bar{\psi}_2)\,\mathrm{d}\bar{z}^2$$

and we see that the variations agree to first order as before. Since the first variation of E vanishes, the second variation of E is independent of higher-order terms, implying again that it suffices to study

$$\frac{\mathrm{d}^2}{\mathrm{d}s\,\mathrm{d}t} E(u(g,\gamma)\circ\zeta_{st}^{-1}, g_{st}, \gamma)|_{s=t=0}$$

where g_{st} is the variation of g induced by $s\psi_1 + t\psi_2$, and

$$\zeta_{s,t}\colon \left(H, \frac{1}{y^2}\mathrm{d}z\,\mathrm{d}\bar{z}\right) \to \left(H, \frac{1}{y^2}|\mathrm{d}z + (s\alpha_1 + t\alpha_2)\,\mathrm{d}\bar{z}|^2\right)$$

is a two-parameter family of conformal diffeomorphisms with $\zeta_{00} = \mathrm{id}$ as before.

Likewise, we put

$$\frac{\partial}{\partial s}\zeta_{st}|_{s=t=0} = \mu$$

$$\frac{\partial}{\partial t}\zeta_{st}|_{s=t=0} = \nu$$

i.e.

$$\begin{aligned} \mu_{\bar{z}} &= \alpha_1 = \bar{\psi}_1 y^2 \\ \nu_{\bar{z}} &= \alpha_2 = \bar{\psi}_2 y^2. \end{aligned} \tag{6.4.13}$$

Using the formula for $(\mathrm{d}/\mathrm{d}t)E(u\circ\zeta_t^{-1}, g_t, \gamma)|_{t=0}$ employed for the derivation of (6.4.3) we compute the second variation as

$$\begin{aligned} &\frac{1}{2}\int \rho^2\{(\mu_z\bar{\nu}_{\bar{z}} + \nu_z\bar{\mu}_{\bar{z}} + \bar{\mu}_z\nu_{\bar{z}} + \bar{\nu}_z\mu_{\bar{z}}) \\ &\quad - (\mu_z + \bar{\mu}_{\bar{z}})(\nu_z + \bar{\nu}_{\bar{z}}) \\ &\quad - (\mu_z\bar{\nu}_{\bar{z}} + \nu_z\bar{\mu}_{\bar{z}} - \mu_{\bar{z}}\bar{\nu}_z - \bar{\mu}_z\nu_{\bar{z}}) \\ &\quad - (\mu_z + \bar{\mu}_{\bar{z}})(\nu_z + \bar{\nu}_{\bar{z}}) + 2(\mu_z + \bar{\mu}_{\bar{z}})(\nu_z + \bar{\nu}_{\bar{z}})\}\,\mathrm{d}z\,\mathrm{d}\bar{z} \\ &= \int \rho^2(\mu_{\bar{z}}\bar{\nu}_z + \bar{\mu}_z\nu_{\bar{z}})\,\mathrm{d}z\,\mathrm{d}\bar{z}. \end{aligned}$$

Therefore, we obtain at $g = \gamma$, noting (6.4.13) and $\rho^2 = 1/y^2$,

$$\frac{\mathrm{d}^2}{\mathrm{d}g^2}E(u(g,\gamma), g, \gamma)(\psi_1, \psi_2) = 2\,\mathrm{Re}\int \psi_1\bar{\psi}_2 y^2\,\mathrm{d}z\,\mathrm{d}\bar{z}. \tag{6.4.14}$$

6.5. The metric structure. The Weil–Petersson metric. Kähler property. The curvature

We now introduce a Riemannian metric on Teichmüller space as follows. We identify the tangent space to T_p at γ with $H(g)$, the space of harmonic Beltrami differentials on (Σ, g).

By Theorem 6.2.1 and the duality between $Q(g)$ and $H(g)$, we have a diffeomorphism between T_p and $H(\gamma)$; consequently, $H(\gamma)$ can be used as a coordinate chart.

For $\alpha, \beta \in H(g)$, we let u^t be the harmonic map from the metric given by $0 \in H(g)$ into the metric given by $t_1\alpha + t_2\beta, t = (t_1, t_2)$.

Definition 6.5.1: The Weil–Petersson metric on T_p at g is given by

$$\langle \alpha, \beta \rangle := \frac{\partial^2}{\partial t_1 \partial t_2} E(u^t)|_{t=0} = 2\,\mathrm{Re} \int_{(\Sigma,g)} (\alpha\bar{\beta})\lambda^2 \, dz \, d\bar{z} \tag{6.5.1}$$

cf. (6.2.26).

If we consider $H(g)$ as a complex vector space, we obtain an induced Hermitian metric by

$$g_{\alpha\bar{\beta}} := \langle \alpha, \beta \rangle_{\mathbb{C}} := E_{\alpha\bar{\beta}} = \int_{(\Sigma,g)} \alpha\bar{\beta}\lambda^2 \, dz \, d\bar{z}. \tag{6.5.2}$$

Likewise, we get an induced metric on $Q(g)$, e.g. for $\varphi, \psi \in Q(g)$

$$\langle \varphi, \psi \rangle_{\mathbb{C}} = \int_{(\Sigma,g)} \varphi\bar{\psi} \frac{1}{\lambda^2} \, dz \, d\bar{z}. \tag{6.5.3}$$

We now want to study derivatives of the Weil–Petersson metric at the origin of our coordinate charts.

In the following, we shall use the fact that infinitesimal diffeomorphisms are L^2-orthogonal to the harmonic Beltrami differentials in a similar way as Ahlfors [A1] and Wolpert [Wp]; otherwise, since we work with harmonic maps, our considerations will be conceptually different from theirs.

We start by deriving that fact. Let $\eta_t : (\Sigma, g) \to (\Sigma, g)$ be a smooth family of diffeomorphisms, with $\eta(0) = \mathrm{id}$, and let $\delta \in H(g)$. We recall the transformation behaviour of η:

$$\eta(\Gamma z) = \Gamma \eta(z) \qquad \text{for } \Gamma \in \mathrm{T}, (\Sigma, g) = H/\mathrm{T}, \text{ and } z \in H.$$

Hence

$$\frac{\partial}{\partial t}\eta_t(\Gamma z) = \Gamma_\eta \frac{\partial}{\partial t}\eta_t(z).$$

Consequently, we may integrate by parts, letting $\lambda^2(z)\,dz\,d\bar{z}$ denote the hyperbolic metric on (Σ, g):

$$\int_{H/\mathrm{T}} \delta(z) \frac{\partial}{\partial z}\left(\frac{\partial}{\partial t}\bar{\eta}_t(z)\right)\lambda^2(z)\,dz\,d\bar{z} = -\int_{H/\mathrm{T}} \frac{\partial}{\partial z}(\lambda^2(z)\delta(z))\frac{\partial}{\partial t}\bar{\eta}_t(z)\,dz\,d\bar{z} = 0$$

since

$$\frac{\partial}{\partial z}(\lambda^2(z)\delta(z)) = 0 \tag{6.5.4}$$

as δ is a harmonic Beltrami differential.

The equation

$$\int_{H/T} \delta(z)\frac{\partial}{\partial z}\left(\frac{\partial}{\partial t}\bar{\eta}_t(z)\right)\lambda^2(z)\,dz\,d\bar{z} = 0 \qquad \text{for } \delta\in H(g) \tag{6.5.5}$$

expresses the fact that the harmonic Beltrami differentials are L^2-orthogonal to the infinitesimal diffeomorphisms of the base surface.

Studying derivatives of $g_{\alpha\bar{\beta}}$ involves changing the base surface (Σ, g). For $\mu, \nu \in H(g)$, we let

$$w: \mu \to \mu + \nu$$
$$\zeta: 0 \to \mu$$
$$u: 0 \to \mu + \nu$$

be harmonic maps. We also put

$$v := u \circ \zeta^{-1}$$

and z will denote a complex parameter on 0; as always, $0 \in H(g)$ corresponds to (Σ, g).

For obtaining $g_{\alpha\bar{\beta}}(\mu)$, we have to differentiate the ν dependence at $\nu = 0$ w.r.t. α and $\bar{\beta}$. For obtaining $g_{\alpha\bar{\beta},\gamma}(0)$, we have to differentiate the μ dependence at $\mu = 0$ w.r.t. γ.

If $h(\rho, \sigma): \rho \to \sigma$ is a map, we shall denote differentiation w.r.t. the image variable σ by $'$, and w.r.t. the domain variable by $^{\bullet}$. The corresponding tangent vectors will be given in square brackets, the image direction always coming first. For example,

$$\dot{h}'[\bar{\beta}][\gamma]$$

denotes differentiation of the σ dependence in the direction $\bar{\beta}$ and of the ρ dependence in the direction γ.

We denote the hyperbolic metric on μ by

$$l^2(\zeta)\,d\zeta\,d\bar{\zeta}.$$

In this notation

$$g_{\alpha\bar{\beta}}(\mu) = \int\left[\frac{\rho^2(w)}{l^2(\zeta)}w_{\bar{\zeta}}\bar{w}_{\zeta}\right]''[\alpha,\bar{\beta}]\,l^2(\zeta)\,d\zeta\,d\bar{\zeta} \tag{6.5.6}$$
$$= \int w'_{\bar{\zeta}}[\alpha]\bar{w}'_{\zeta}[\bar{\beta}]\,l^2(\zeta)\,d\zeta\,d\bar{\zeta}$$

where all quantities are evaluated on the surface corresponding to μ.

Since $w'_{\bar{\zeta}}[\alpha]$ is a harmonic Beltrami differential ($\alpha \in H(g)$), now for the surface corresponding to μ and a family $\eta(v): \mu \to \mu$ of diffeomorphisms depending on v, satisfying $\eta(0) = \text{id}$, equation (6.5.5) yields

$$\int w'_{\bar{\zeta}}[\alpha]\bar{\eta}'_{\zeta}[\bar{\beta}]l^2(\zeta)\,\mathrm{d}\zeta\,\mathrm{d}\bar{\zeta} = 0 \qquad \text{for } \alpha, \beta \in H(g). \tag{6.5.7}$$

Thus, in the above notations also

$$\int w'_{\bar{\zeta}}[\alpha]\overline{w \circ \eta'_{\zeta}}[\bar{\beta}]l^2(\zeta)\,\mathrm{d}\zeta\,\mathrm{d}\bar{\zeta} = \int w'_{\bar{\zeta}}[\alpha]\bar{w}'_{\zeta}[\bar{\beta}]l^2(\zeta)\,\mathrm{d}\zeta\,\mathrm{d}\bar{\zeta} \tag{6.5.8}$$

and in particular

$$\int w'_{\bar{\zeta}}[\alpha]\bar{w}'_{\zeta}[\bar{\beta}]l^2(\zeta)\,\mathrm{d}\zeta\,\mathrm{d}\bar{\zeta} = \int w'_{\bar{\zeta}}[\alpha]\bar{v}'_{\zeta}[\bar{\beta}]l^2(\zeta)\,\mathrm{d}\zeta\,\mathrm{d}\bar{\zeta} \tag{6.5.9}$$

with $v = u \circ \zeta^{-1}$ as above.

We now come to the main point of our approach, namely exploiting the harmonicity of $w: \mu \to \mu + v$.

As a harmonic map, w satisfies the differential equation

$$w_{\zeta\bar{\zeta}} - \frac{2}{(w - \bar{w})} w_{\zeta} w_{\bar{\zeta}} = 0. \tag{6.5.10}$$

We shall differentiate this equation w.r.t. μ and obtain certain explicit formulae at $\mu = v = 0$.

We first have to write $w(\zeta)$ as a composition $w(\zeta) = \omega(z(\zeta))$, with

$$z: \mu \to 0 \qquad \omega: 0 \to \mu + v.$$

We point out that $z(\zeta)$ and $\omega(z)$ are no longer harmonic for $\mu, v \neq 0$. Nevertheless, decomposing w in this manner will be useful for the computations.

Equation (6.5.10) becomes

$$\begin{aligned}\omega_z z_{\zeta\bar{\zeta}} + \omega_{\bar{z}}\bar{z}_{\zeta\bar{\zeta}} &+ \left(\omega_{zz} - \frac{2}{\omega - \bar{\omega}}\omega_z\omega_z\right) z_{\zeta} z_{\bar{\zeta}} + \left(\omega_{\bar{z}\bar{z}} - \frac{2}{\omega - \bar{\omega}}\omega_{\bar{z}}\omega_{\bar{z}}\right)\bar{z}_{\zeta}\bar{z}_{\bar{\zeta}} \\ &+ \left(\omega_{z\bar{z}} - \frac{2}{\omega - \bar{\omega}}\omega_z\omega_{\bar{z}}\right)(z_{\zeta}\bar{z}_{\bar{\zeta}} + \bar{z}_{\zeta} z_{\bar{\zeta}}) = 0.\end{aligned} \tag{6.5.11}$$

From the chain rule, one computes

$$\begin{aligned} z_{\zeta\bar{\zeta}} = \frac{1}{(\zeta_z\bar{\zeta}_{\bar{z}} - \zeta_{\bar{z}}\bar{\zeta}_z)^3}&[\zeta_{zz}\bar{\zeta}_{\bar{z}}\bar{\zeta}_{\bar{z}}\zeta_{\bar{z}} + \zeta_{\bar{z}\bar{z}}\bar{\zeta}_{\bar{z}}\bar{\zeta}_z\zeta_z - \bar{\zeta}_{zz}\zeta_{\bar{z}}\zeta_{\bar{z}}\bar{\zeta}_{\bar{z}} - \bar{\zeta}_{\bar{z}\bar{z}}\zeta_{\bar{z}}\bar{\zeta}_z\zeta_z \\ &+ (\bar{\zeta}_{z\bar{z}}\zeta_{\bar{z}} - \zeta_{\bar{z}z}\bar{\zeta}_{\bar{z}})(\zeta_z\bar{\zeta}_{\bar{z}} + \zeta_{\bar{z}}\bar{\zeta}_z)]\end{aligned}$$

and (6.5.11) becomes

$$\begin{aligned} 0 = \omega_z \frac{1}{\zeta_z\bar{\zeta}_{\bar{z}} - \zeta_{\bar{z}}\bar{\zeta}_z}&[\zeta_{zz}\bar{\zeta}_{\bar{z}}\bar{\zeta}_{\bar{z}}\zeta_{\bar{z}} + \zeta_{\bar{z}\bar{z}}\bar{\zeta}_{\bar{z}}\bar{\zeta}_z\zeta_z - \bar{\zeta}_{zz}\zeta_{\bar{z}}\zeta_{\bar{z}}\bar{\zeta}_{\bar{z}} - \bar{\zeta}_{\bar{z}\bar{z}}\zeta_{\bar{z}}\bar{\zeta}_z\zeta_z \\ &+ (\bar{\zeta}_{z\bar{z}}\zeta_{\bar{z}} - \zeta_{\bar{z}z}\bar{\zeta}_{\bar{z}})(\zeta_z\bar{\zeta}_{\bar{z}} + \zeta_{\bar{z}}\bar{\zeta}_{\bar{z}})]\end{aligned}$$

$$+\omega_{\bar z}\frac{1}{\zeta_z\bar\zeta_{\bar z}-\zeta_{\bar z}\bar\zeta_z}[\bar\zeta_{\bar z\bar z}\zeta_z\zeta_z\bar\zeta_z+\bar\zeta_{zz}\zeta_z\zeta_{\bar z}\bar\zeta_{\bar z}-\zeta_{\bar z\bar z}\bar\zeta_z\bar\zeta_z\zeta_z-\zeta_{zz}\bar\zeta_z\zeta_{\bar z}\bar\zeta_{\bar z}$$
$$+(\zeta_{\bar z z}\bar\zeta_z-\bar\zeta_{z\bar z}\zeta_z)(\zeta_z\bar\zeta_{\bar z}+\zeta_{\bar z}\bar\zeta_z)]$$
$$-\left(\omega_{zz}-\frac{2}{\omega-\bar\omega}\omega_z\omega_z\right)\zeta_{\bar z}\bar\zeta_{\bar z}-\left(\omega_{\bar z\bar z}-\frac{2}{\omega-\bar\omega}\omega_{\bar z}\omega_{\bar z}\right)\zeta_z\bar\zeta_z$$
$$+\left(\omega_{z\bar z}-\frac{2}{\omega-\bar\omega}\omega_z\omega_{\bar z}\right)(\zeta_z\bar\zeta_{\bar z}-\zeta_{\bar z}\bar\zeta_z). \tag{6.5.12}$$

We recall that w goes from μ to $\mu+v$; therefore the μ dependence of w and hence also of ω is twofold.

We now take the total μ derivative of (6.5.12) and evaluate at $\mu=v=0$. We first compute derivatives of $\omega_{\bar z}$ w.r.t. changes of the domain. Equation (6.4.12) tells us already that first derivatives vanish (at $\mu=v=0$); here we give a slightly different derivation in order to get a method for also computing second derivatives.

We recall

$$\zeta_{\bar z}=\mu\zeta_z$$

hence, at $\mu=0$,

$$\zeta'_{\bar z}[\gamma]=\gamma\zeta_z=\gamma$$

and

$$\bar\zeta'_z[\gamma]=0.$$

We then obtain, using that $\omega(z)=z$ at $\mu=v=0$,

$$0=-\omega_z\gamma_z+\frac{2}{\omega-\bar\omega}\omega_z\omega_z\gamma+\dot\omega_{\bar z z}[\gamma]-\frac{2}{\omega-\bar\omega}\omega_z\dot\omega_{\bar z}[\gamma]$$
$$+\omega'_{\bar z z}[\gamma]-\frac{2}{\omega-\bar\omega}\omega_z\omega'_{\bar z}[\gamma].$$

Now $\omega'_{\bar z}[\gamma]=\gamma$, and since γ is a harmonic Beltrami differential,

$$\gamma_z-\frac{2}{z-\bar z}\gamma=0.$$

Hence

$$0=\dot\omega_{\bar z z}[\gamma]-\frac{2}{z-\bar z}\dot\omega_{\bar z}[\gamma] \tag{6.5.13}$$

i.e. $\dot\omega_{\bar z}[\gamma]$ is a harmonic Beltrami differential.

On the other hand, $\dot\omega[\gamma]$ represents an infinitesimal diffeomorphism of our base surface 0, and $\dot\omega_{\bar z}[\gamma]$ is hence orthogonal to all harmonic Beltrami differentials; cf. (6.5.5). Consequently,

$$\dot\omega_{\bar z}[\gamma]=0. \tag{6.5.14}$$

Likewise, we obtain at $\mu = \nu = 0$, by differentiating (6.5.12) w.r.t. $\bar{\delta}$,

$$0 = \dot{\omega}_{\bar{z}z}[\bar{\delta}] - \frac{2}{z-\bar{z}}\dot{\omega}_{\bar{z}}[\bar{\delta}] + \omega'_{\bar{z}z}[\bar{\delta}] - \frac{2}{z-\bar{z}}\omega'_{\bar{z}}[\bar{\delta}].$$

Since $\omega'_{\bar{z}}[\bar{\delta}] = 0$ as $\omega_{\bar{z}}$ is holomorphic in μ ($\omega_{\bar{z}} = \mu\omega_z$) at $\mu = 0$, $\dot{\omega}_{\bar{z}}[\bar{\delta}]$ is harmonic; hence as before

$$\dot{\omega}_{\bar{z}}[\bar{\delta}] = 0. \tag{6.5.15}$$

Of course, (6.5.14) and (6.5.15) are equivalent to (6.4.12). Namely, $\dot{\omega}[\gamma]$ and $\dot{\omega}[\bar{\delta}]$ are holomorphic infinitesimal diffeomorphisms; they vanish. This can be seen as follows. An infinitesimal diffeomorphism is nothing but a vector field. Since $p > 1$, our surface has no non-vanishing holomorphic vector fields. This is a consequence of the Riemann–Roch theorem; it can also be seen elementarily, as follows.

Let X be a holomorphic vector field, and let ρ be a holomorphic 1-form. Then $X(\rho)$ is a holomorphic function, hence constant. Thus, either $X \equiv 0$, or X has no zeroes. In the latter case, we consider the norm of X:

$$|X|^2 = \lambda^2(z)X(z)\bar{X}(z) > 0$$

where $\lambda^2(z)$ is the hyperbolic metric. Using $X_{\bar{z}} = 0$, and the fact that the hyperbolic metric has curvature -1, we compute

$$\Delta \log |X|^2 = -2.$$

This contradicts the divergence theorem, and thus $X \equiv 0$. Thus, if we have a object X that transforms as a vector field and satisfies $X_{\bar{z}} \equiv 0$, then we already have $X \equiv 0$.

We now want to compute mixed second derivatives of $\omega_{\bar{z}}$, namely $\dot{\omega}'_{\bar{z}}[\alpha][\bar{\delta}]$ and $\dot{\omega}'_{\bar{z}}[\bar{\beta}][\bar{\delta}]$ at $\mu = \nu = 0$. An essential point will be that $\dot{\omega}'$ also transforms as an infinitesimal diffeomorphisms, because of $\dot{\omega} = 0$. Corresponding differentiation of (6.5.12) yields, using $\dot{\omega}[\bar{\delta}] = 0$,

$$\begin{aligned} 0 &= -\omega'_{\bar{z}}[\alpha]\bar{\delta}_{\bar{z}} - \omega'_{\bar{z}z}[\alpha]\bar{\delta} + \dot{\omega}'_{\bar{z}z}[\alpha][\bar{\delta}] - \frac{2}{z-\bar{z}}\dot{\omega}'_{\bar{z}}[\alpha][\bar{\delta}] \\ &= -(\alpha\bar{\delta})_{\bar{z}} + \dot{\omega}'_{\bar{z}z}[\alpha][\bar{\delta}] - \frac{2}{z-\bar{z}}\dot{\omega}'_{\bar{z}}[\alpha][\bar{\delta}] \qquad \text{since } \omega'_{\bar{z}}[\alpha] = \alpha, \end{aligned} \tag{6.5.16}$$

and, using $\omega'_{\bar{z}}[\bar{\beta}] = 0$,

$$0 = \dot{\omega}'_{\bar{z}z}[\bar{\beta}][\bar{\delta}] - \frac{2}{z-\bar{z}}\dot{\omega}'_{\bar{z}}[\bar{\beta}][\bar{\delta}]. \tag{6.5.17}$$

We now investigate the transformation behaviour of $\dot{\omega}'$; we have, for arbitrary μ, ν,

$$\omega(\Gamma z) = \Gamma^\nu \omega(z) \tag{6.5.18}$$

where Γ^ν is an element of the group T^ν, the group of automorphisms of H

corresponding to ν, as in Section 6.4. Hence, at $\mu = \nu = 0$

$$\dot{\omega}'(\Gamma z) = \Gamma'\dot{\omega}(z) + \Gamma\dot{\omega}'(z) = \Gamma\dot{\omega}'(z) \qquad \text{since } \dot{\omega} = 0. \tag{6.5.19}$$

Therefore, $\dot{\omega}'$ transforms as an infinitesimal diffeomorphism, and, by (6.5.7), $\dot{\omega}'_{\bar{z}}$ then is L^2-orthogonal to the harmonic Beltrami differentials. (6.5.17) therefore implies

$$\dot{\omega}'_{\bar{z}}[\bar{\beta}][\bar{\delta}] = 0. \tag{6.5.20}$$

Similarly, we compute

$$\begin{aligned}
0 &= -\omega'_z[\alpha]\gamma_z - \omega'_{zz}[\alpha]\gamma + \frac{4}{z-\bar{z}}\omega'_z[\alpha]\gamma - \frac{2}{(z-\bar{z})^2}(\omega'[\alpha] - \bar{\omega}'[\alpha])\gamma \\
&\quad + \dot{\omega}'_{\bar{z}z}[\alpha][\gamma] - \frac{2}{z-\bar{z}}\dot{\omega}'_{\bar{z}}[\alpha][\gamma] \\
&= -\omega'_{zz}[\alpha]\gamma + \frac{2}{z-\bar{z}}\omega'_z[\alpha]\gamma - \frac{2}{(z-\bar{z})^2}(\omega'[\alpha] - \bar{\omega}'[\alpha])\gamma \\
&\quad + \dot{\omega}'_{\bar{z}z}[\alpha][\gamma] - \frac{2}{z-\bar{z}}\dot{\omega}'_{\bar{z}}[\alpha][\gamma]
\end{aligned} \tag{6.5.21}$$

since γ is harmonic, i.e. $\gamma_z = 2\gamma/(z-\bar{z})$.

Moreover, at $\mu = \nu = 0$, from (6.2.20)

$$0 = H_\alpha = \left[\frac{(z-\bar{z})^2}{(\omega-\bar{\omega})^2}\omega_z\bar{\omega}_{\bar{z}}\right]'[\alpha]$$

yielding

$$\omega'_z[\alpha] + \bar{\omega}'_{\bar{z}}[\alpha] - \frac{2}{(z-\bar{z})}(\omega'[\alpha] - \bar{\omega}'[\alpha]) = 0$$

and upon differentiation, since $\bar{\omega}'_z[\alpha] = 0$,

$$\omega'_{zz}[\alpha] - \frac{2}{(z-\bar{z})}\omega'_z[\alpha] + \frac{2}{(z-\bar{z})^2}(\omega'[\alpha] - \bar{\omega}'[\alpha]) = 0$$

so that (6.5.21) gives

$$\dot{\omega}'_{\bar{z}z}[\alpha][\gamma] - \frac{2}{z-\bar{z}}\dot{\omega}'_{\bar{z}}[\alpha][\gamma] = 0$$

and then as before

$$\dot{\omega}'_{\bar{z}}[\alpha][\gamma] = 0 \tag{6.5.22}$$

and with a similar derivation, using in addition $\bar{\omega}'_z[\bar{\beta}] = \bar{\beta}$ and

$$\bar{\beta}_{\bar{z}} + \frac{2}{z-\bar{z}}\bar{\beta} = 0$$

since β is harmonic, we also obtain

$$\dot{\omega}'_{\bar{z}}[\bar{\beta}][\gamma]=0. \tag{6.5.23}$$

The next step consists in computing $\dot{\omega}'_{\bar{z}}[\alpha][\bar{\delta}]$ from (6.5.16), which is equivalent to

$$(z-\bar{z})^2\dot{\omega}'_{\bar{z}z}[\alpha][\bar{\delta}]-2(z-\bar{z})\dot{\omega}'_{\bar{z}}[\alpha][\bar{\delta}]=(z-\bar{z})^2(\alpha\bar{\delta})_{\bar{z}}. \tag{6.5.24}$$

But

$$(z-\bar{z})^2\frac{\partial}{\partial\bar{z}}\left(-(z-\bar{z})^2\frac{\partial^2}{\partial z\partial\bar{z}}-2\right)=-\left((z-\bar{z})^2\frac{\partial}{\partial z}-2(z-\bar{z})\right)\left(\frac{\partial}{\partial\bar{z}}\left((z-\bar{z})^2\frac{\partial}{\partial\bar{z}}\right)\right) \tag{6.5.25}$$

and $\Delta=-(z-\bar{z})^2(\partial^2/\partial_z\partial_{\bar{z}})$ is the Laplace–Beltrami operator so that (6.5.24) is equivalent to

$$\left((z-\bar{z})^2\frac{\partial}{\partial z}-2(z-\bar{z})\right)\dot{\omega}'_{\bar{z}}[\alpha][\bar{\delta}]=\left((z-\bar{z})^2\frac{\partial}{\partial z}-2(z-\bar{z})\right)$$
$$\cdot\left(\frac{\partial}{\partial\bar{z}}\left((z-\bar{z})^2\frac{\partial}{\partial\bar{z}}((\Delta-2)^{-1}(-\alpha\bar{\delta}))\right)\right). \tag{6.5.26}$$

Now $(z-\bar{z})^2(\partial/\partial\bar{z})(\Delta-2)^{-1}(-\alpha\bar{\delta})$ again transforms as a vector field, namely with factor Γ_z, because $\alpha\bar{\delta}$ transforms as a function and Δ and hence also $(\Delta-2)^{-1}$ is a zero-order operator. Therefore,

$$\frac{\partial}{\partial\bar{z}}\left((z-\bar{z})^2\frac{\partial}{\partial\bar{z}}((\Delta-2)^{-1}(-\alpha\bar{\delta}))\right)$$

again is L^2-orthogonal to the harmonic Beltrami differentials.

Therefore, as both expressions occurring in (6.5.26) are orthogonal to the harmonic Beltrami differentials, we finally conclude

$$\dot{\omega}'_{\bar{z}}[\alpha][\bar{\delta}]=-\frac{\partial}{\partial\bar{z}}\left((z-\bar{z})^2\frac{\partial}{\partial\bar{z}}(\Delta-2)^{-1}(\alpha\bar{\delta})\right). \tag{6.5.27}$$

We now start to compute derivatives of $g_{\alpha\bar{\beta}}(\mu)$ at $\mu=0$. We pull formula (6.5.6) back to the base surface 0, as in Section 6.4:

$$g_{\alpha\bar{\beta}}(\mu)=\int\frac{\mathrm{d}^2}{\mathrm{d}_\alpha\mathrm{d}_{\bar{\beta}}}(\rho^2(\omega)\omega_z\bar{\omega}_{\bar{z}}\zeta_{\bar{z}}\bar{\zeta}_z+\rho^2(\omega)\omega_{\bar{z}}\bar{\omega}_z\zeta_z\bar{\zeta}_{\bar{z}}-\rho^2(\omega)\omega_z\bar{\omega}_z\bar{\zeta}_{\bar{z}}\zeta_{\bar{z}}-\rho^2(\omega)\bar{\omega}_{\bar{z}}\omega_{\bar{z}}\zeta_z\bar{\zeta}_z)$$
$$\cdot\frac{1}{\zeta_z\bar{\zeta}_{\bar{z}}-\bar{\zeta}_z\zeta_{\bar{z}}}\,\mathrm{d}z\,\mathrm{d}\bar{z}. \tag{6.5.28}$$

Since at $\mu=0$, $\zeta(z)=z$, and

$$\left(\frac{\zeta_z\zeta_{\bar{z}}}{\zeta_z\bar{\zeta}_{\bar{z}}-\zeta_{\bar{z}}\bar{\zeta}_z}\right)'[\gamma]=\left(\frac{\zeta_{\bar{z}}\bar{\zeta}_z}{\zeta_z\bar{\zeta}_{\bar{z}}-\zeta_{\bar{z}}\bar{\zeta}_z}\right)'[\gamma]=0$$

we compute

$$g_{\alpha\bar{\beta},\gamma}(0) = \frac{\mathrm{d}}{\mathrm{d}\gamma} g_{\alpha\bar{\beta}}(0)$$

$$= \int (\dot{\omega}'_{\bar{z}}[\alpha][\gamma]\bar{\omega}'_z[\bar{\beta}] + \omega'_{\bar{z}}[\alpha]\dot{\bar{\omega}}'_z[\bar{\beta}][\gamma])\lambda^2(z)\mathrm{d}z\,\mathrm{d}\bar{z}$$

$$= 0 \tag{6.5.29}$$

as $\dot{\omega}'$ transforms as an infinitesimal diffeomorphism (cf. (6.5.19)), and $\dot{\omega}'_{\bar{z}}$ is hence orthogonal to the harmonic Beltrami differentials (cf. (6.5.5)).

Similarly

$$g_{\alpha\bar{\beta},\bar{\delta}}(0) = 0. \tag{6.5.30}$$

This yields the following result of Ahlfors and Weil; cf. [A1].

Theorem 6.5.1: *The Weil–Petersson metric is a Kähler metric.*

The Kähler property of a Hermitian metric follows from (6.5.29), (6.5.30); namely, our coordinates are holomorphic at 0 (cf. 6.3), and (6.5.29), (6.5.30) show that they are also normal, and the existence of such coordinates is equivalent to the Kähler property.

Combining the previous results with the formulae of Section 6.2 and using again the method of Section 6.4, we can now compute the second derivatives. We let, as in Section 6.2.,

$$H = \frac{\rho^2(u)}{\lambda^2(z)} u_z \bar{u}_{\bar{z}} \qquad L = \frac{\rho^2(u)}{\lambda^2(z)} u_{\bar{z}} \bar{u}_z$$

for a harmonic map u, and denote complex derivatives at $\mu = 0$ by corresponding subscripts:

$$H'[\alpha, \bar{\beta}](0) = H_{\alpha\bar{\beta}} \qquad \text{etc.}$$

We compute first, using

$$\left(\frac{\zeta_z \bar{\zeta}_{\bar{z}}}{\zeta_z \bar{\zeta}_{\bar{z}} - \zeta_{\bar{z}} \bar{\zeta}_z}\right)''[\gamma, \bar{\delta}] = \left(\frac{\zeta_{\bar{z}} \bar{\zeta}_z}{\zeta_z \bar{\zeta}_{\bar{z}} - \zeta_{\bar{z}} \bar{\zeta}_z}\right)''[\gamma, \bar{\delta}] = \gamma\bar{\delta}$$

and (6.5.20), (6.5.22), (6.5.23) from (6.5.28),

$$g_{\alpha\bar{\beta},\gamma\bar{\delta}}(0) = \frac{\mathrm{d}^2}{\mathrm{d}\gamma \mathrm{d}\bar{\delta}} g_{\alpha\bar{\beta}}(0)$$

$$= \int (L_{\alpha\bar{\beta}\gamma\bar{\delta}} + H_{\alpha\bar{\beta}}\gamma\bar{\delta} + L_{\alpha\bar{\beta}}\gamma\bar{\delta} + \ddot{\omega}'_{\bar{z}}[\alpha][\gamma, \bar{\delta}]\bar{\beta}$$

$$+ \dot{\omega}'_{\bar{z}}[\alpha][\bar{\delta}]\dot{\bar{\omega}}'_z[\bar{\beta}][\gamma] + \alpha\ddot{\bar{\omega}}'_z[\bar{\beta}][\gamma, \bar{\delta}])\lambda^2(z)\,\mathrm{d}z\,\mathrm{d}\bar{z}. \tag{6.5.31}$$

On the other hand, pulling (6.5.9) back to the surface 0 by a change of variables, we also have

$$g_{\alpha\bar\beta}(\mu)=\int\frac{d^2}{d\alpha d\bar\beta}(\rho^2(\omega)\omega_z\bar u_{\bar z}\zeta_{\bar z}\bar\zeta_z+\rho^2(\omega)\omega_{\bar z}\bar u_z\zeta_z\bar\zeta_{\bar z}-\rho^2(\omega)\omega_z\bar u_z\zeta_{\bar z}\bar\zeta_{\bar z}$$

$$-\rho^2(\omega)\omega_{\bar z}\bar u_{\bar z}\zeta_z\zeta_{\bar z})\frac{1}{\zeta_z\bar\zeta_{\bar z}-\zeta_{\bar z}\bar\zeta_z}\,dz\,d\bar z \tag{6.5.32}$$

where, as before, $u: 0\to\mu+v$ is harmonic. Differentiating (6.5.32), we get

$$g_{\alpha\bar\beta,\gamma\bar\delta}(0)=\int(L_{\alpha\bar\beta\gamma\bar\delta}+H_{\alpha\bar\beta}\gamma\bar\delta+L_{\alpha\bar\beta}\gamma\bar\delta+\ddot\omega'_z[\alpha][\gamma,\bar\delta]\bar\beta)\lambda^2(z)\,dz\,d\bar z. \tag{6.5.33}$$

Comparing (6.5.31) and (6.5.33) yields

$$\int(\dot\omega'_{\bar z}[\alpha][\bar\delta]\dot{\bar\omega}'_z[\bar\beta][\gamma]+\alpha\ddot{\bar\omega}'_z[\bar\beta][\gamma,\bar\delta])\lambda^2(z)\,dz\,d\bar z=0 \tag{6.5.34}$$

hence also

$$\int\ddot\omega'_{\bar z}[\alpha][\gamma,\bar\delta]\bar\beta\lambda^2(z)\,dz\,d\bar z=-\int\dot\omega'_{\bar z}[\alpha][\bar\delta]\dot{\bar\omega}'_z[\bar\beta][\gamma]\lambda^2(z)\,dz\,d\bar z, \tag{6.5.35}$$

and inserting this into (6.5.33) and using (6.5.27), we obtain

$$g_{\alpha\bar\beta,\gamma\bar\delta}(0)=\int\Big\{L_{\alpha\bar\beta\gamma\bar\delta}+H_{\alpha\bar\beta}\gamma\bar\delta+L_{\alpha\bar\beta}\gamma\bar\delta-\frac{\partial}{\partial\bar z}\Big((z-\bar z)^2\frac{\partial}{\partial\bar z}(\Delta-2)^{-1}(\alpha\bar\delta)\Big)$$

$$\cdot\frac{\partial}{\partial z}\Big((z-\bar z)^2\frac{\partial}{\partial z}(\Delta-2)^{-1}(\bar\beta\gamma)\Big)\Big\}\lambda^2(z)\,dz\,d\bar z. \tag{6.5.36}$$

In order to evaluate (6.5.36), we recall from Section 6.2, that

$$\begin{aligned}L_{\alpha\bar\beta}&=\alpha\bar\beta\\ H_{\alpha\bar\beta}&=-2(\Delta-2)^{-1}(\alpha\bar\beta)\\ L_{\alpha\bar\beta\gamma\bar\delta}&=-H_{\alpha\bar\beta}L_{\gamma\bar\delta}-H_{\gamma\bar\delta}L_{\alpha\bar\beta}-H_{\alpha\bar\delta}L_{\gamma\bar\beta}-H_{\gamma\bar\beta}L_{\alpha\bar\delta}.\end{aligned}$$

Also, when integrating the last term in (6.5.36) by parts, we obtain, since $\lambda^2(z)=-2/(z-\bar z)^2$, the operator

$$\begin{aligned}&-\frac{\partial}{\partial\bar z}\Big\{(z-\bar z)^2\frac{\partial}{\partial\bar z}\Big(-\frac{2}{(z-\bar z)^2}\frac{\partial}{\partial z}\Big((z-\bar z)^2\frac{\partial}{\partial z}\Big)\Big)\Big\}\\ &=2(z-\bar z)^2\frac{\partial^4}{\partial z^2\partial\bar z^2}-4(z-\bar z)\frac{\partial^3}{\partial z^2\partial\bar z}+4(z-\bar z)\frac{\partial^3}{\partial\bar z^2\partial z}\\ &=\frac{2}{(z-\bar z)^2}(\Delta\cdot\Delta-2\Delta)\qquad\text{since }\Delta=-(z-\bar z)^2\,\partial^2/\partial z\partial\bar z\\ &=-\lambda^2(z)((\Delta-2)^2+2(\Delta-2)).\end{aligned} \tag{6.5.37}$$

Since $(\Delta-2)$ is self-adjoint, we obtain altogether

$$g_{\alpha\bar{\beta},\gamma\bar{\delta}}(0)=\int\{(2(\Delta-2)^{-1}(\gamma\bar{\delta}))\alpha\bar{\beta}+(2(\Delta-2)^{-1}(\gamma\bar{\beta}))\alpha\bar{\delta}\}\lambda^2(z)\,\mathrm{d}z\,\mathrm{d}\bar{z} \quad (6.5.38)$$

and also

$$g_{\alpha\bar{\delta},\gamma\bar{\beta}}(0)=g_{\alpha\bar{\beta},\gamma\bar{\delta}}(0). \quad (6.5.39)$$

Since

$$H_{\alpha\gamma}=0=L_{\alpha\gamma} \qquad \text{etc.}$$
$$L_{\alpha\gamma\bar{\beta}\bar{\delta}}=L_{\alpha\bar{\beta}\gamma\bar{\delta}}$$

we obtain likewise

$$\begin{aligned} g_{\alpha\gamma,\bar{\beta}\bar{\delta}}(0)&=\int L_{\alpha\bar{\beta}\gamma\bar{\delta}}\lambda^2(z)\,\mathrm{d}z\,\mathrm{d}\bar{z}\\ &=\int\{(4(\Delta-2)^{-1}(\gamma\bar{\delta}))\alpha\bar{\beta}+(4(\Delta-2)^{-1}(\gamma\bar{\beta}))\alpha\bar{\delta}\}\lambda^2(z)\,\mathrm{d}z\,\mathrm{d}\bar{z} \end{aligned} \quad (6.5.40)$$

and

$$g_{\bar{\beta}\bar{\delta},\alpha\gamma}(0)=g_{\alpha\gamma,\bar{\beta}\bar{\delta}}(0). \quad (6.5.41)$$

This enables us to evaluate the curvature tensor $R_{\alpha\bar{\beta}\gamma\bar{\delta}}$ of the Weil–Petersson metric. The curvature of the Weil–Petersson metric has been computed by Tromba [Tr3], Wolpert (cf. [Wp] where such results are also independently attributed to Royden), and Siu [Si], by different methods. Here, we can recover all their results.

Since, by (6.5.29), (6.5.30), first derivatives of the metric vanish, we have

$$R_{\alpha\bar{\beta}\gamma\bar{\delta}}(0)=\tfrac{1}{2}(g_{\alpha\bar{\beta},\gamma\bar{\delta}}(0)+g_{\alpha\bar{\delta},\gamma\bar{\beta}}(0)-g_{\alpha\gamma,\bar{\beta}\bar{\delta}}(0)-g_{\bar{\beta}\bar{\delta},\alpha\gamma}(0)). \quad (6.5.42)$$

(Note that the quantities $g_{\alpha\gamma,\bar{\beta}\bar{\delta}}$ and $g_{\bar{\beta}\bar{\delta},\alpha\gamma}$ appear because our coordinate system is not holomorphic.)

From (6.5.38)–(6.5.42)

$$R_{\alpha\bar{\beta}\gamma\bar{\delta}}(0)=\int\{(-2(\Delta-2)^{-1}(\gamma\bar{\delta}))\alpha\bar{\beta}+(-2(\Delta-2)^{-1}(\gamma\bar{\beta}))\alpha\bar{\delta}\}\lambda^2(z)\,\mathrm{d}z\,\mathrm{d}\bar{z}. \quad (6.5.43)$$

In real notation with

$$L_{ij}=2\,\mathrm{Re}\,\alpha_i\bar{\alpha}_j \quad (6.5.44)$$

for α_i, α_j harmonic Beltrami differentials, and

$$H_{ij}=-2(\Delta-2)^{-1}L_{ij} \quad (6.5.45)$$

we have

$$R_{ijkl}=\int(H_{il}L_{kj}+H_{kj}L_{il}-H_{ik}L_{lj}-H_{jl}L_{ik})\lambda^2(z)\,\mathrm{d}z\,\mathrm{d}\bar{z}. \quad (6.5.46)$$

Remark: The formula for the curvature agrees with the formula for the fourth variation of the energy w.r.t. changes of the image structure; cf. (6.2.16), (6.2.24), (6.2.21), (6.2.22). This observation is due to Wolf [Wf].

Equation (6.5.46) implies:

Theorem 6.5.2: *The sectional curvature of the Weil–Petersson metric is negative.*

Proof: We have to evaluate

$$R_{ijij} = \int (2H_{ij}L_{ij} - H_{ii}L_{jj} - H_{jj}L_{ii}). \tag{6.5.47}$$

We shall show that the integrand is pointwise non-positive.

From (6.5.44) we have

$$|L_{ij}| \leqslant |L_{ii}|^{1/2} |L_{jj}|^{1/2}. \tag{6.5.48}$$

If $i \neq j$, we actually have strict inequality somewhere.

Since Green's function for Δ on a fundamental domain H/T (as which we can represent our surface (Σ, γ); cf. Theorem 3.3.1) is negative, the same is, of course, true for the fundamental solution of $\Delta - 2$. (This is, of course, a simple consequence of the maximum principle.) Hence Hölder's inequality yields, using (6.5.48),

$$\begin{aligned} |H_{ij}| &\leqslant 2|(\Delta-2)^{-1} L_{ii}|^{1/2} |(\Delta-2)^{-1} L_{jj}|^{1/2} \\ &= |H_{ii}|^{1/2} |H_{jj}|^{1/2}. \end{aligned} \tag{6.5.49}$$

For $i \neq j$, we have strict inequality somewhere. This remark, together with (6.5.47), (6.5.48), (6.5.49) implies that, for $i \neq j$

$$R_{ijij} < 0$$

q.e.d.

Another consequence of (6.5.43) is:

Theorem 6.5.3: *The holomorphic sectional curvature $H(\varphi)$ of the Weil–Petersson metric is bounded by*

$$H(\varphi) \leqslant \frac{-1}{2\pi(p-1)} \tag{6.5.50}$$

where p is the genus of Σ, and φ is a unit vector.

Proof: We have to evaluate (6.5.46) with $\varphi_i = \varphi$ and $\varphi_j = J_{\varphi_i} = \sqrt{-1}\,\varphi_i$. In this case

$$L_{ij} = \alpha_i \bar{\alpha}_j + \bar{\alpha}_i \alpha_j = 0 \tag{6.5.51}$$

hence also (cf. (6.2.8))

$$H_{ij}=0. \tag{6.5.52}$$

Also

$$L_{ii}=2\alpha_i\bar{\alpha}_i \geqslant 0 \tag{6.5.53}$$

hence also

$$H_{ii} \geqslant 0, \tag{6.5.54}$$

by the maximum principle (cf. (6.2.8)).
By normalization

$$\int L_{ii}\lambda^2 \, dz \, d\bar{z} = 2\int \alpha_i\bar{\alpha}_i \, dz \, d\bar{z} = 1.$$

We put

$$f = -H_{ii} = 2(\Delta-2)^{-1}L_{ii}.$$

Then

$$\int(\Delta f - 2f) = 2\int L_{ii} = 2$$

and hence

$$\int f = -1.$$

From this last relation, we deduce

$$-\int H_{ii}L_{ii} = \tfrac{1}{2}\int f(\Delta f - 2f) = -\int f^2 + \tfrac{1}{2}|\nabla f|^2 \leqslant -\frac{1}{\operatorname{vol}\Sigma} = \frac{-1}{4\pi(p-1)} \tag{6.5.55}$$

and (6.5.50) follows from (6.5.47) and (6.5.51)–(6.5.55)

q.e.d.

Finally:

Theorem 6.5.4: *The Ricci curvature of the Weil–Petersson metric is strictly negative. In fact, for a unit vector*

$$\operatorname{Ric}(\varphi) \leqslant -\frac{1}{2\pi(p-1)}.$$

(The negativity of the Ricci curvature was already shown in [A2].)

Proof: Since all sectional curvatures are negative by Theorem 6.5.2, the Ricci curvature is bounded from above by the holomorphic sectional curvature, and the result follows from Theorem 6.5.3.

q.e.d.

Remark: Further applications of the technique presented in this section can be found in [J18].

APPENDIX. REMARKS ON NOTATION AND TERMINOLOGY

N is a complete Riemannian manifold of dimension d, usually compact or of bounded geometry; the latter means that the absolute value of the sectional curvature is uniformly bounded and the injectivity radius is uniformly bounded away from zero. Although one can weaken this assumption sometimes, i.e. in Section 4.7, we have chosen this as a unifying underlying assumption, as it is geometrically and conceptually natural.

We denote by $d(\cdot,\cdot)$ the distance function on N, and by R its curvature tensor. N will always be equipped with the Levi–Civita connection ∇. For $p \in N$, $r > 0$, we put $B(p,r) := \{q \in N : d(p,q) \leqslant r\}$. $\langle\cdot,\cdot\rangle$ usually is the scalar product in the tangent bundle TN. We sometimes use local coordinates $u^1, \ldots, u^d$ on N. In these coordinates, we write the metric tensor as $(g_{ij}(u))_{i,j=1,\ldots,d}$, we put $(g^{ij}) := (g_{ij})^{-1}$ and $g := \det(g_{ij})$ and we denote the Christoffel symbols by Γ^i_{jk} $(= \frac{1}{2}[g^{il}(g_{jl,k} + g_{kl,j} - g_{kj,l})])$. We always employ the standard summation convention that an index occurring twice in a product has to be summed over its range; usually, the index is a latin postscript and is summed from 1 to d.

We do not distinguish notationally between the d-dimensional vector space $\mathbb{R}^d$ and this space equipped with its standard Euclidean structure, i.e. d-dimensional Euclidean space.

$D := \{(x,y) \in \mathbb{R}^2 : x^2 + y^2 \leqslant 1\}$ is the closed unit disk. We often use the complex notation, namely

$$z = x + \mathrm{i}y$$

$$\frac{\partial}{\partial z} = \frac{1}{2}\left(\frac{\partial}{\partial x} - \mathrm{i}\frac{\partial}{\partial y}\right) \qquad \frac{\partial}{\partial \bar{z}} = \frac{1}{2}\left(\frac{\partial}{\partial x} + i\frac{\partial}{\partial y}\right)$$

$$\mathrm{d}z = \mathrm{d}x + \mathrm{i}\mathrm{d}y \qquad \mathrm{d}\bar{z} = \mathrm{d}x - \mathrm{i}\mathrm{d}y.$$

For abbreviation, we also put $u_x := \partial u/\partial x$, etc.

Σ is a compact surface, equipped with a conformal structure; it turns out that for the study of our two-dimensional conformally invariant variational problems, an introduction of a metric on Σ is not necessary. Sometimes, however, we have to use hyperbolic metrics for auxiliary purposes.

Σ may or may not have a boundary. If $\partial\Sigma \neq \emptyset$, then $\partial\Sigma$ is considered to be contained in Σ so that Σ is compact in any case. We also assume $\partial\Sigma \in C^1$, which can always be achieved by conformal mapping (cf. chapter 3). We note that the assumption $\partial\Sigma \subset \Sigma$ excludes isolated boundary points.

We study maps $u: \Sigma \to N$. They usually belong to certain mapping spaces, for example to $C^{k,\alpha}$ ($k \in \mathbb{N}$, $0 < \alpha < 1$), the space of k-times differentiable mappings, the kth derivatives of which satisfy a Hölder condition with exponent α. Usually, we write C^α instead of $C^{0,\alpha}$.

We also need Sobolev spaces of maps; for this purpose we embed N differentiably (by Whitney's Theorem) or isometrically (by Nash's Theorem) into some Euclidean space $\mathbb{R}^k$ and put

$$W^{1,2}(\Sigma, N) := \{v \in W^{1,2}(\Sigma, \mathbb{R}^k) : v(x) \in N \text{ for almost all } x \in \Sigma\}.$$

(Here, $W^{1,2}(\Sigma, \mathbb{R}^k)$ is the standard Sobolev space of maps that are once weakly differentiable and which together with all their weak first derivatives have finite L^2-norm.)

We also define $H^{1,2}(\Sigma, N)$ as the $W^{1,2}$ closure of $C^\infty(\Sigma, N)$. The following result of Schoen and Uhlenbeck [SU] depends on the assumption that Σ is two-dimensional:

Lemma A.1:

$$H^{1,2}(\Sigma, N) = W^{1,2}(\Sigma, N)$$

Proof: By a standard extension theorem (cf. [GT, Theorem 7.25]), we can assume that Σ is compactly contained in Σ_1 and that any given $u \in W^{1,2}(\Sigma, N)$ is the restriction of some $u_1 \in W^{1,2}(\Sigma_1, \mathbb{R}^k)$ to Σ. We embed Σ_1 differentiably into some $\mathbb{R}^l$.

We let U be a normal neighbourhood of Σ_1 in $\mathbb{R}^l$, V a normal neighbourhood of N in $\mathbb{R}^k$, and

$$\mu: U \to \Sigma_1$$
$$\pi: V \to N$$

be the nearest point projections.

Furthermore, for $x \in \Sigma_1$, we put (for $r > 0$)

$$B^2(x, r) := \{y \in \Sigma_1 : |x - y| \leq r\}$$

and

$$B^l(x, r) := \{y \in \mathbb{R}^l : |x - y| \leq r\}.$$

For $0 < \epsilon < \operatorname{dist}(\Sigma, \partial\Sigma_1)$,

$$G_\epsilon(x) := \int_{B^2(x,\epsilon)} |\nabla u_1|^2$$

is a continuous function of $x \in \Sigma$.

$G_\epsilon(x)$ is an increasing function of ϵ, and $\lim_{\epsilon \to 0^+} G_\epsilon(x) = 0$ for all $x \in \Sigma$. Consequently, $G_\epsilon(x)$ converges uniformly to zero as $\epsilon \to 0$. We then put, for $x \in U$,

$$u_2(x) := u_1(\mu(x)).$$

Since the metric on U is uniformly equivalent to the product metric on $\Sigma x B^{l-2}(0,1)$, for $x \in \mu^{-1}(\Sigma)$

$$\int_{B^l(x,\epsilon)} |\nabla u_2|^2 \leqslant c\epsilon^{l-2} \int_{B^l(\mu(x),\epsilon)} |\nabla u_1|^2 = c\epsilon^{l-2} G_\epsilon(x). \tag{A.1}$$

Also $u_2(x) \in N$ for almost all $x \in \mu^{-1}(\Sigma)$.

Choosing a mollifier $\varphi(x)$ on $B^l(0,1)$, with $\varphi_\epsilon(x) := \epsilon^{-l}\varphi(x/\epsilon)$, then

$$u_{2,\epsilon}(x) := \int \varphi_\epsilon(x-y) u_2(y)\,\mathrm{d}y \qquad (x \in \Sigma).$$

Equation (A.1) and the Poincaré inequality give

$$\begin{aligned}\epsilon^{-l} \int_{B^l(x,\epsilon)} |u_2(y) - u_{2,\epsilon}(x)|^2\,\mathrm{d}y &\leqslant c\epsilon^{2-l} \int_{B^l(x,\epsilon)} |\nabla u_2|^2\,\mathrm{d}y \\ &\leqslant cG_\epsilon(x).\end{aligned} \tag{A.2}$$

Equation (A.2) and $u_2(\mu^{-1}(\Sigma)) \subset N$ imply for all $x \in \Sigma$ that

$$\operatorname{dist}(u_{2,\epsilon}(x), N) \leqslant cG_\epsilon^{1/2}(x). \tag{A.3}$$

We put, for $x \in \Sigma$,

$$v_\epsilon(x) := \pi \circ u_{2,\epsilon}(x).$$

By (A.3) and since G_ϵ converges uniformly to zero, v_ϵ is well defined and smooth for sufficiently small $\epsilon > 0$. Now $u_{2,\epsilon}$ converges to u_2 in $W^{1,2}$ as $\epsilon \to 0$ by standard properties of mollifications, and therefore also $v_\epsilon = \pi \circ u_{2,\epsilon}$ converges in $W^{1,2}$ to $\pi \circ u_2$. Since $\pi \circ u_2 = u_2$, then v_ϵ converges in $W^{1,2}$ to $\pi \circ u_2$. So far, this convergence takes place in $W^{1,2}(U)$, but it is clear from the construction that this also yields convergence in $W^{1,2}(\Sigma)$. This means that v_ϵ converges in $W^{1,2}(\Sigma)$ to u, proving the claim.

q.e.d.

The next lemma, which is well known, shows that weak solutions of variational problems of the type considered in this book extend as weak solutions through isolated singularities if their $H^{1,2}$-norm is finite.

By embedding the target manifold again into some $\mathbb{R}^k$ we may assume that the image is $\mathbb{R}^k$, and since the question is local it suffices to consider the case of the punctured unit disk $D' := \{(x,y) \in \mathbb{R}^2 : 0 < x^2 + y^2 \leqslant 1\}$.

Lemma A.2: Suppose $u \in H^{1,2}(D', \mathbb{R}^k)$ satisfies

$$\int_{D'} \nabla u(z) \nabla \varphi(z) \, dz = \int_{D'} f(z, u(z), \nabla u(z)) \varphi(z) \, dz \tag{A.4}$$

for all $\varphi \in H_0^{1,2} \cap L^\infty(D', \mathbb{R}^k)$, where f satisfies

$$|f(z, u, p)| \leqslant a + b|p|^2 \tag{A.5}$$

with constants a, b, for all $(z, u, p) \in D' \times \mathbb{R}^k \times \mathbb{R}^{2k}$.

Then also

$$\int_D \nabla u(z) \nabla \eta(z) \, dz = \int_D f(z, u(z), \nabla u(z)) \eta(z) \, dz \tag{A.6}$$

for all $\eta \in H_0^{1,2} \cap L^\infty(D, \mathbb{R}^k)$.

Proof. For $n \geqslant 2$, we define

$$\sigma_n(z) := \begin{cases} 1 & \text{for } r \leqslant (1/n)^2 \\ \log(1/nr)/\log n & \text{for } (1/n)^2 \leqslant r \leqslant 1/n \\ 0 & \text{for } r \geqslant 1/n \end{cases}$$

and for $\eta \in H_0^{1,2} \cap L^\infty(D, \mathbb{R}^k)$

$$\varphi_n(z) := (1 - \sigma_n|z|)) \eta(z) \in H_0^{1,2} \cap L^\infty(D', \mathbb{R}^k).$$

In fact,

$$\int_D |\nabla \sigma_n(|z|)|^2 \, dz = 2\pi \int_{(1/n)^2}^{1/n} \left(\frac{d\sigma_n}{dr} \right)^2 r \, dr = \frac{1}{\log n} \tag{A.7}$$

tends to zero as $n \to \infty$.

By assumption

$$\int_{D'} \nabla u(z) \nabla \varphi_n(z) \, dz = \int_{D'} f(z, u(z), \nabla u(z)) \varphi_n(z) \, dz \tag{A.8}$$

As $n \to \infty$, the right-hand side of (A.8) tends to

$$\int_D f(z, u(z), \nabla u(z)) \eta(z) \, dz$$

by Lebesgue's theorem on dominated convergence, because $|f(z, u(z), \nabla u(z))|$ is in L^1 by (A.5) and since $u \in H^{1,2}$, and $|\varphi_n| \leqslant |\eta| \in L^\infty$, and φ_n converges to η pointwise almost everywhere.

The left-hand side tends to

$$\int_D \nabla u(z) \nabla \eta(z) \, dz$$

because

$$\int_D \nabla u(z) \nabla (\sigma_n(z) \eta(z)) \, dz$$

tends to zero, by using (A.8), $\eta\in H^{1,2}\cap L^{\infty}$, $u\in H^{1,2}$, and dominated convergence again.

q.e.d.

For applications of Lemma A.2 in this book, we note that the result holds and the argument goes through also in the following situation:

$$D_{+} := \{(x,y)\in\mathbb{R}^2 : x^2+y^2\leqslant 1, y\geqslant 0\},\ D'_{+} := D_{+}\backslash\{0\}$$

$u\in H^{1,2}(D'_{+},\mathbb{R}^k)$ satisfies

$$\int_{D'_{+}} \nabla u(z)\nabla\varphi(z)\,\mathrm{d}z = \int_{D'_{+}} f(z,u(z),\nabla u(z))\varphi(z)\,\mathrm{d}z$$

with f as before, for all $\varphi\in H^{1,2}\cap L^{\infty}(D'_{+},\mathbb{R}^k)$ which are limits of C^{∞}-functions with support in $\{(x,y)\in\mathbb{R}^2 : 0<x^2+y^2<1, y\geqslant 0\}$ and which satisfy a free boundary constraint on $\{y=0\}$, i.e. $\varphi(z)$ is required to be tangential to some given submanifold M of $\mathbb{R}^k$ for almost all points in $\{y=0\}$. In the proof, we have only to observe that this free boundary constraint is preserved by multiplication with a scalar valued function ($(1-\sigma_n)$ in the proof).

Then (A.6) holds with D_{+} instead of D'_{+} and for those η that satisfy the free boundary constraint on $\{y=0\}$.

REFERENCES

[Ab] Abresch, U. (1987) Constant mean curvature tori in terms of elliptic functions, *J. Reine Angew. Math.* **374**, 169–192.

[Ad] Adams, J. F. (1982) Maps from a surface to the projective plane, *Bull. London. Math. Soc.* **14**, 533–534.

[A1] Ahlfors, L. (1961) Some remarks on Teichmüller's space of Riemann surfaces, *Ann. Math.* **74**, 171–191.

[A2] Ahlfors, L. (1961) Curvature properties of Teichmüller's space, *J. d'Anal. Math.* **9**, 161–176.

[AB] Ahlfors, L., and Bers, L. (1960) Riemann's mapping theorem for variable metrics, *Ann. Math.* **72**, 385–404.

[Alb1] Al'ber, S. I. (1964) On *n*-dimensional problems in the calculus of variations in the large, *Sov. Math. Dokl.* **5**, 700–704.

[Alb2] Al'ber, S. I. (1967) Spaces of mappings into a manifold with negative curvature, *Sov. Math. Dokl.* **9**, 6–9.

[Al] Almgren, F. (1966) Some interior regularity theorems and an extension of Bernstein's theorem, *Ann. Math.* **84**, 277–292.

[AlS] Almgren, F., and Simon, L. (1979) Existence of embedded solutions of Plateau's problem, *Ann. Sc. Norm. Sup. Pisa* **6**, 447–495.

[Alt] Alt, H. W. (1972) Verzweigungspunkte von H-Flächen, I, *Math. Z.* **127**, 333–362; Verzweigungspunkte von H-Flächen, II, *Math. Ann.* **201**, 33–55.

[Ba] Baldes, A. (1982) Harmonic mappings with a partially free boundary, *Manuscripta Math.* **40**, 255–275.

[Be1] Bers, L. (1960) Spaces of Riemann surfaces, *Proc. Int. Congr. Math. 1958*, Cambridge University Press, Cambridge, pp. 349–361.

[Be2] Bers, L. (1965) Automorphic forms and general Teichmüller spaces, *Proc. Conf. Complex Analysis, Minneapolis 1964*, Springer, Berlin, pp. 109–113.

[BJS] Bers, L., John, F. and Schechter, M. (1964) *Partial Differential Equations*, Interscience, New York.

[Bl] Blaschke, W. (1945) *Vorlesungen über Differentialgeometrie*, part I, Springer, Berlin, 2nd edn.

[Bö] Böhme, R. (1987) The conformal structure of Riemann surfaces with boundary parametrizing minimal surfaces.

[BC] Brézis, H., and Coron, J. M. (1983) Large solutions for harmonic maps in two dimensions, *Commun. Math. Phys.* **92**, 203–215.

[BK] Buser, P., and Karcher, H. (1981) Gromov's almost flat manifolds, *Astérisque* **81**.

[Ca] Calabi, E. (1967) Minimal immersions of surfaces into Euclidean spheres, *J. Diff. Geom.* **1**, 111–125.

[Ch] Chavel, I. (1984) *Eigenvalues in Riemannian Geometry*, Academic Press, Orlando, FL.

[Cq] Choquet, G. (1945) Sur un type de transformation analytique généralisant la representation conforme et définie au moyen de fonctions harmoniques, *Bull. Sci. Math* (2) **69**, 156–165.

[Cr] Coron, J. M. (1984) Topologie et cas limite des injections de Sobolev. *C.R. Acad. Sci. Paris, Ser. I*, **299**, 209–212.

[CH] Coron, J. M., and Helein, F. (1989) Harmonic diffeomorphisms, minimizing harmonic maps and rotational symmetry. *Compositio Math.* **69**, 175–228.

[C1] Courant, R. (1937) Plateau's problem and Dirichlet's principle, *Ann. Math.* (2) **38**, 679–724.

[C2] Courant, R. (1940) The existence of minimal surfaces of given topological structure under prescribed boundary conditions, *Acta Math.* **72**, 51–98.

[C3] Courant, R. (1941) Critical points and unstable minimal surfaces, *Proc. Natl. Acad. Sci. USA* **27**, 51–57.

[C4] Courant, R. (1945) On Plateau's problem with free boundaries, *Proc. Natl. Acad. Sci. USA* **31**, 242–246.

[C5] Courant, R. (1950) *Dirichlet's Principle, Conformal Mapping, and Minimal Surfaces*, Interscience, New York.

[Db] Darboux, G. (1894) *Théorie Générale des Surfaces*, tome III, Gauthier–Villars, Paris.

[Di] Ding, W. (1986) Lusternik–Schnirelmann theory for harmonic maps *Acta Math. Sinica* **2**, 105–122.

[D1] Douglas, J. (1931) Solution of the problem of Plateau, *Trans. Am. Math. Soc.* **33**, 263–321.

[D2] Douglas, J. (1939) Minimal surfaces of higher topological structure, *Ann. Math.* (2) **40**, 205–298.

[Dz1] Dziuk, G. (1981) Über die Stetigkeit teilweise freier Minimalflächen, *Manuscripta Math.* **36**, 241–251.

[Dz2] Dziuk, G. (1985) C^2-regularity for partially free minimal surfaces, *Math. Z.* **189**, 71–79.

[EE] Earle, C. J., and Eells, J. (1969) A fibre bundle description of Teichmüller theory, *J. Diff. Geom.* **3**, 19–43.

[Ed] Edmonds, A. (1979) Deformations of maps to branched coverings in dimension two, *Ann. Math.* **110**, 113–125.

[EL1] Eells, J., and Lemaire, L. (1978) A report on harmonic maps, *Bull. London Math. Soc.* **10**, 1–68.

[EL2] Eells, J., and Lemaire, L. (1980) Deformations of metrics and associated harmonic maps, *Patodi Mem. Vol. Geometry and Analysis* Tata Institute, Bombay, pp. 33–45.

[EL3] Eells, J., and Lemaire, L. (1980) On the construction of harmonic and holomorphic maps between surfaces, *Math. Ann.* **252**, 27–52.

[EL4] Eells, J., and Lemaire, L. (1983) Selected topics in harmonic maps, *CBMS Regional Conf. Ser.* **50**, Am. Math. Soc., Providence, R. I.

[EL5] Eells, J., and Lemaire, L. (1988) Another report on harmonic maps, *Bull. London Math. Soc.* **20**, 385–524.

[EW] Eells, J., and Wood, J. C. (1976) Restrictions on harmonic maps of surfaces, *Topology* **15**, 263–266.

[F] Federer, H. (1969) *Geometric Measure Theory.* Grundlehren vol. 163, Springer, New York.

[Fe] Fenchel, W. (1932) Elementare Beweise und Anwendungen einiger Fixpunktsätze, *Mat. Tidsskr.* B, 66–87.

[FHS] Freedman, M., Hass, J. and Scott, P. (1983) Least area incompressible surfaces in 3-manifolds, *Inv. Math.* **71**, 609–642.

[FK] Fricke, R., and Klein, F. (1926) *Vorlesungen über die Theorie der automorphen Funktionen*, Teubner, Leipzig.

[GR] Gerstenhaber, M., and Rauch, H. E. (1954) On Extremal Quasiconformal Mappings I, II, *Proc. Natl. Acad. Sci. USA* **40**, 808–812, 991–994.

[GT] Gilbarg, D., and Trudinger, N. S. (1983) *Elliptic Partial Differential Equations of Second Order*, Grundlehren vol. 224, Springer, Berlin, 2nd edn.

[GKM] Gromoll, D., Klingenberg, W., and Meyer, W. (1975) *Riemannsche Geometrie im Grossen* (*Lecture Notes in Mathematics 55*), Springer, Berlin.

[Gr1] Grüter, M. (1981) Regularity of weak H-surfaces, *J. Reine Angew. Math.* **329**, 1–15.

[Gr2] Grüter, M. (1984) Conformally invariant variational integrals and the removability of isolated singularities, *Manuscripta Math.* **47**, 85–104.

[Gr3] Grüter, M. (1986) Eine Bemerkung zur Regularität stationärer Punkte von konform invarianten Variationsintegralen, *Manuscripta Math.* **55**, 451–453.

[GHN1] Grüter, M., Hildebrandt, S., and Nitsche, J. (1981) On the boundary behaviour of minimal surfaces with a free boundary which are not minima of the area, *Manuscripta Math.* **35**, 387–410.

[GHN2] Grüter, M., Hildebrandt, S., and Nitsche, J. (1986) Regularity for stationary surfaces of constant mean curvature with free boundaries, *Acta Math.* **156**, 119–152.

[G] Gulliver, R. (1973) Regularity of minimizing surfaces of prescribed mean curvature, *Ann. Math.* **97**, 275–305.

[GJ] Gulliver, R., and Jost, J. (1987) Harmonic maps which solve a free boundary problem. *J. Reine Angew. Math.* **381**, 61–89.

[GL] Gulliver, R., and Leslie, J. (1973) On boundary branch points of minimizing surfaces, *Arch. Rat. Mech. Anal.* **52**, 20–25.

[GOR] Gulliver, R., Osserman, R., and Royden, H. (1973) A theory of branched immersion of surfaces, *Am. J. Math.* **95**, 750–812.

[GS] Gulliver, R., and Spruck, J. (1976) On embedded minimal surfaces, *Ann. Math.* **103**, 331–347 (erratum 1979 **109**, 407–412).

[Hp] Halpern, N. (1981) A proof of the collar lemma, *Bull. London Math. Soc.* **13**, 141–144.

[HS] Hardt, R., and Simon, L. (1979) Boundary regularity and embedded solutions for the oriented Plateau problem, *Ann. Math.* **110**, 439–486.

[Ht] Hartman, P. (1967) On homotopic harmonic maps, *Can. J. Math.* **19**, 673–687.

[HtW] Hartman, P., and Winter, A. (1953) On the local behaviour of solutions of nonparabolic partial differential equations, *Am. J. Math.* **75**, 449–476.

[H1] Heinz, E. (1956/57) On certain nonlinear elliptic differential equations and univalent mappings, *J. d'Anal.* **5**, 197–272.

[H2] Heinz, E. (1960) Neue a-priori Abschätzungen für den Ortsvektor einer Fläche positiver Gausscher Krümmung durch ihr Linienelement, *Math. Z.* **74**, 129–157.

[H3] Heinz, E. (1965) Existence theorems for one-to-one mappings associated with elliptic systems of second order, I, *J. d'Anal.* **15**, 325–353; (1967) II, *J. d'Anal.* **17**, 145–184.

[H4] Heinz, E. (1968) Über das Nichtverschwinden der Funktionaldeterminante bei einer Klasse eineindeutiger Abbildungen, *Math. Z.* **105**, 87–89.

[H5] Heinz, E. (1968) Zur Abschätzung der Funktionaldeterminante bei einer Klasse topologischer Abbildungen, *Nachr. Akad. Wiss. Gött.*, 183–197.

[H6] Heinz, E. (1970) Über das Randverhalten quasilinearer elliptischer Systeme mit isothermen Parametern, *Math. Z.* **113**, 99–105.

[HH] Heinz, E., and Hildebrandt, S. (1970) Some remarks on minimal surfaces in Riemannian manifolds, *Commun. Pure Appl. Math.* **23**, 371–377.

[HT] Heinz, E., and Tomi, F. (1969) Zu einem Satz von Hildebrandt über das Randverhalten von Minimalflächen, *Math. Z.* **111**, 372–386.

[Hl1] Hélein, F. (1988) Homéomorphismes quasiconformes entre surfaces riemanniennes, *C.R. Acad. Sci. Paris* **307**, 725–730.

[Hl2] Hélein, F. (1990) Regularité des applications faiblement harmoniques entre une surface et une sphère, *C.R. Acad. Sci. Paris Series* I, to appear

[Hl3] Hélein, F. (1990) Regularity of weakly harmonic maps from a surface into a manifold with symmetries, preprint.

[Hi1] Hildebrandt, S. (1969) Boundary behaviour of minimal surfaces, *Arch. Rat. Mech. Anal.* **35**, 47–82.

[Hi2] Hildebrandt, S. (1970) On the Plateau problem for surfaces of constant mean curvature, *Commun. Pure Appl. Math.* **23**, 97–114.

[Hi3] Hildebrandt, S. (1982) Nonlinear elliptic systems and harmonic mappings, *Proc. Beijing Symp. on Differential Geometry and Differential Equations (1980)*, Science Press, Beijing (also in SFB 72, Vorlesungsreihe No. 4, Bonn: (1980)).

[HKW1] Hildebrandt, S., Kaul, H., and Widman, K.-O. (1975) Harmonic mappings into Riemannian manifolds with non-positive sectional curvature, *Math. Scand.* **37**, 257–263.

[HKW2] Hildebrandt, S., Kaul, H., and Widman, K.-O. (1977) An existence theorem for harmonic mappings of Riemannian manifolds, *Acta Math.* **138**, 1–16.

[HN] Hildebrandt, S., and Nitsche, J. (1979) Minimal surfaces with free boundaries, *Acta Math.* **143**, 251–272.

[HW] Hildebrandt, S., and Widman, K.-O. (1975) Some regularity results for quasilinear elliptic systems of second order, *Math. Z.* **142**, 67–86.

[Ho] Hopf, H. (1950/51) Über Flächen mit einer Relation zwischen den Hauptkrümmungen, *Math. Nachr.* **4**, 232–249.

[Jä] Jäger, W. (1970) Behaviour of minimal surfaces with free boundaries, *Commun. Pure Appl. Math.* **23**, 803–818.

[JäK] Jäger, W., and Kaul, H. (1979) Uniqueness and stability of harmonic maps and their Jacobi fields, *Manuscr. Math.* **28**, 269–291.

[J1] Jost, J. (1981) Univalency of harmonic mappings between surfaces, *J. Reine Angew. Math.* **342**, 141–153.

[J2] Jost, J. (1983) Existence proofs for harmonic mappings with the help of a maximum principle, *Math. Z.* **184**, 489–496.

[J3] Jost, J. (1984) The Dirichlet problem for harmonic maps from a surface with boundary onto a 2-sphere with non-constant boundary values, *J. Diff. Geom.* **19**, 393–401.

[J4] Jost, J. (1984) *Harmonic Maps between Surfaces (Lecture Notes in Mathematics 1062)*, Springer, Berlin.

[J5] Jost J. (1985) Conformal mappings and the Plateau–Douglas problem in Riemannian manifolds, *J. Reine Angew. Math.* **359**, 37–54.

[J6] Jost, J. (1984) *Harmonic mapping between Riemannian manifolds, Proc. Centre for Mathematical Analysis*, vol. 4, Australian National University Press, Canberra.

[J7] Jost, J. (1985) A note on harmonic maps between surfaces, *Ann. Inst. H. Poincaré (Anal. Nonlinéaire)* **2**, 397–405.

[J8] Jost, J. (1986) On the existence of harmonic maps into the real projective plane, *Compositio Math.* **59**, 15–19.

[J9] Jost, J. (1986/87) Existence results for embedded minimal surfaces of controlled topological type, Parts I, II, III, *Ann. Sc. Norm. Sup. Pisa (ser. IV)* Part I, **13** (1986), 15–50; Part II, **13** (1986), 401–426; Part III, **14** (1987), 165–167.

[J10] Jost, J. (1986) On the regularity of minimal surfaces with free boundaries in Riemannian manifolds, *Manuscripta Math.* **56**, 279–291.

[J11] Jost, J. (1987) Continuity of minimal surfaces with piecewise smooth free boundaries, *Math. Ann.* **276**, 599–614.

[J12] Jost, J. (1987) On the existence of embedded minimal surfaces of higher genus with free boundaries in Riemannian manifolds, in: *Variational Methods for Free Surface Interfaces*, ed. P. Concus and R. Finn, Springer, New York, pp. 65–75.

[J13] Jost, J. (1988) Das Existenzproblem for Minimalflächen *Jber. Deutsch. Math. Ver.* **91**, 1–32.

[J14] Jost, J. (1988) Harmonic maps—analytic theory and geometric significance, in: *Partial Differential Equations and Calculus of Variations* (*Lecture Notes in Mathematics 1357*) ed. S. Hildebrandt and R. Leis, Springer, Berlin, pp. 264–296.

[J15] Jost, J. (1987) Two-dimensional geometric variational problems, in: *Proc. Int. Congr. Math. 1986, Berkeley*, American Mathematical Society, Providence, RI, pp. 1094–1100.

[J16] Jost, J. (1989) Embedded minimal surfaces in manifolds diffeomorphic to the three-dimensional ball or sphere, *J. Diff. Geom.* **30**, 555–577.

[J17] Jost, J. (1988) Embedded minimal disks with a free boundary on a polyhedron in $\mathbb{R}^3$, *Math. Z.* **199**, 311–320.

[J18] Jost, J. (1990) Harmonic maps and curvature computations in Teichmüller theory.

[JK] Jost, J., and Karcher, H. (1982) Geometrische Methoden zur Gewinnung von a-priori-Schranken für harmonische Abbildungen, *Manuscripta Math.* **40**, 27–77.

[JS] Jost, J., and Schoen, R. (1982) On the existence of harmonic diffeomorphisms between surfaces, *Inv. Math.* **66**, 353–359.

[JSt] Jost, J., and Struwe, M. (1990) Morse-Conley theory for minimal surfaces of varying topological type, *Inv. Math.*

[KPS] Karcher, H., Pinkall, U., and Sterling, I. (1988) New minimal surfaces in S^3, *J. Diff. Geom.* **28**, 109–185.

[Ke] Keen, L. (1974) Collars on Riemann surfaces, *Ann. Math. Studies* **79**, 263–268.

[Kp] Kilpeläinen, T. (1985) Homogeneous and conformally invariant variational integrals, *Ann. Acad. Sci. Fenn.* A, *I. Mat. Diss.* **57**, Helsinki.

[Ki] Kinderlehrer, D. (1969) The boundary regularity of minimal surfaces, *Ann. Sc. Norm. Sup. Pisa* **23**, 711–744.

[Kl] Klingenberg, W. (1982) *Riemannian Geometry*, de Gruyter, Berlin.

[Kn1] Kneser, H. (1926) Lösung der Aufgabe 41, *Jber. Deutsch. Math. Ver.* **35**, 123–124.

[Kn2] Kneser, H. (1930) Die kleinste Bedeckungszahl innerhalb einer Klasse von Flächenabbildungen, *Math. Ann.* **103**, 347–358.

[LU] Ladyženskaja, O. A., and Ural'ceva, N. N. (1968) *Équations aux Dérivées Partielles de Type Elliptique*, Dunod, Paris.

[LV] Lavrent'ev, M. A. (1935) Sur une classe des réprésentations continues, *Mat. Sb.*, 407–434.

[La] Lawson, H. B. (1970) Complete minimal surfaces in S^3, *Ann. Math.* **92**, 335–374.

[L1] Lemaire, L. (1978) Applications harmoniques de surfaces Riemanniennes, *J. Diff. Geom.* **13**, 51–78.

[L2] Lemaire, L. (1982) Boundary value problems for harmonic and minimal maps of surfaces into manifolds, *Ann. Sc. Norm. Sup. Pisa* (*4*) **8**, 91–103.

[Lw] Lewy, H. (1934) On the existence of a closed surface realizing a given Riemannian metric, *Proc. Natl. Acad. Sci. USA* **24**, 104–106.

[Li] Lichtenstein, L. (1916) Zur Theorie der konformen Abbildung. Konforme Abbildung nichtanalytischer singularitätenfreier Flächenstücke auf ebene Gebiete, *Bull. Acad. Sci. Cracovie, Cl. Sci. Mat. Nat.* A, 192–217.

[Lu] Luckhaus, S. (1978) The Douglas-problem for surfaces of prescribed mean curvature, *Preprint* 234, SFB 72, Universität Bonn.

[Ma] Matelski, J. (1970) A compactness theorem for Fuchsian groups of the second kind, *Duke Math. J.* **43**, 829–840.

[Mc] McShane, E. (1933) Parametrization of saddle surfaces with applications to the problem of Plateau, *Trans. Am. Math. Soc.* **35**, 716–733.

[MSY] Meeks, W., Simon, L., and Yau, S. T. (1982) Embedded minimal surfaces, exotic spheres, and manifolds with positive Ricci curvature, *Ann. Math.* **116**, 621–659.

[MY1] Meeks, W., and Yau, S. T. (1982) The classical Plateau problem and the topology of three-dimensional manifolds, *Topology* **21**, 409–442.

[MY2] Meeks, W., and Yau, S. T. (1980) Topology of three dimensional manifolds and the embedding problems in minimal surface theory, *Ann. Math.* **112**, 441–484.

[MY3] Meeks, W., and Yau, S. T. (1982) The existence of embedded minimal surfaces and the problem of uniqueness, *Math. Z.* **179**, 151–168.

[M1] Morrey, C. B. (1938) On the solution of quasi-linear elliptic partial differential equations, *Trans. Am. Math. Soc.* **43**, 126–166.

[M2] Morrey, C. B. (1948) The problem of Plateau on a Riemannian manifold, *Ann. Math.* **49**, 807–851.

[M3] Morrey, C. B. (1966) *Multiple Integrals in the Calculus of Variations*, Springer, Berlin.

[MT] Morse, M., and Tompkins, C. (1939/41) The existence of minimal surfaces of general critical types, *Ann. Math.* (2) **40**, 443–472; Erratum **42** (1941), 331.

[Mu] Mumford, D. (1971) A remark on Mahler's compactness theorem, *Proc. Am. Math. Soc.* **28**, 288–294.

[NN] Newlander, A., and Nirenberg, L. (1957) Complex analytic coordinates in almost complex manifolds, *Ann. Math.* **65**, 391–404.

[Ni1] Nitsche, J. (1969) The boundary behaviour of minimal surfaces, Kellogg's Theorem and branch points on the boundary, *Inv. Math.* **8**, 313–333.

[Ni2] Nitsche, J. (1985) Stationary partitioning of convex bodies, *Arch. Rat. Mech. Anal.* **89**, 1–19.

[Ol1] Olum, P. (1953) On mappings into spaces in which homotopy groups vanish, *Ann. Math.* **57**, 561–574.

[Ol2] Olum, P. (1953) Mappings of manifolds and the notion of degree, *Ann. Math.* **58**, 458–480.

[O] Osserman, R. (1970) A proof of the regularity everywhere of the classical solution of Plateau's problem, *Ann. Math.* **91**, 550–569.

[P] Pitts, J. (1981) *Existence and Regularity of Minimal Surfaces on Riemannian Manifolds*, Princeton University Press, Princeton, N.J.

[PR] Pitts, J., and Rubinstein, H. (1986) Existence of minimal surfaces of bounded topological type in three-manifolds, in: *Proc. Miniconf. on Geometry and Partial Differential Equations (Canberra, 1985)*, Proc. Centre for Mathematical Analysis vol. 10, Australian National University Press, Canberra, pp. 163–176.

[R1] Radó, T. (1926) Aufgabe 41, *Jber. Deutsch. Math. Ver.* **35**, 49.

[R2] Radó, T. (1930) On Plateau's problem. *Ann. Math.* (2) **31**, 457–469.

[R3] Radó, T. (1930) The problem of the least area and the problem of Plateau, *Math. Z.* **32**, 763–796.

[Rn] Randol, B. (1979) Cylinders in Riemann surfaces, *Comm. Math. Helv.* **54**, 1–5.

[Re] Reich, E. (1985) On the variational principle of Gerstenhaber and Rauch, *Ann. Acad. Sci. Fenn.* **10**, 469–475.

[RS] Reich, E., and Strebel, K. (1987) On the Gerstenhaber–Rauch principle, *Israel J. Math.* **57**, 89–100.

[RV] Ruh, E., and Vilms, J. (1970) The tension field of the Gauss map, *Trans. Am. Math. Soc.* **149**, 569–573.

[SkU1] Sacks, J., and Uhlenbeck, K. (1981) The existence of minimal immersions of 2-spheres, *Ann. Math.* **113**, 1–24.

[SkU2] Sacks, J., and Uhlenbeck, K. (1982) Minimal immersion of closed Riemann surfaces, *Trans. Am. Math. Soc.* **271**, 639–652.

[Sa] Sampson, J. H. (1978) Some properties and applications of harmonic mappings, *Am. Sci. Ecole Normale Superieure* **11**, 211–228.

[S] Schoen, R. (1984) Analytic aspects of the harmonic map problem, in: *Math. Sci. Res. Inst. Publ. 2, (Seminar on nonlinear partial differential equations, Berkeley, CA 1983)*, ed. S. S. Chern, Springer, New York, pp. 321–358.

[SS] Schiffer, M., and Spencer, D. (1954) *Functionals of Finite Riemann Surfaces*, Princeton University Press, Princeton, N.J.

[SU] Schoen, R., and Uhlenbeck, K. (1983) Boundary regularity and miscellaneous results on harmonic maps, *J. Diff. Geom.* **18**, 253–268.

[SY1] Schoen, R., and Yau, S. T. (1978) On univalent harmonic maps between surfaces, *Inv. Math.* **44**, 265–278.

[SY2] Schoen, R., and Yau, S. T. (1979) Existence of incompressible minimal surfaces and the topology of three-dimensional manifolds with non-negative scalar curvature, *Ann. Math.* **110**, 127–142.

[SY3] Schoen, R., and Yau, S. T. (1979) Compact group actions and the topology of manifolds with non-positive curvature, *Topology* **18**, 361–380.

[Sz] Schulz, F. (1989) Univalent solutions of elliptic systems of Heinz–Lewy type, *Ann. Inst. H. Poincaré* (Anal. Nonlinéaire) **6**, 347–361.

[Se] Sealey, H. (1980) Some properties of harmonic mappings, *Thesis*, University of Warwick.

[Sh] Shibata, K. (1963) On the existence of a harmonic mapping, *Osaka J. Math.* **15**, 173–211.

[Sf1] Shiffman, M. (1939) The Plateau problem for minimal surfaces of arbitrary topological structure, *Am. J. Math.* **61**, 853–883.

[Sf2] Shiffman, M. (1939) The Plateau problem for non-relative minima, *Ann. Math.* (2) **40**, 834–854.

[Sf3] Shiffman, M. (1942) Unstable minimal surfaces with several boundaries, *Ann. Math.* (2) **43**, 197–222.

[SSm] Simon, L., and Smith, F. On the existence of embedded minimal 2-spheres in the 3-sphere, endowed with an arbitrary metric, to appear.

[Si] Siu, Y.-T. (1986) Curvature of the Weil–Petersson metric in the moduli space of compact Kähler–Einstein manifolds of negative first Chern class, *Aspects of Mathematics* vol. 9, ed. K. Diederich, Vieweg, Braunschweig, Wiesbaden, pp. 261–298.

[Sp] Spruck, J., The elliptic sinh Gordon equation and the construction of toroidal soap bubbles, to appear.

[Sy] Smyth, B. (1984) Stationary minimal surfaces with boundary on a simplex, *Inv. Math.* **76**, 411–420.

[Sö1] Ströhmer, G. (1980) Instabile Minimalflächen in Riemannschen Mannigfaltigkeiten nichtpositiver Schnittkrümmung, *J. Reine Angew. Math.* **315**, 16–39.

[Sö2] Ströhmer, G. (1982) Instabile Lösungen der Eulerschen Gleichungen gewisser Variationsproblem, *Arch. Rat. Mech. Anal.* **79**, 219–239.

[St1] Struwe, M. (1985) On the evolution of harmonic mappings, *Comm. Math. Helv.* **60**, 558–581.

[St2] Struwe, M. (1984) On a free boundary problem for minimal surfaces, *Inv. Math.* **75**, 547–560.

[St3] Struwe, M. (1984) On a critical point theory for minimal surfaces spanning a wire in $\mathbb{R}^N$, *J. Reine Angew. Math.* **349**, 1–23.

[St4] Struwe, M. (1986) A Morse theory for annulus-type minimal surfaces. *J. Reine Angew. Math.* **368**, 1–27.

[St5] Struwe, M., *Plateau's problem and the calculus of variations*, Princeton University Press, Princeton, N.J.

[Ty] Taylor, J. (1976) The structures of singularities in soap-bubble-like and soap-film-like minimal surfaces, *Ann. Math.* **103**, 486–539.

[Tm1] Teichmüller, O. (1943) Extremale quasikonforme Abbildungen und quadratische Differentiale, *Abh. Preuss. Akad. Wiss., Math.-Naturw. Klasse* **4**, 1–197 (also in [Tm3], pp. 335–531).

[Tm2] Teichmüller, O. (1944) Veränderliche Riemannsche Flächen, *Deutsche Math.* **7**, 344–359 (also in: [Tm3], pp. 712–727).

[Tm3] Teichmüller, O. (1982) *Gesammelte Abhandlungen—Collected Papers*, ed. L. Ahlfors and F. Gehring, Springer, Berlin.

[To1] Tolksdorf, P. (1984) A parametric variational principle for minimal surfaces of varying topological type, *J. Reine Angew. Math.* **354**, 16–49.

[To2] Tolksdorf, P. (1985) On minimal surfaces with free boundaries in given homotopy classes, *Ann. Inst. H. Poincaré* (*Anal. Nonlinéaire*) **2**, 157–165.

[TT1] Tomi, F., and Tromba, A. (1978) Extreme curves bound embedded minimal surfaces of the type of the disc, *Math. Am.* **158**, 137–145.

[TT2] Tomi, F., and Tromba, A. (1985) On Plateau's problem for minimal surfaces of higher genus in $\mathbb{R}^3$, *Bull. Am. Math. Soc.* **13**, 169–171.

[TT3] Tomi, F., and Tromba, A. (1988) Existence theorems for minimal surfaces of non-zero genus spanning a contour, *Memoirs Am. Math. Soc.* **71**, no 382.

[Tr1] Tromba, A. (1990) On Plateau's problem for minimal surfaces of higher genus in $\mathbb{R}^n$, to appear.

[Tr2] Tromba, A. (1900) A new proof that Teichmüller space is a cell, *Trans. Am. Math. Soc.* **303**, 257–262.

[Tr3] Tromba, A. (1986) On a natural algebraic affine connection on the space of almost complex structures and the curvature of Teichmüller space with respect to its Weil–Petersson metric, *Manuscripta Math.* **56**, 475–497.

[Tr4] Tromba, A. (1987) Global analysis and Teichmüller theory, in: *Seminar on New Results in Nonlinear PDE*, ed. A. Tromba, Vieweg, Braunschweig, Wiesbaden.

[Tr5] Tromba, A. (1987) On an energy function for the Weil–Petersson metric on Teichmüller space, *Manuscripta Math.* **59**, 249–260.

[Wa] Walter, R. (1984) Explicit examples to the *H*-problem of Heinz Hopf.

[W1] Wente, H. (1975) The differential equation $\Delta x = 2Hx_u \Lambda x_v$ with vanishing boundary values, *Proc. Am. Math. Soc.* **50**, 131–137.

[W2] Wente, H. (1980) Large solutions to the volume constrained Plateau problem, *Arch. Rat. Mech. Anal.* **75**, 59–77.

[W3] Wente, H. (1986) Counterexample to a question of H. Hopf, *Pacific J. Math.* **121**, 193–243.

[W4] Wente, H. (1987) Twisted tori of constant mean curvature, *Aspects Math. E* **10**, 1–36.

[Wf] Wolf, M. (1989) The Teichmüller theory of harmonic maps, *J. Diff. Geom.* **29**, 449–479.

[Wp] Wolpert, S. (1986) Chern forms and the Riemann tensor for the moduli space of curves, *Inv. Math.* **85**, 119–145.

[Wo] Wood, J. C. (1977) Singularities of harmonic maps and applications of the Gauss–Bonnet Formula, *Am. J. Math.* **99**, 1329–1344.

[Ye1] Ye, R. (1984) Randregularität von Minimalflächen, *Diplomarbeit*, Universität Bonn.

[Ye2] Ye, R. (1991) On the existence of area-minimizing surfaces with free boundary, *Math. Z.*, to appear.

[Z] Zieschang, H. (1981) *Finite Groups of Mapping Classes of Surfaces*, (*Lecture Notes in Mathematics 875*) Springer, Berlin.

INDEX